Audio in the Pendulum

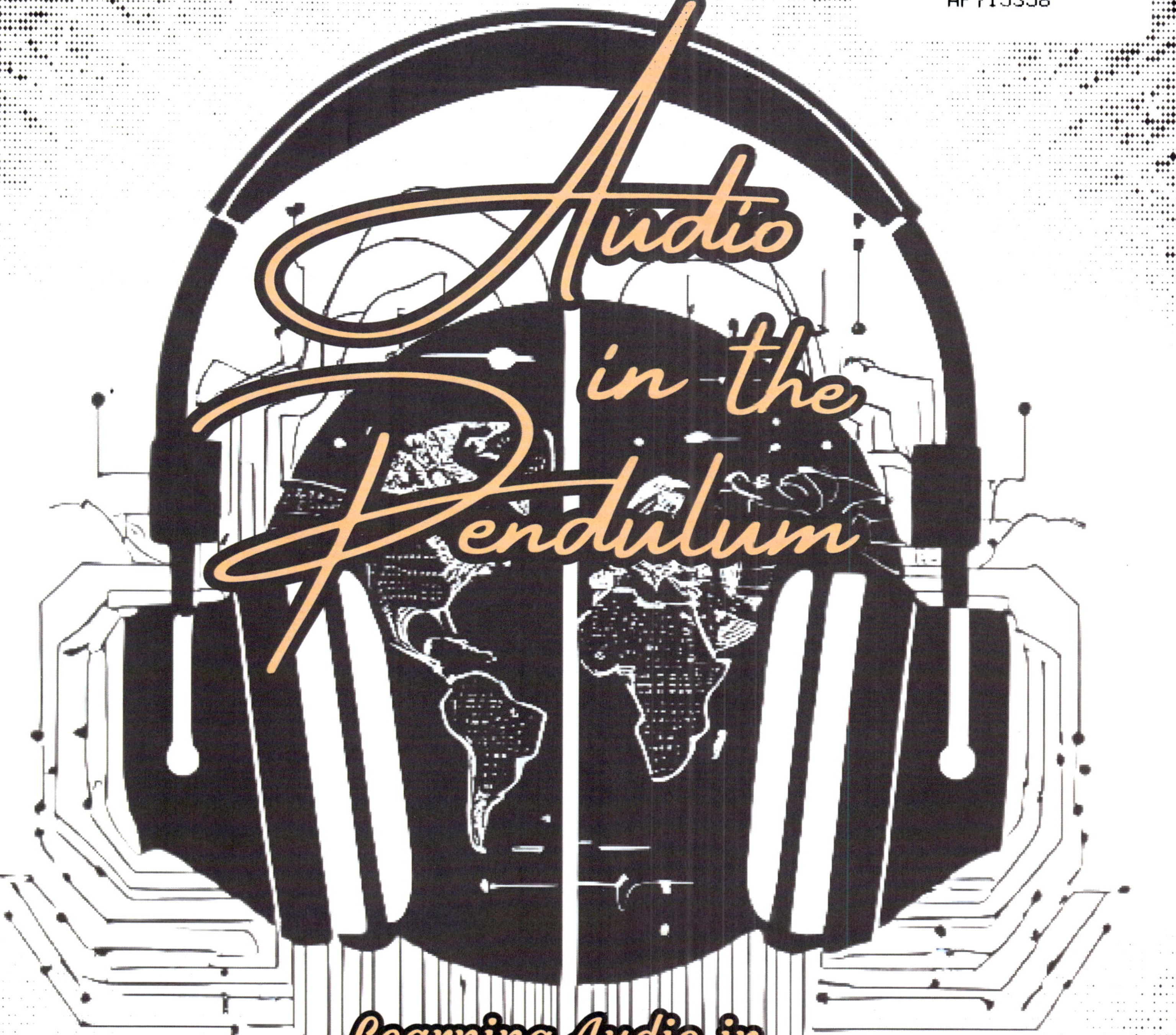

Learning Audio in The Pendulum Process

by Samuel Friend

PALMETTO
PUBLISHING
Charleston, SC
www.PalmettoPublishing.com

Paperback ISBN: 979-8-9926137-1-1
eBook ISBN: 979-8-9926137-3-5

Please note that much of this handbook is based on personal experience and anecdotal evidence. All concepts and analogies were created with the utmost honesty, and with the intent to inform and entertain the reader. None of the information in this handbook is intended to replace common sense, legal, medical or other professional advice. A reader is always encouraged to seek further information to gain a conclusive understanding of the subject matter. Any trademarks, service marks, product names, or named features are assumed to be the property of their respective owners, and are used only for reference. There is no implied endorsement if we use one of these terms.

Table of Contents

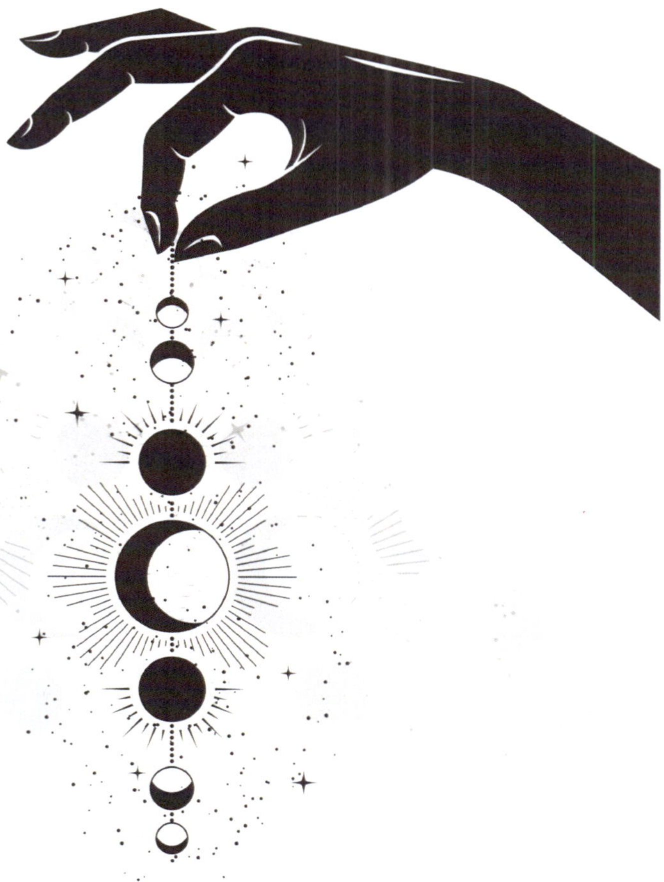

Welcome to The Pendulum Process...

The Pendulum Process is the balancing act and intuitive force that allows an audio engineer to make tasteful decisions. To make tasteful decisions, one must grasp the concepts of too much, not enough, and balance within a given context. The Pendulum Process enlightens the "myopic" or "short-sighted" audio engineer. Remember, your goal as an audio engineer is to preserve the "greatness" in great ideas.

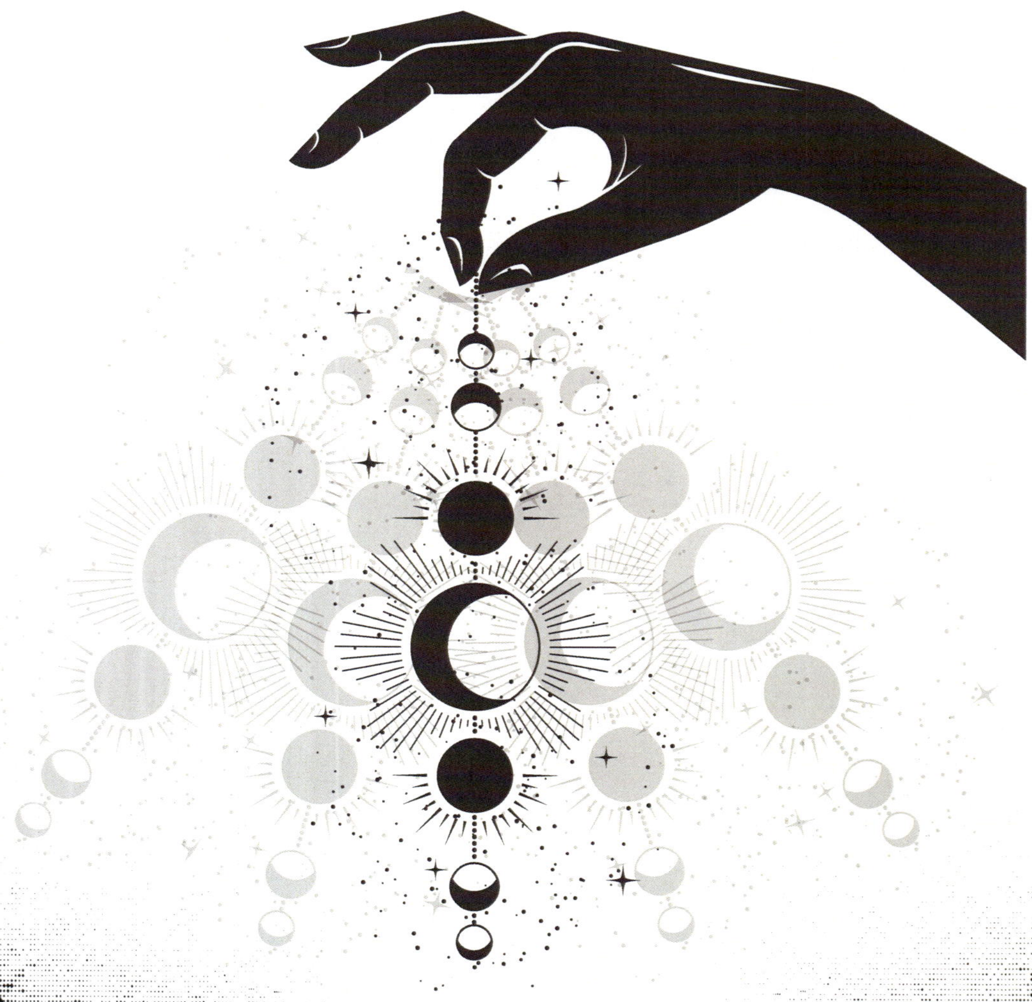

Learning the Map of Music Production

The method of learning Music Production is similar to navigating a theme park map. The more you explore the park, the better you'll be at creating the right experience. The sequence in which you explore is flexible, but ensure that you delve into and understand each area.

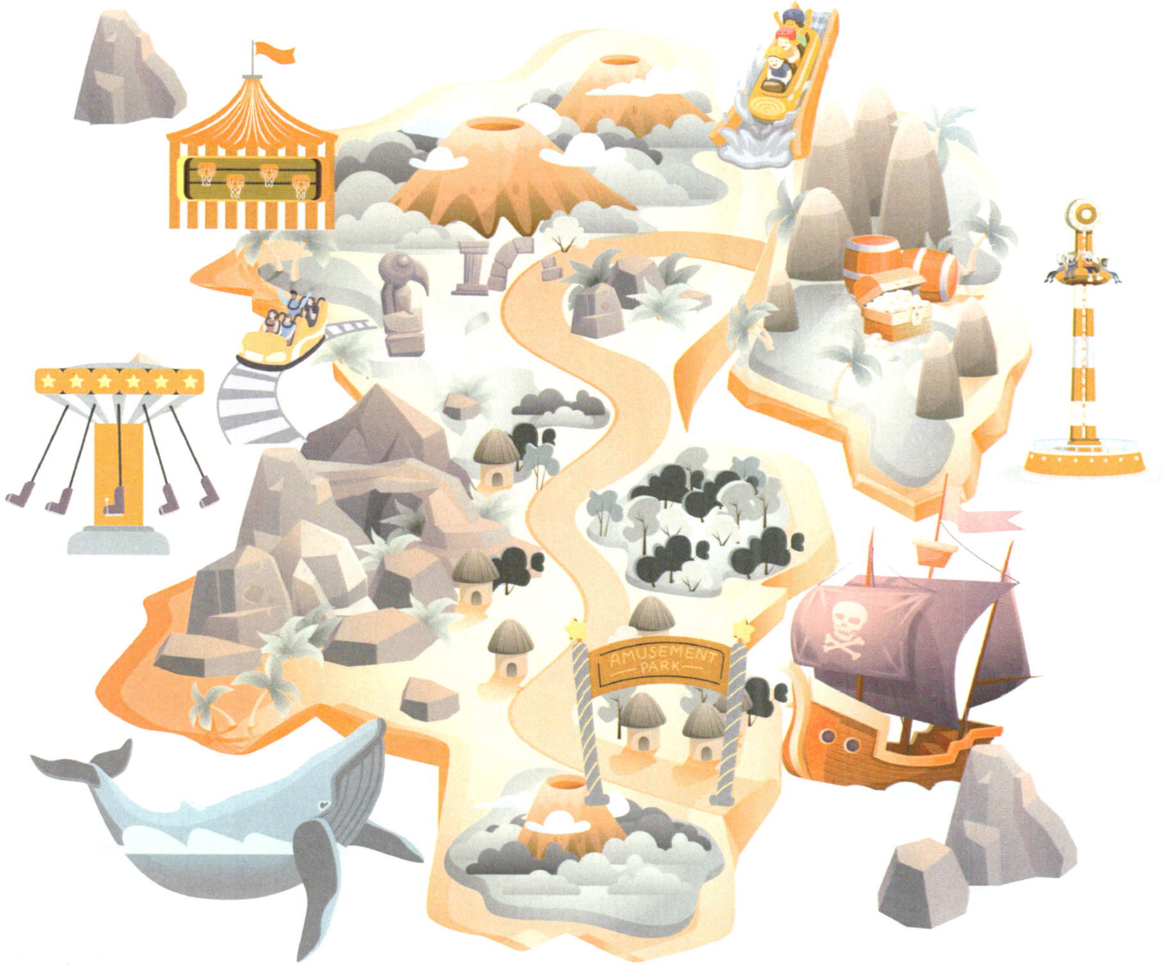

The "Big 4" Skillsets in the Pendulum Process

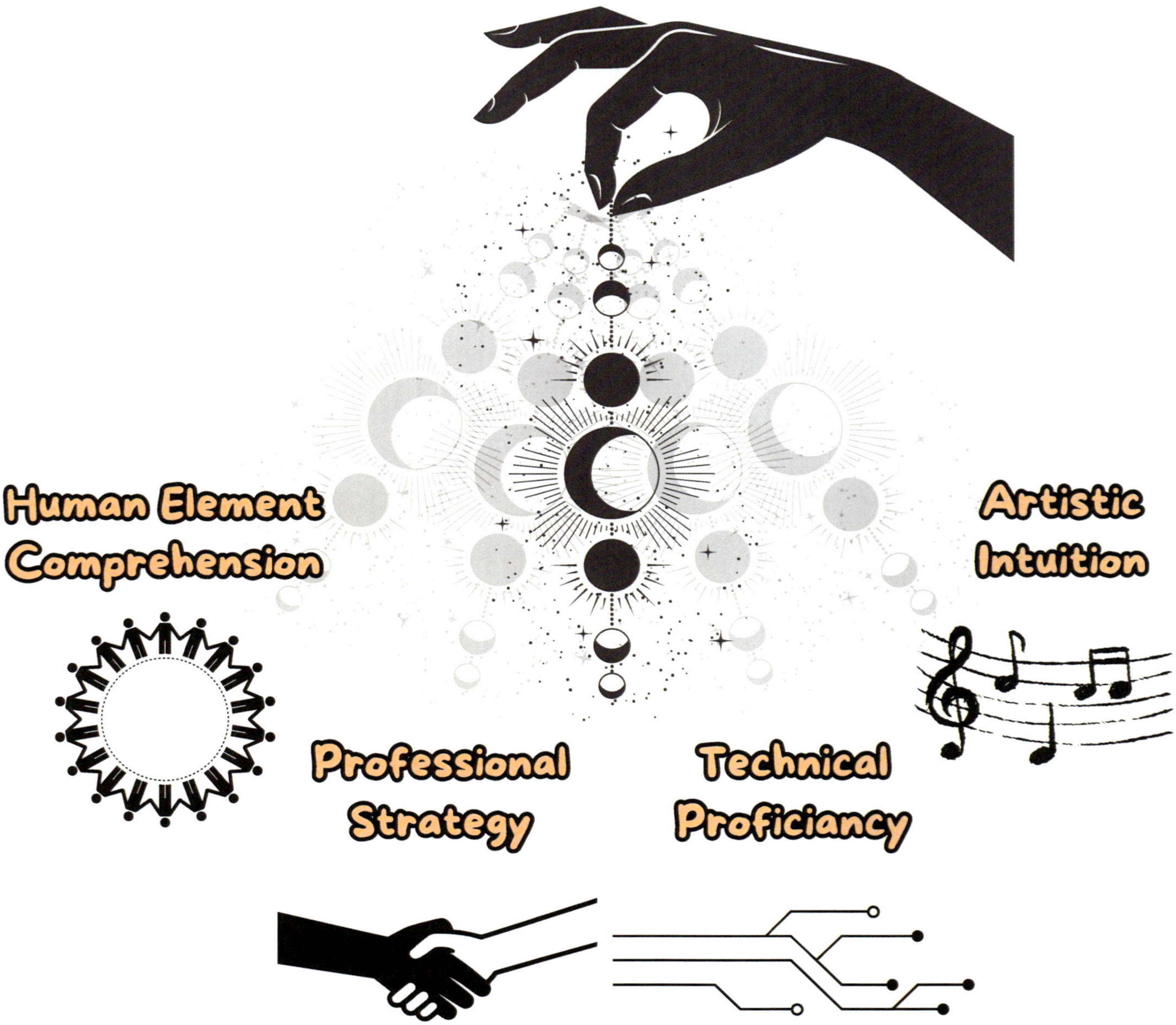

A balance of the "Big 4" will help a producer make tasteful decisions...

Feel the Pendulum Process

Ask yourself these 3 questions...

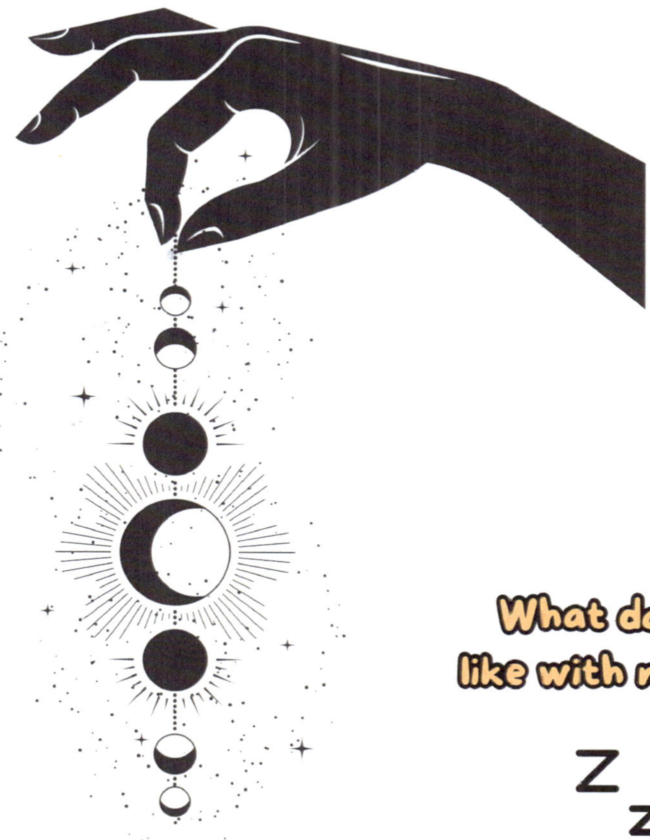

What does it feel like with too much?

What does it feel like with not enough?

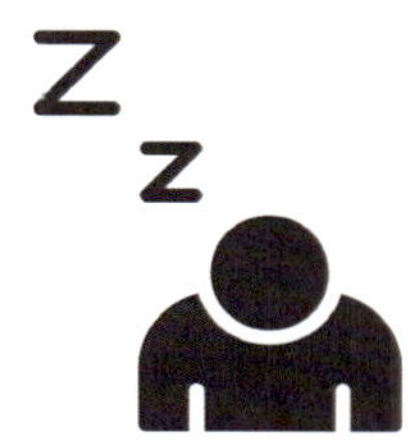

What does it feel like with a balance?

Too much of any 1 feeling can be boring or create tension...
A predictably balanced song can also become uninspiring...
A tasteful balance of all 3 will resolve this tension...

Awareness in the Pendulum Process

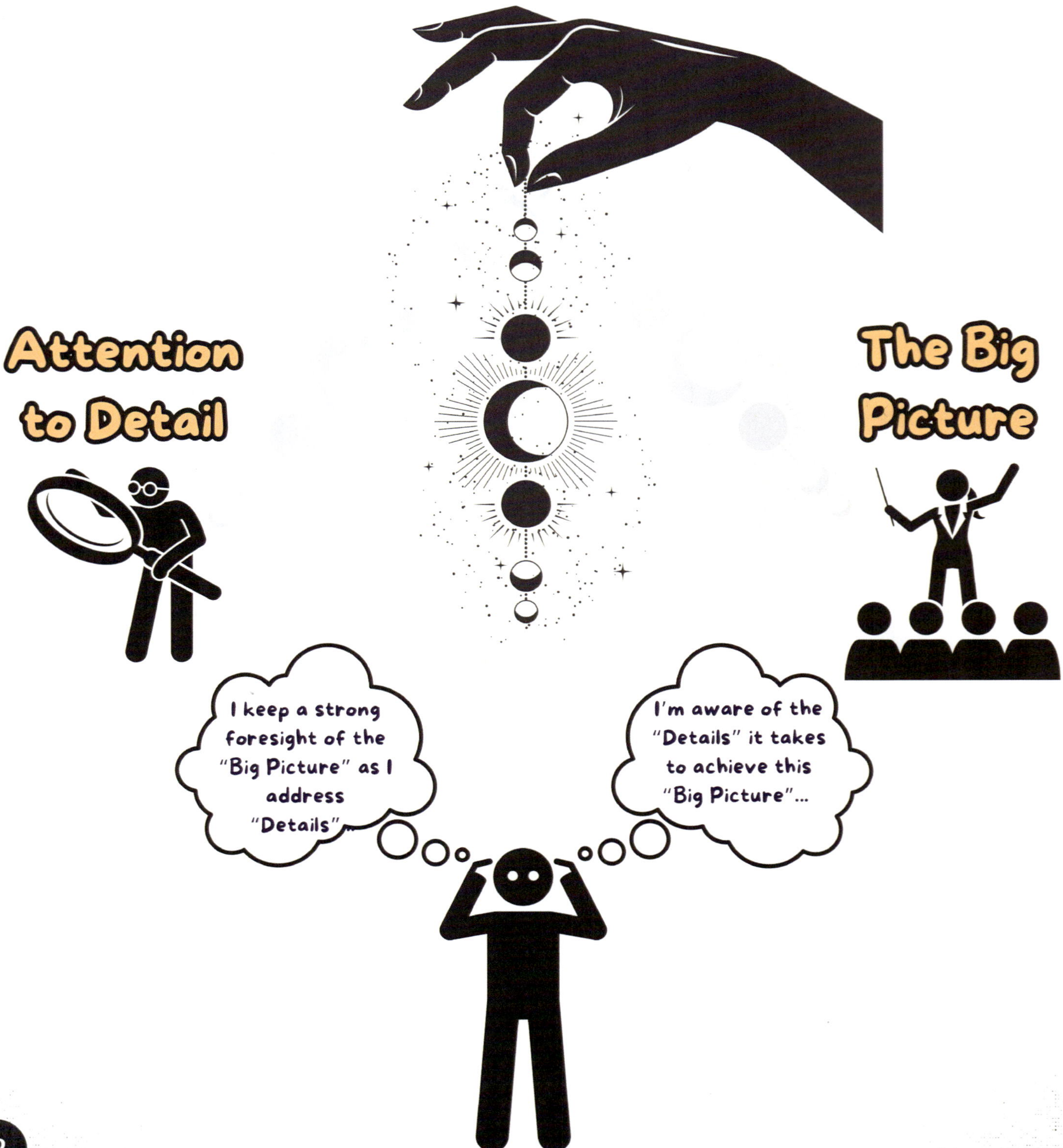

Getting Started with your Audio Setup

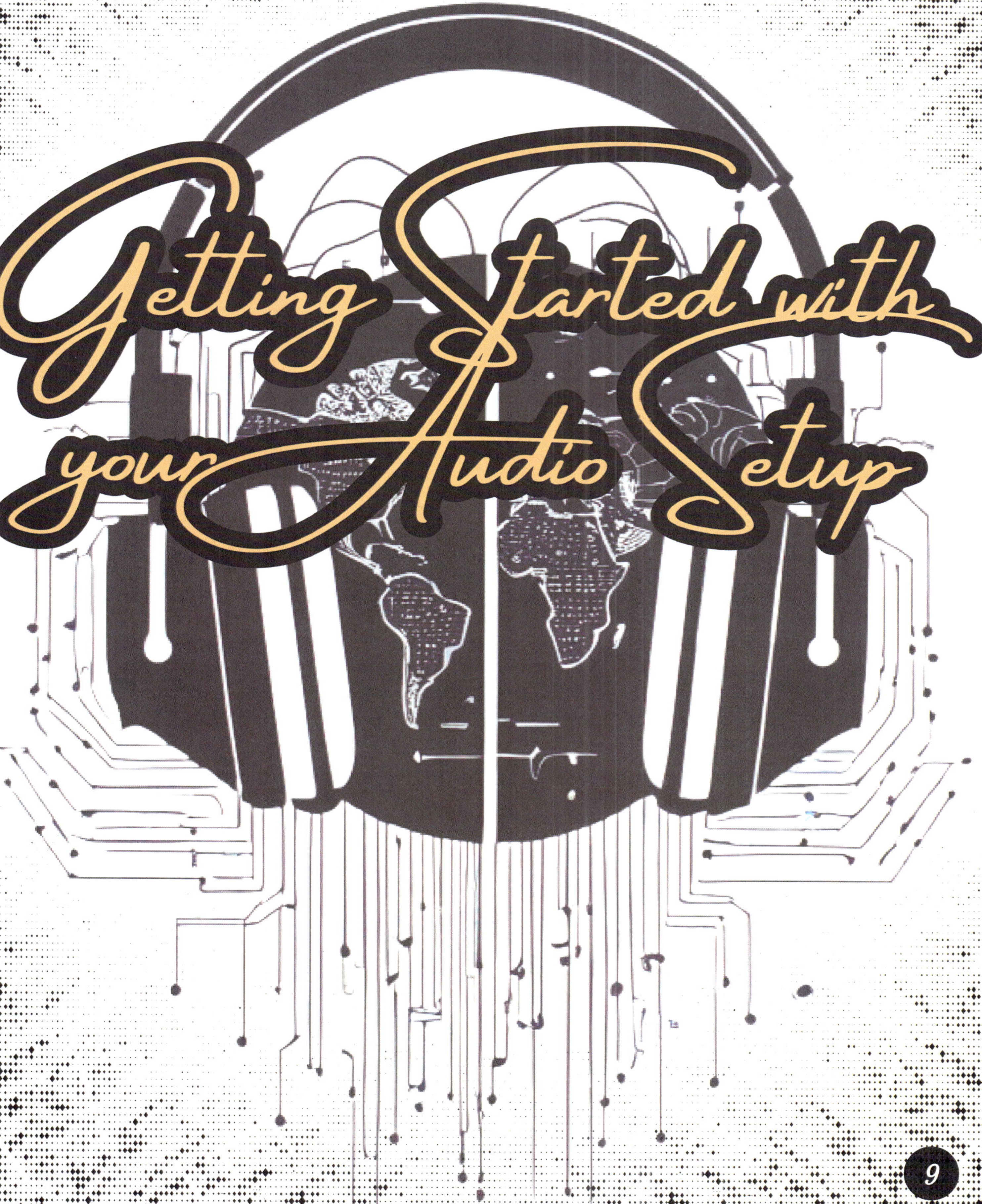

Basic Audio Production Setup

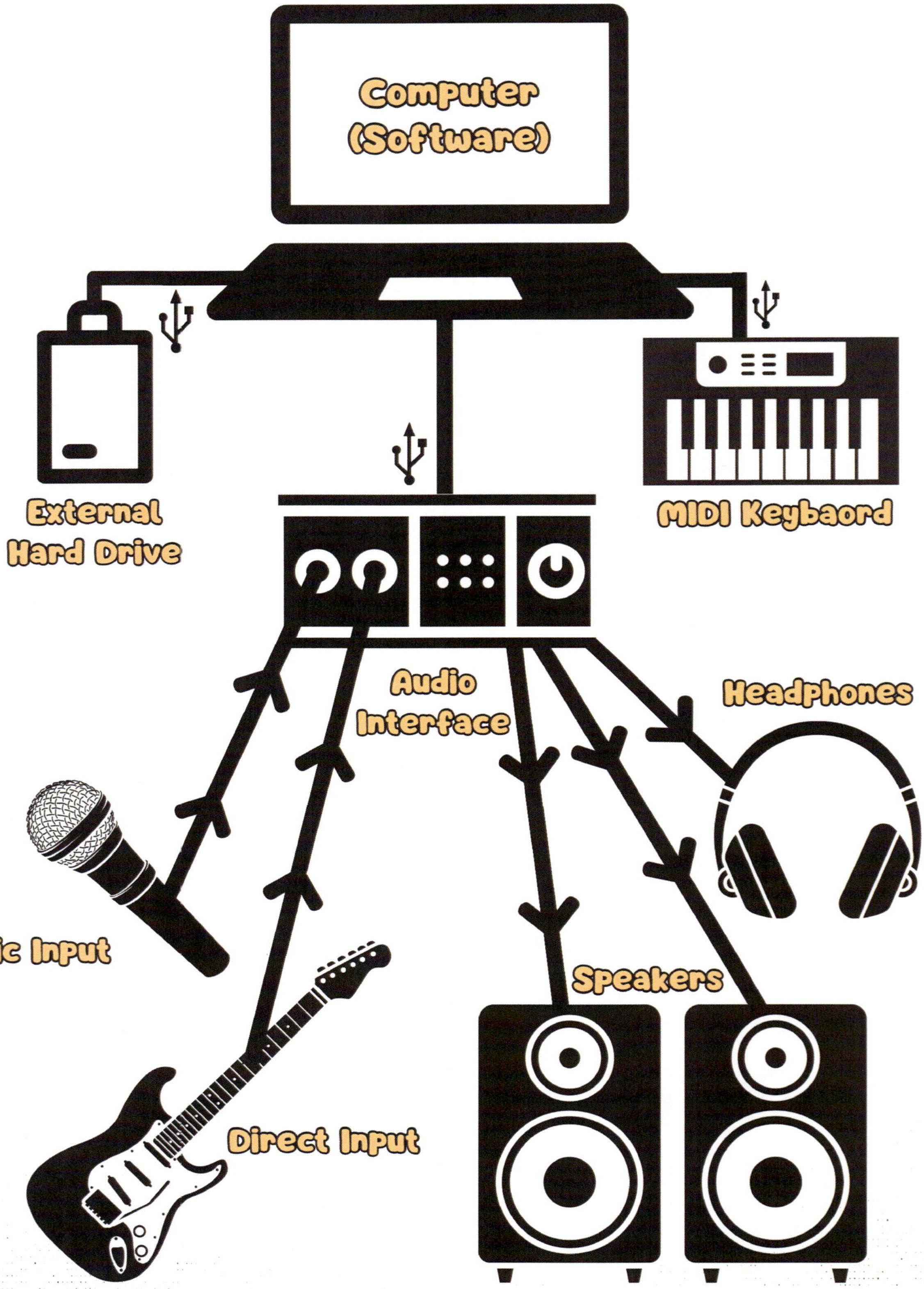

Basic Gear List

Okay, the hard part... buying gear!

To start with the bare minimum, you'll need a reliable computer with the proper specifications and a good pair of headphones.

This is my recommended basic gear list...

Next, purchase an audio interface, external hard drive, and MIDI Keyboard. If you want to record audio, buy a microphone (dynamic mic if you don't have a nice sounding acoustic room). Don't forget a good XLR cable!!

Finally, buy a pair of good speakers to allow a more comfortable workflow. Don't forget 2 TRS cables!!

Buy it Nice or Buy it Twice!

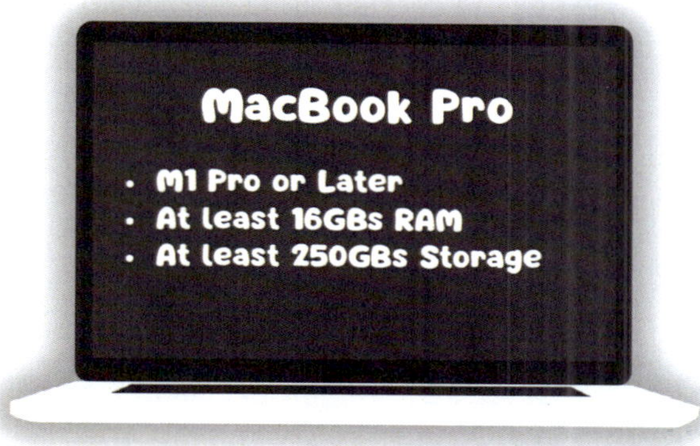

External Hard Drive
Samsung
1 TB T5 SSD

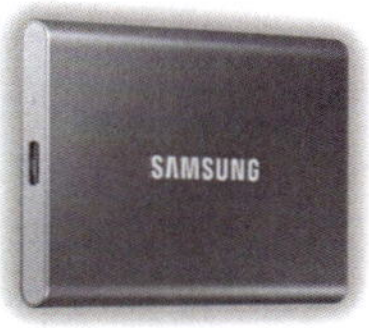

Dynamic Mic
Shure
SM-7B

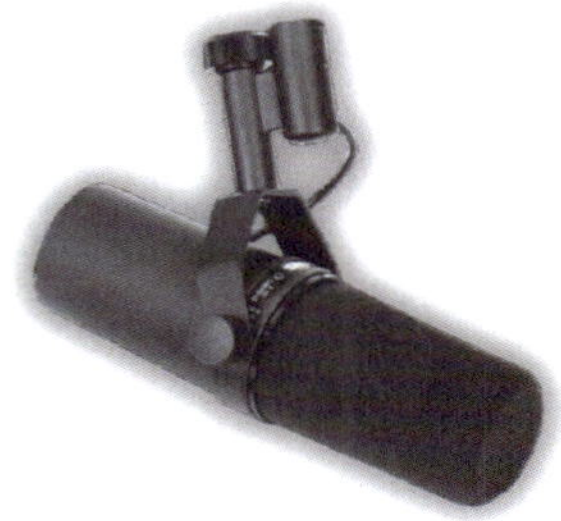

MIDI Keyboard
Native Instruments
Komplete Kontrol A49

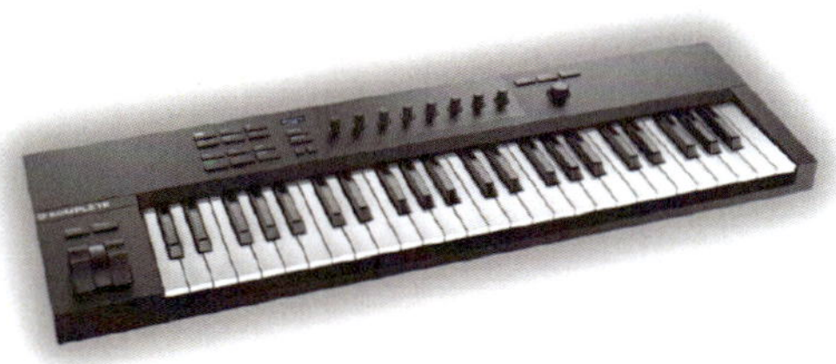

Audio Interface
Universal Audio Volt-276

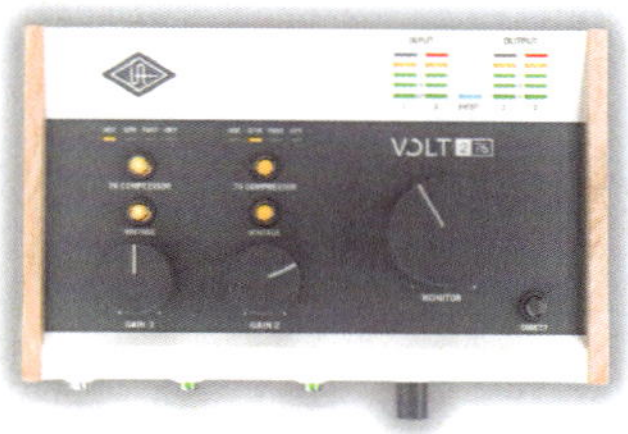

Headphones
Audio-Technica
M40x

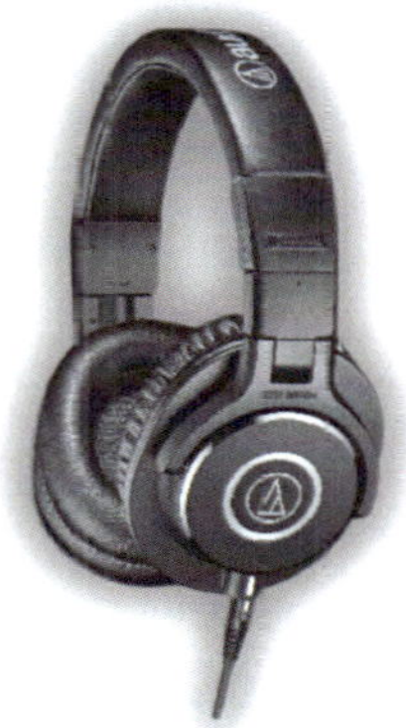

Or

Condenser Mic
Lewitt
LCT 440 Pure

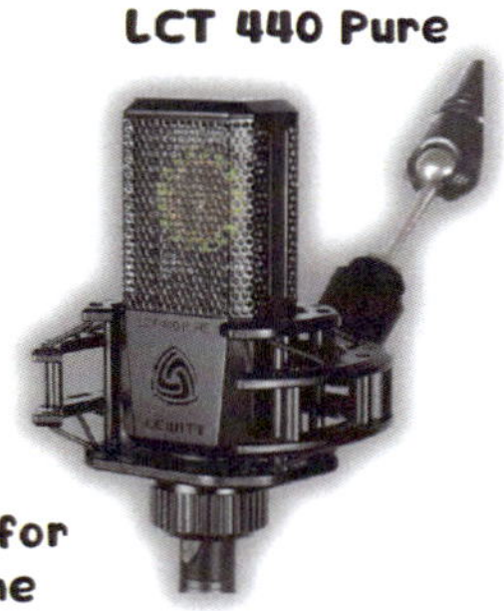

Pair of TRS Cables for Speakers

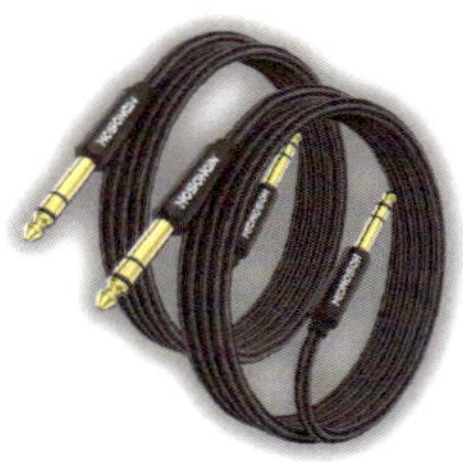

Speakers
IK Multimedia
iLoud MTM

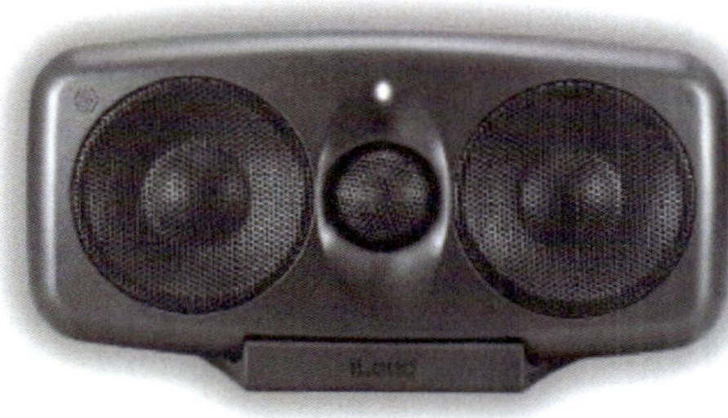

XLR Cable for Microphone

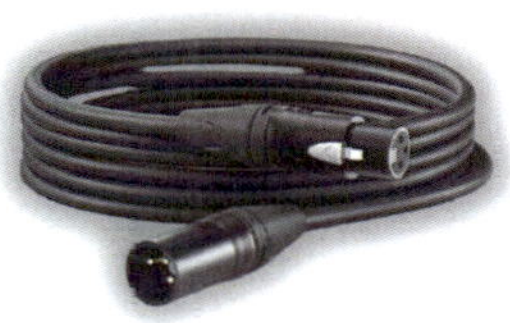

With proper care, all of this equipment will last a long time, and remain an asset throughout your music production journey. As you grow, so will your equipment. Keep in mind that nothing is truly "cheap" and making an "upfront" investment is essential. It's an investment in your growth, and you deserve to acquire the tools you need.

Music Production Software

DAW (Digital Audio Workstation)

Pro Tools

Great DAW for Recording, Mixing, and Editing Audio (Industry Standard)

Logic Pro

Great "All in One" package for creating and mixing.

Ableton Live

Great for interactive "In the Box" creative workflow.

Cubase

Great "All In One" DAW with strong MIDI Composition Features.

FL Studio

Great creative DAW optimized for "beat making" and "sample based" music production.

3rd Party Apps & Plugins

Native Instruments

Great virtual instrument host.

Fabfilter

Great audio processing plugins A top choice by many professionals

Antares

Great Auto-Tune

Waves

Great audio processing plugins A top choice by many professionals

Universal Audio

Great plugin emulations of analog equipmen

Connectivity

Think of a "Highway" when you connect to a USB or Thunderbolt port on your computer. As the amount of Data or number of "Cars" increase, so does the need for more Bandwidth or "Lanes". There are many protocols for connectivity, but be aware of the protocols recommended for audio.

USB 3.0

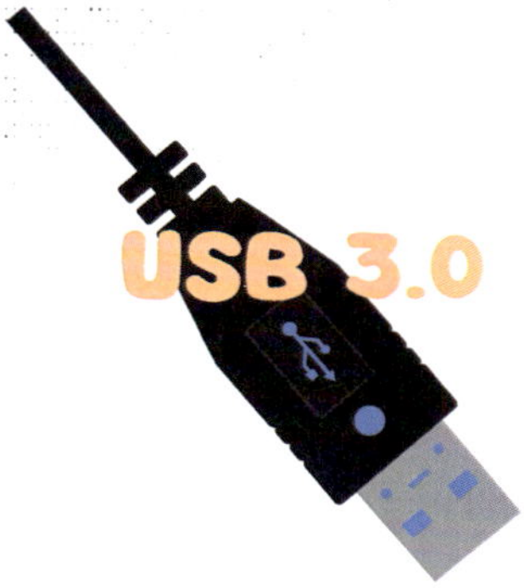

USB 3.0 offers high bandwidth sufficient for most audio playback and recording devices. Depending on the hardware buffer size, number of tracks, and number of plugins in the session, the overall performance of the session may start to suffer.

USB 2.0 (Minimum)

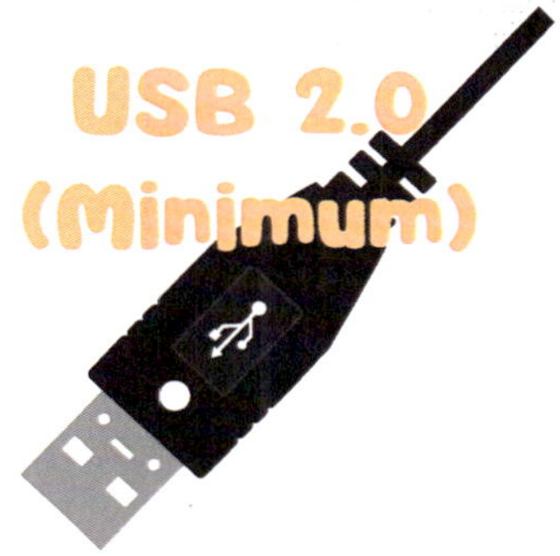

USB 2.0 is still widely utilized for devices that do not require high data transfer speeds such as MIDI Keyboards, License Keys, Trackballs, and simple audio interfaces.

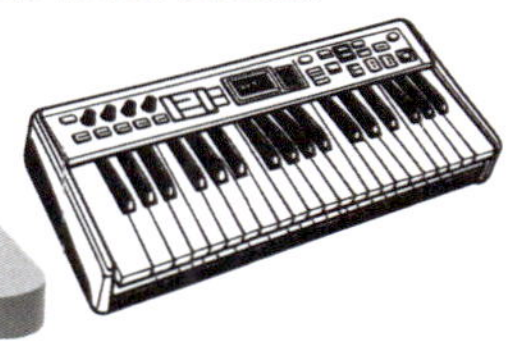

Thunderbolt

Thunderbolt is a proprietary Mac protocol that features extremely high data transfer rates. With large sessions featuring high track count, high numbers of plugins, and low latency playback, a Thunderbolt connection is currently the strongest choice for audio interfaces. Thunderbolt 3 or higher utilize USB-C type connectors, however not all USB-C cables are capable of utilizing Thunderbolt data transfer speeds. Be sure to look for the distinctive thunderbolt icon located on the connector.

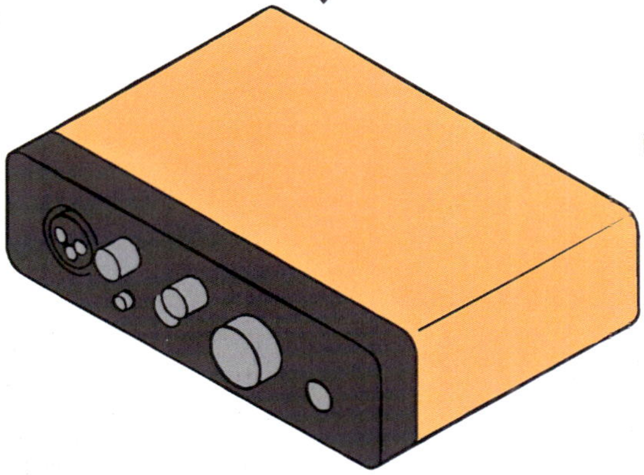

USB-C

USB-C is a connector type that can transmit a number of connectivity protocols, many of which are more than sufficient in the world of audio. Remember, this is simply a connector type and doesn't specifically indicate high transfer rates. Both Mac and PC utilize this type of connector, but the protocol may vary. Lastly, it is important to note that a point-to-point connection is the most reliable connection for devices, but if you need additional ports, choose a "Dock" that utilizes your computer's highest connectivity protocol.

Software Compatibility Hierarchy

When purchasing or updating software, grasping the compatibility hierarchy is crucia At the base level, a sample library must shar compatibility with the host plugin. Following that, the plugin needs to be compatible with the Digital Audio Workstation (DAW). Lastly, t DAW should be compatible with your compute operating system (the host of all software)

One common compatibility issue that severely hinders compute performance is updating an operating system before DAW a plugin companies release compatible updates. To mitigate this, a good rule of thumb is to always stay one or two operatin systems behind the latest versio unless you have thoroughly confirmed compatibility across a applications.

Another common issue can occur when updatin a DAW without first checking compatibility with third-party plugin Launching incompatible plugins is one of the leading reasons for DA session crashes.

DAW must be compatible with Operating System

Computer Operating System
Ex: Mac OS Monterey

Plugin must be compatible with DAW

DAW Version
Ex: Pro Tools 2023.1

Library must be compatible with Plugin

Plugin Version
Ex: Native Instruments or Waves

Sample Libraries
Ex: Spitfire Albion 1

Music Production Software Storage

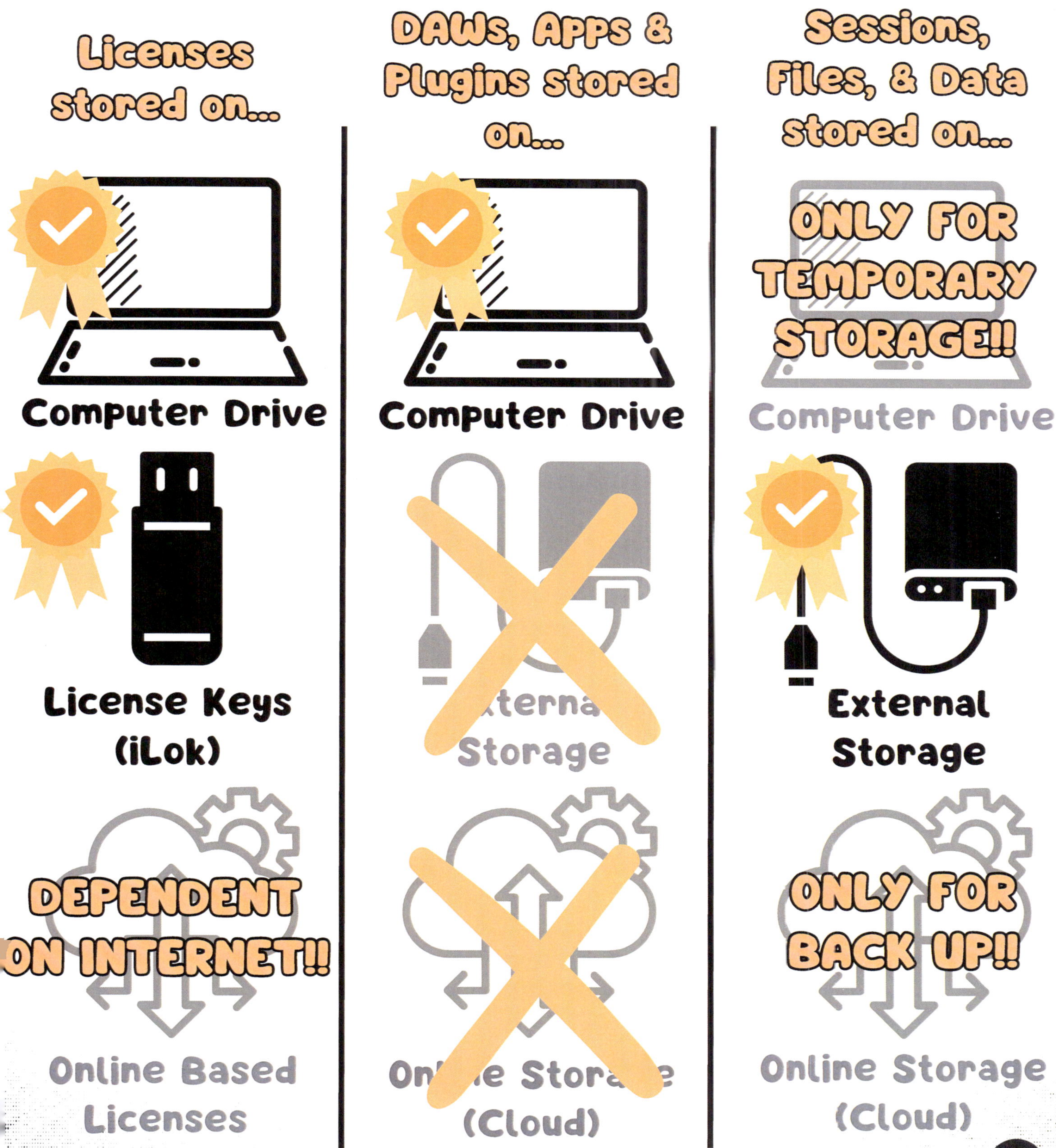

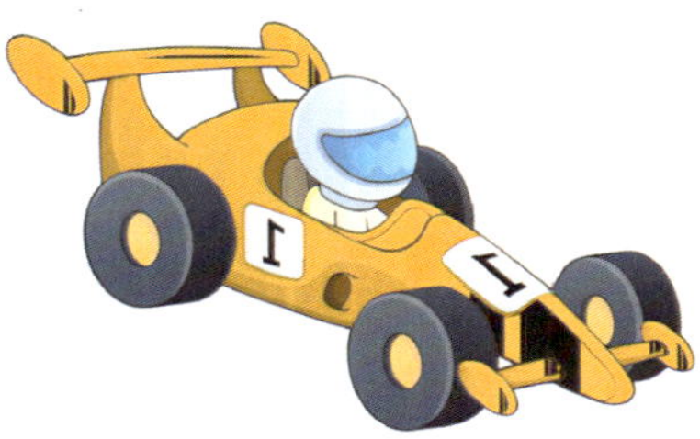

Fast Data Storage

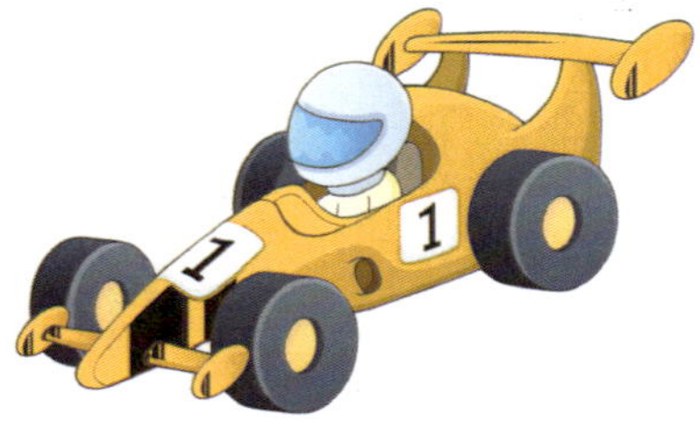

"Session Hard Drive"

These hard drives offer high-performance read and write speeds essential for running DAW sessions. Furthermore, they serve as the storage space for your session and audio files during your production work.

Local Hard Drive

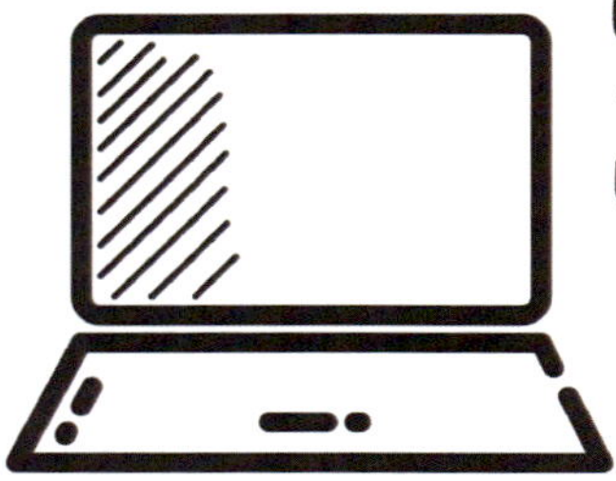

Local hard drives offer high speed and can serve as temporary storage for your DAW sessions. However, local storage can fill up quickly, leading to a decline in overall computer performance.

7200 RPM Hard Disk Drive

HDD's are a great "budget friendly" option for attaining lots of storage. However, make sure the drive is 7200 RPM and establish a direct connection with your computer via USB 3 or higher.

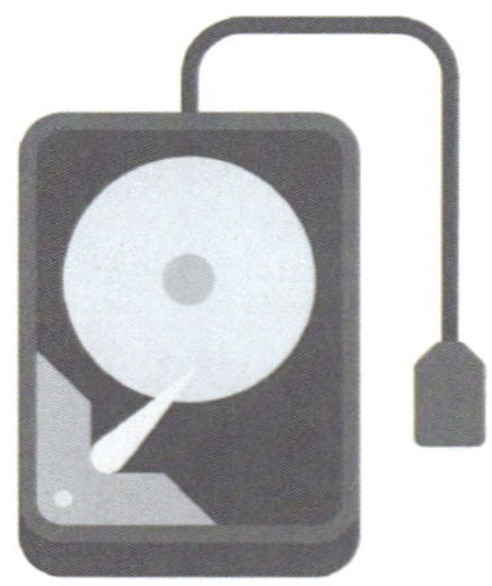

SSD

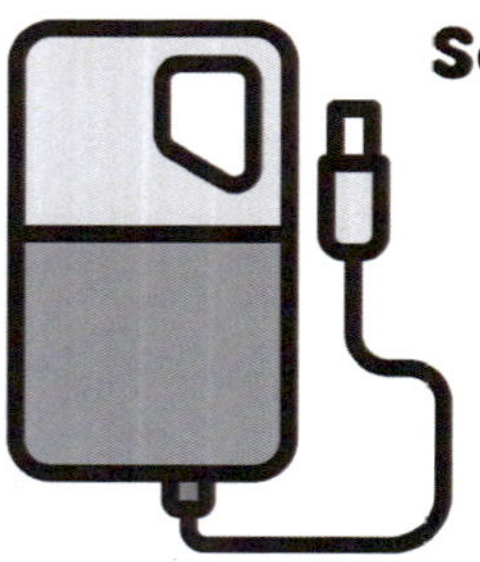

Solid State Drives are the "race car" of the hard drive world. They are extremely fast, but storage can be relatively expensive compared to other hard drives. They are very energy efficient and aren't as susceptible to heat, making them very reliable.

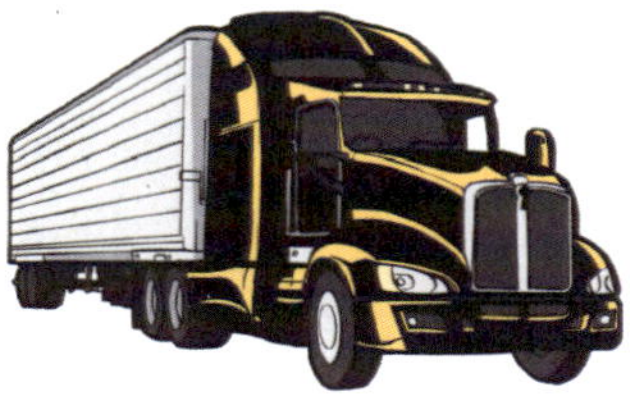

Slow Data Storage

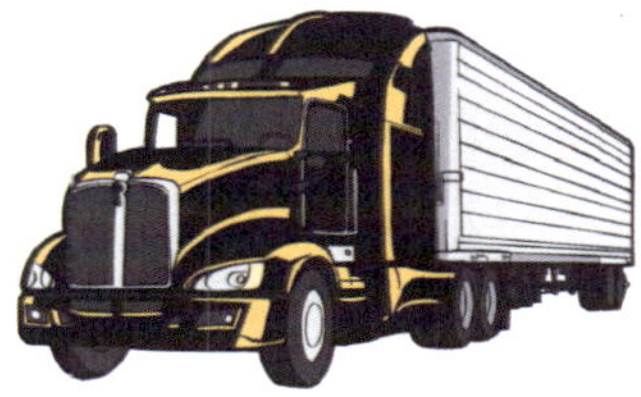

"Backup Hard Drives"

The purpose of a backup drive is for archiving data. While most types of hard drives can store data, backup drives should offer high amounts of dependable storage capacity. Speed is not a top priority. Instead, it is crucial to store your backup drives in a secure, weather-protected location to prevent any potential problems.

Hard Disk Drive

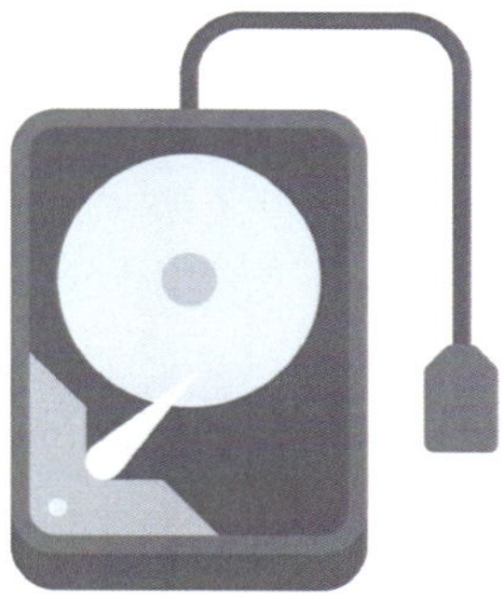

The majority of storage drives are basic Hard Disk Drives, providing a reliable storage capacity ranging from 1 to 10TB. These drives are cost-effective and typically function well for most purposes.

Online Storage

Another excellent choice for backing up your data is an online or "Cloud-Based" storage service. This storage solution is convenient as long as you have basic internet connectivity.

Tip* Always anticipate equipment failures. Redundancy is key. The more locations you have your data backed up, the more secure you are against data loss.

Analog Audio Cables

XLR

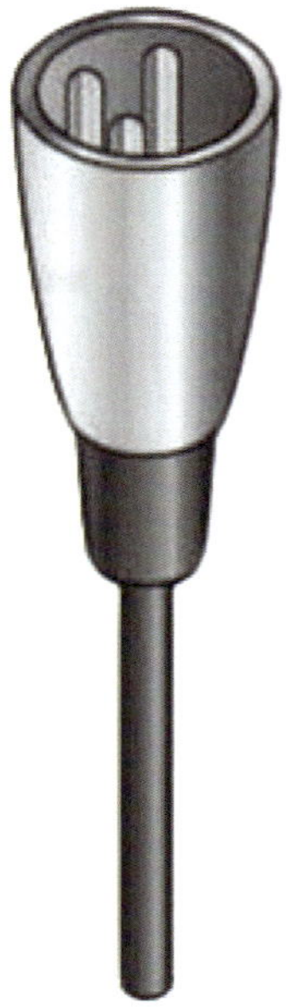

"An XLR cable is a balanced audio cable frequently used for microphones and pro audio devices..."

TS

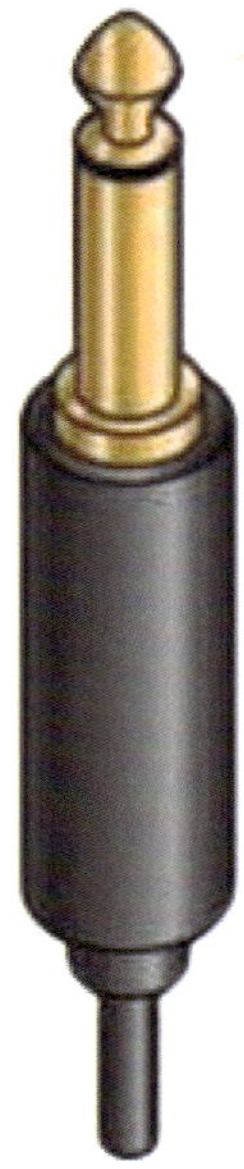

"A Tip/Sleeve or 'TS' cable is an un-balanced audio cable frequently used for instruments such as guitars and keyboards..."

TRS

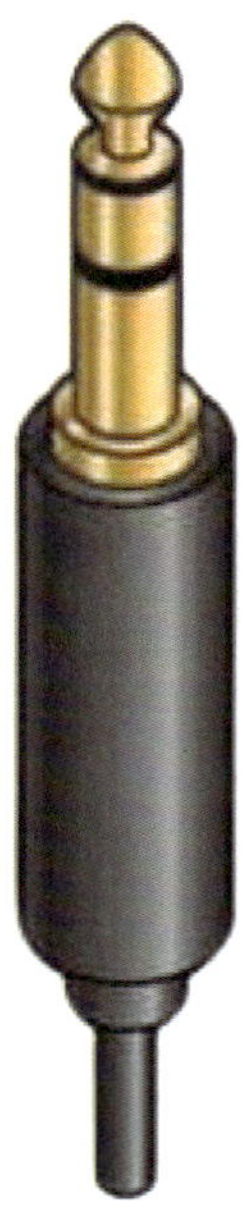

"A Tip/Ring/Sleeve or 'TRS' cable is a balanced audio cable frequently used for line level pro audio devices..."

RCA

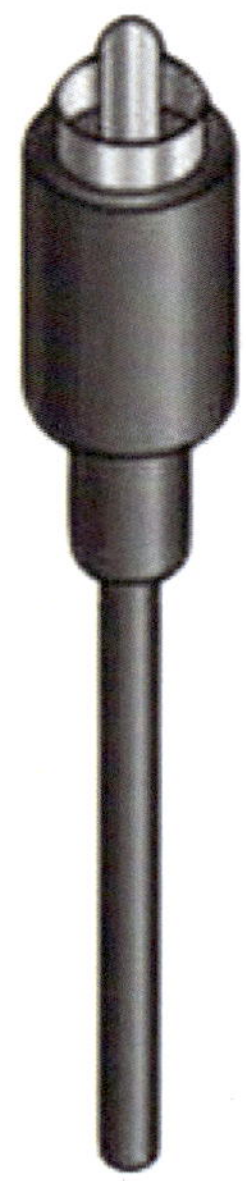

"An RCA cable is an un-balanced audio cable frequently used for consumer audio devices and DJ turntables..."

Un-Balanced Audio Signal

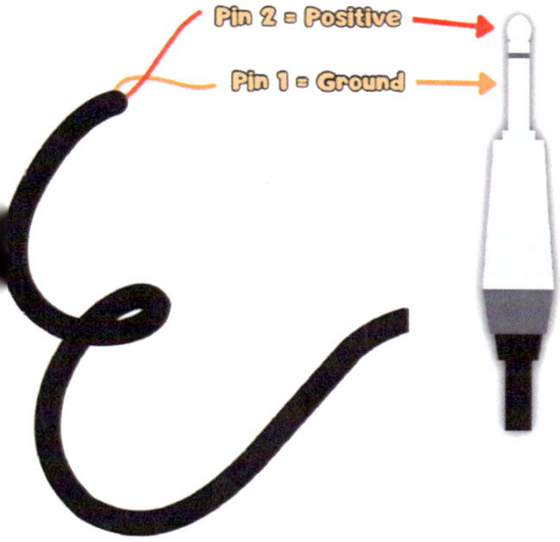

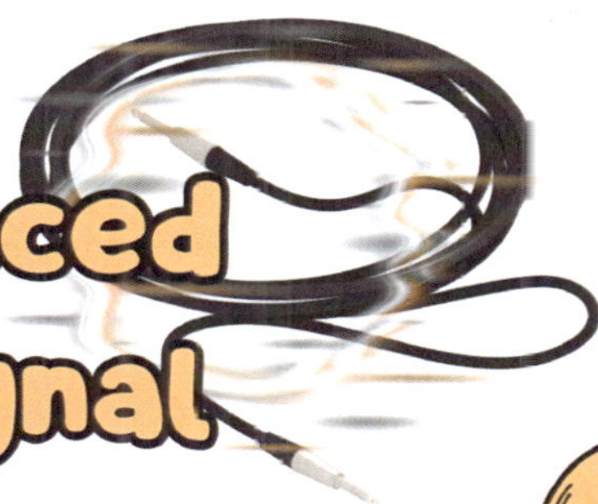

Cables behave similar to a broom sweeping across the floor... just as the broom accumulates dirt as it travels, cables accumulate electronic noise and interference...

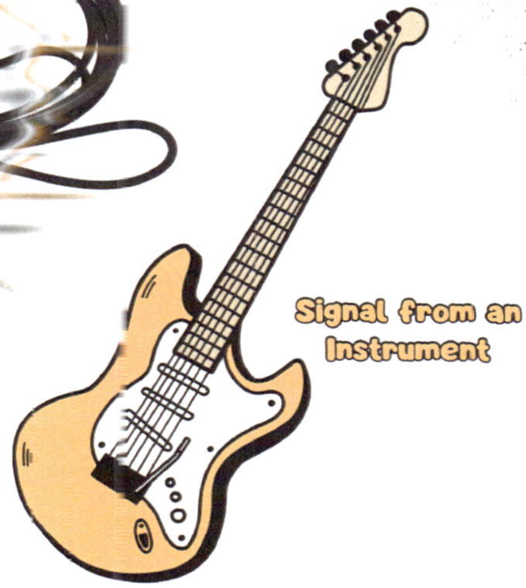

Most often used with instruments and consumer audio equipment, unbalanced cables consist of one conductor and a ground cable. Their design makes them prone to picking up noise and interference. To reduce this unwanted noise it's advisable to keep the length of an unbalanced audio signal to no more than 15 feet.

During Cable Run

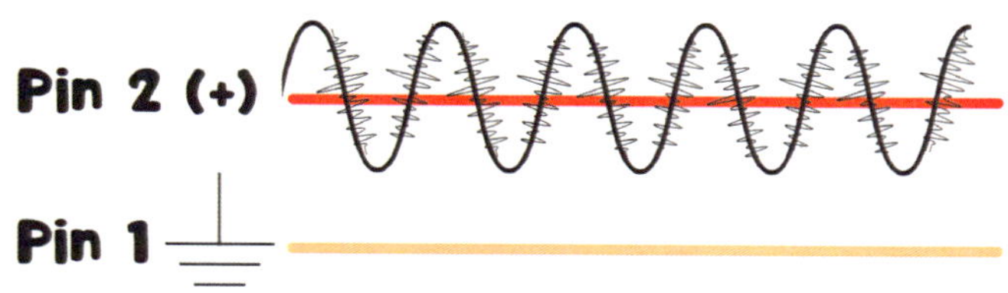

An un-balanced audio signal is prone to picking up electronic interference. The noisier the environment, the greater the amount of unwanted noise that will interfere with the signal.

End of Cable Run

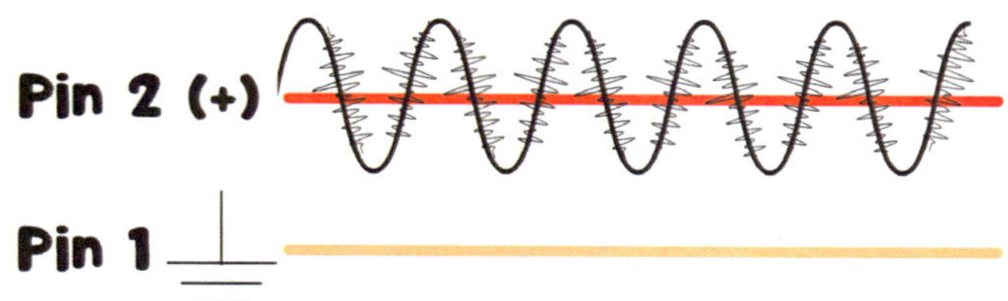

At the termination point of the cable, there is no change to the signal.

Result

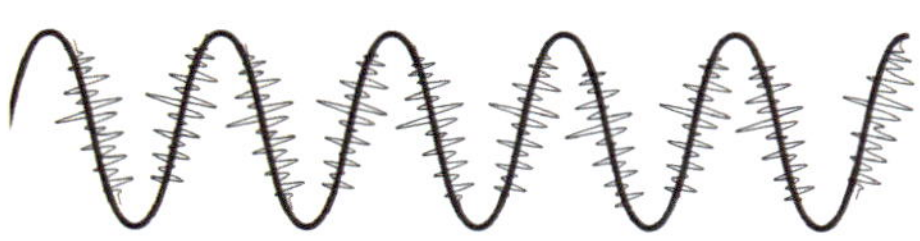

In conclusion, the noise that gathers within the cable run remains in the resulting signal.

Balanced Audio Signal

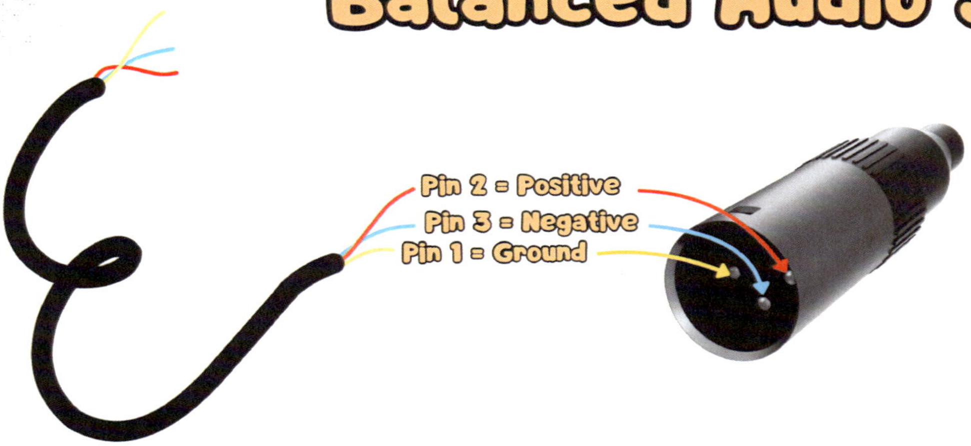

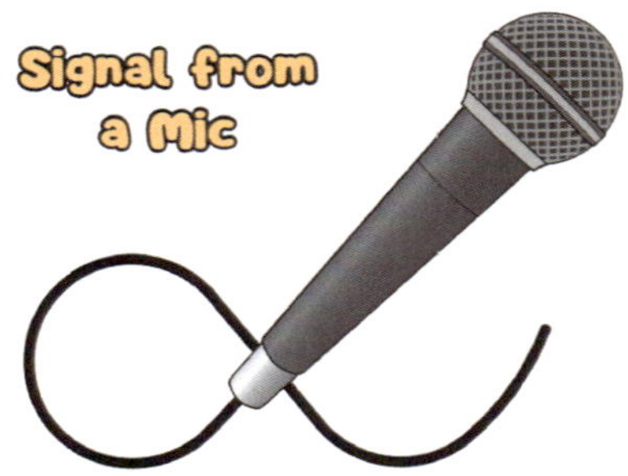

A balanced audio signal is designed to cancel environmental noise accumulated within the wire path of an audio signal. By adding a 2nd conductor wire, a signal can utilize a cleaver cancelation trick when the wire reaches its termination point. In short, a balanced audio approach allows an audio signal to be sent long distances without the risk of accumulating copious amounts of noise within the audio signal.

During Cable Run

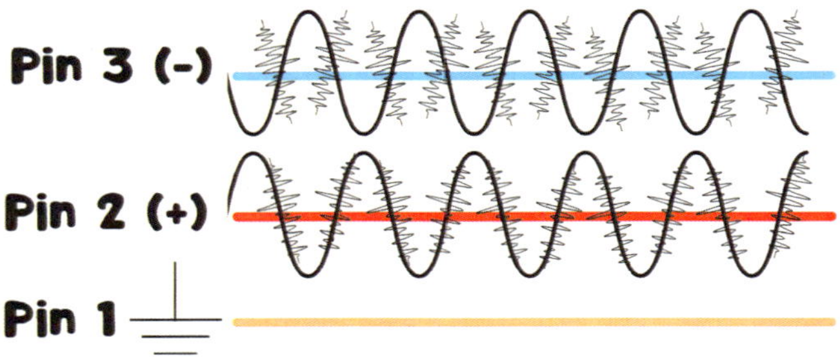

A balanced audio signal transmits two identical audio signals through pins 2 and 3, with an inverted signal on pin 3. Throughout the cable run, the signal remains essentially silent, while it accumulates noise and interference.

End of Cable Run

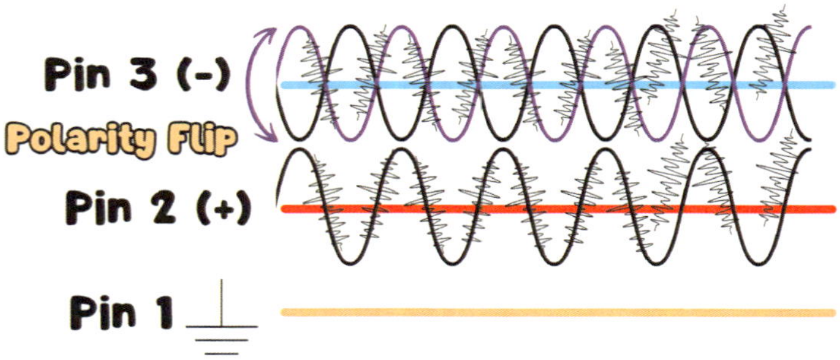

When a balanced cable terminates to a balanced input, a polarity flip is applied to pin 3, allowing the signal to sum "in-phase" whilst flipping the noise "out of phase".

Result

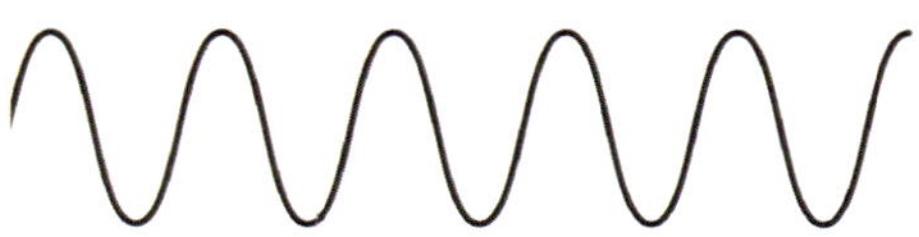

Consequently, the signal cancels interference and receives a +6dB boost to the original signal.

Impedence

Impedance is an essential concept in electrical engineering that plays a crucial role in audio equipment as well. Measured in ohms, impedance refers to the opposition of electrical current within a piece of equipment. In other words, the inputs and outputs of audio equipment can vary in resistance, which creates alterations in volume and tone. It's best to match high impedance outputs with high impedance inputs, while matching low impedance outputs to low impedance inputs. With out the use of a transformer, miss matching impedance can result in signal and tonal loss.

High Impedance
"Hi-Z" Output
(Instruments)

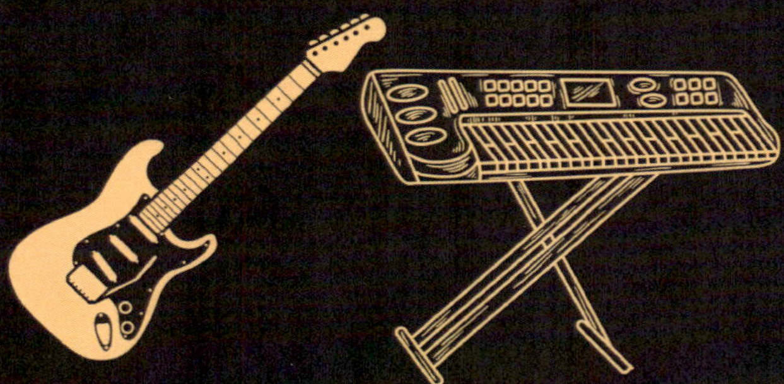

Low Impedance
"Lo-Z" Output
(Pro Audio)

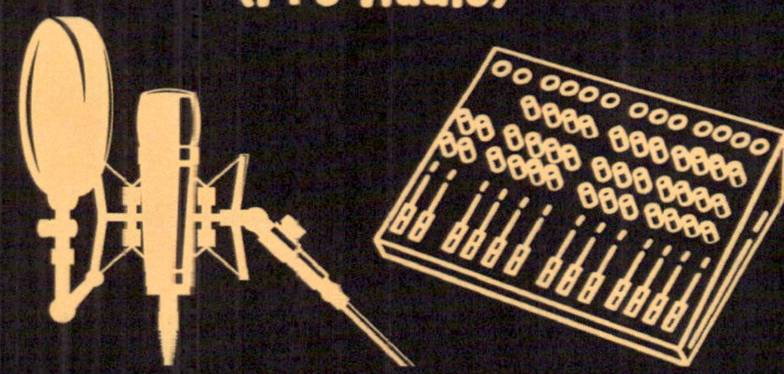

Utilize a "direct box" to transform a Hi-Z signal into a Lo-Z signal...

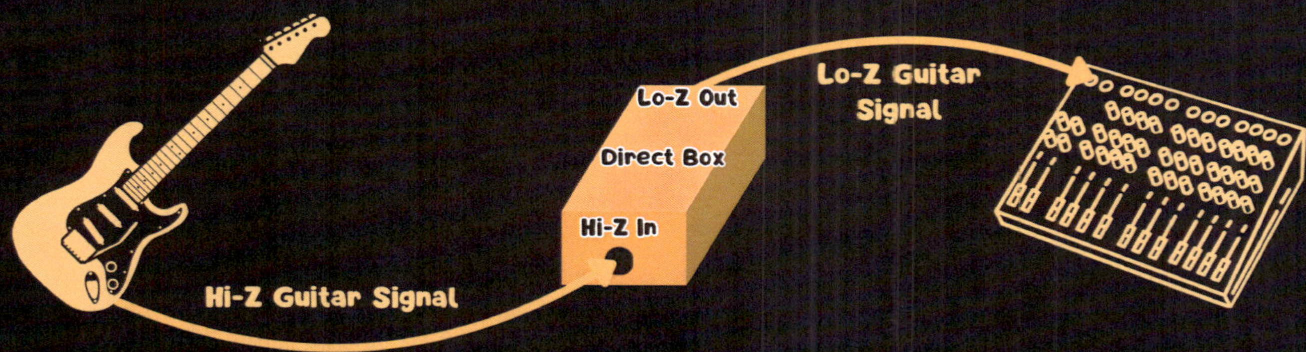

Direct Injection "DI" Box

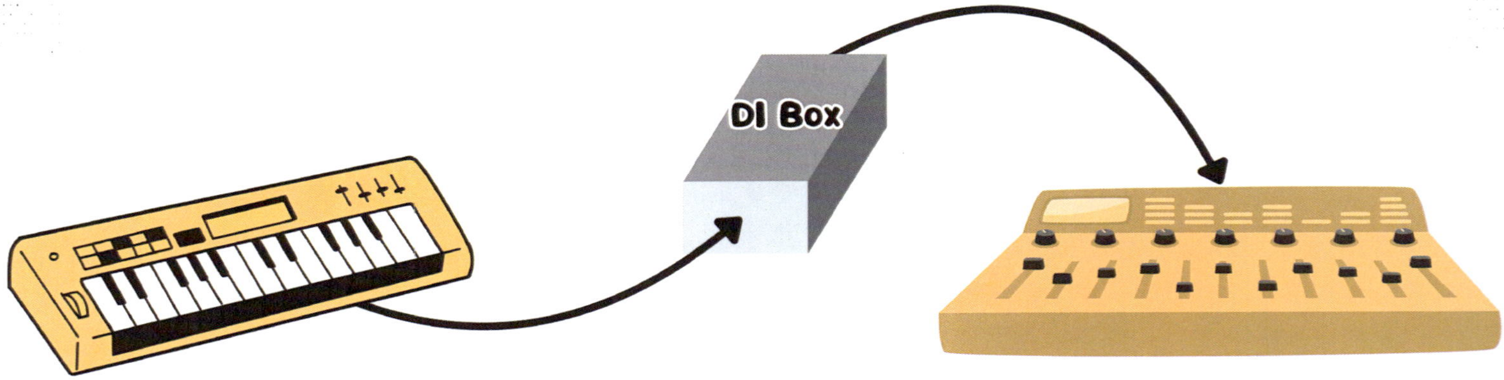

When working with direct audio signals such as keyboards, guitars, or audio from headphone jacks, it's best to use a direct box. There are many variations, but the main function of a DI is to transform the audio signal from high to low impedance while balancing the incoming unbalanced signal. This helps maintain a "clean / full volume" signal while transmitting audio over long distances. In the occasion that a mixer channel has a Hi-Z input, the use of a DI is not required if the cable length is no greater than 15 feet.

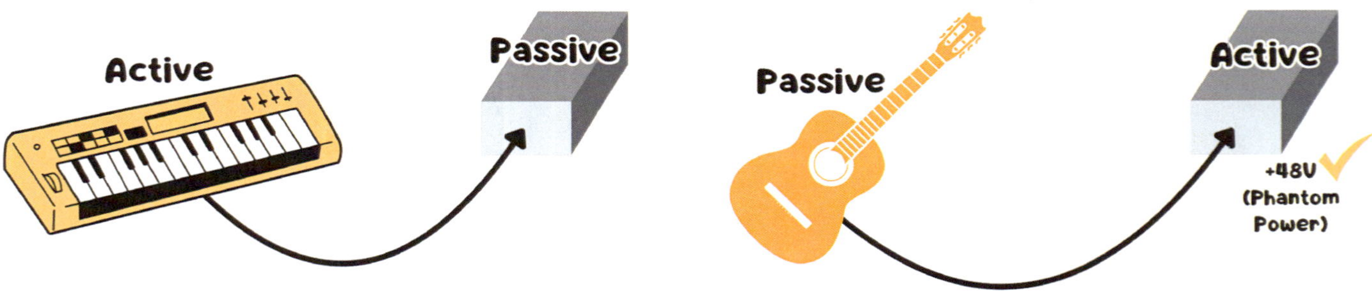

For best results, remember to pair passive DI boxes with active instruments, while pairing active DI boxes (requiring +48V phantom power) with passive instruments. A good way to know if an instrument Is utilizing active electronics is to check for a power supply or battery. This is not to be fooled with a volume knob, as passive instruments may still utilize attenuation knobs for volume.

Powered Vs. Non-Powered Speakers

Regardless of whether the amplifier for a speaker is located within the speaker enclosure or externally, it's essential to distinguish between the two primary methods of delivering amplified audio signals to the speakers. Notice the "powered" speaker only requires a line level signal over a standard XLR or TRS cable. This is because the amplifier is pre-wired and housed within the speaker inclosure. For ease of use, these amps are specifically chosen by the manufacturer, and offer features such as a power button, volume adjustment, and tone controls. On the other hand, amplifiers can exist outside the speaker, creating a "non-powered" speaker configuration. With this approach, speaker wires are connected between the amp and speaker by the user, and a deeper knowledge of proper speaker amplification is required.

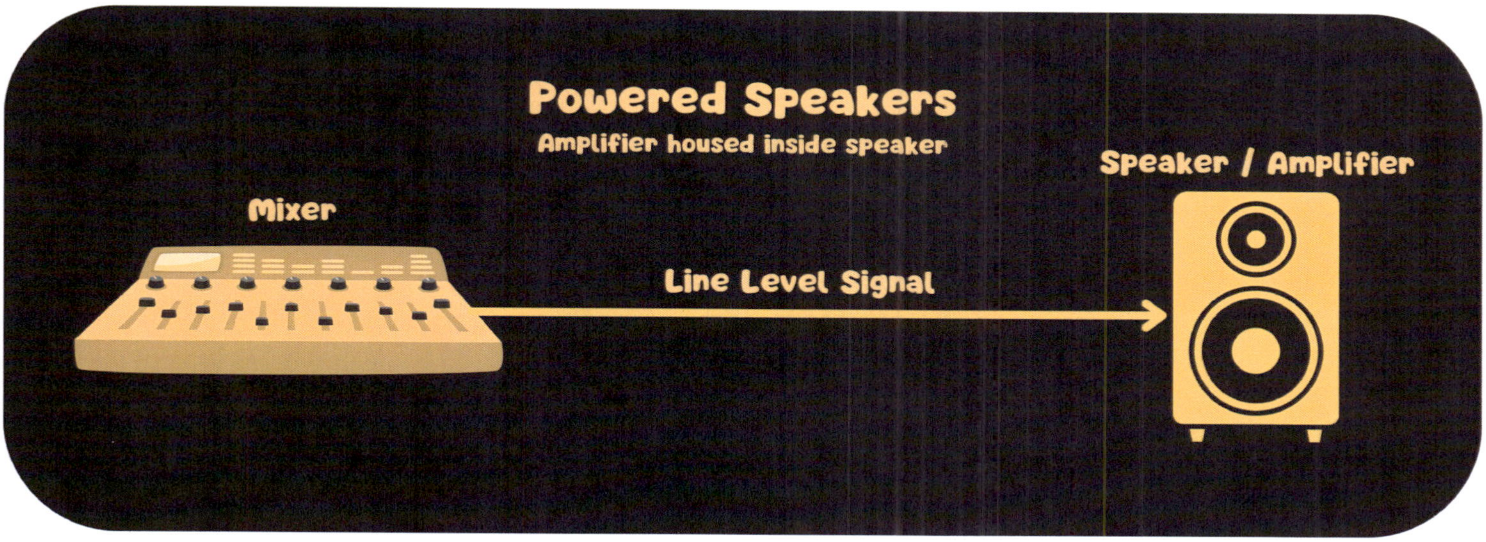

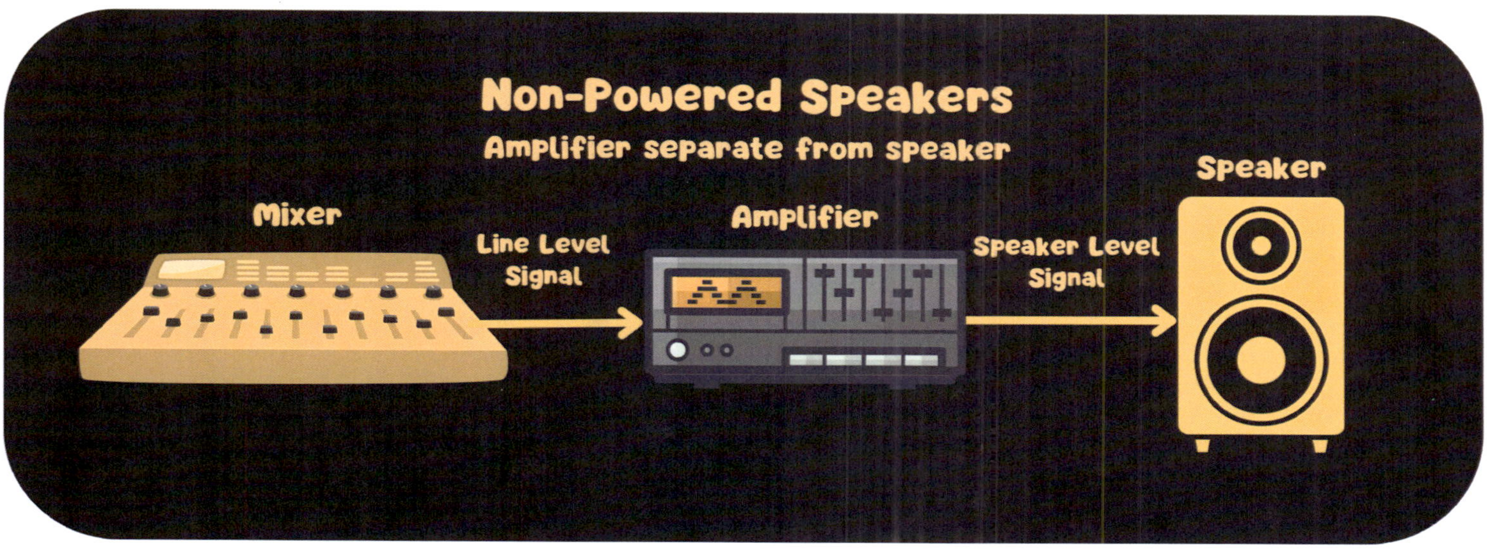

Continue to the next page for a deeper look at speaker amplification...

Matching Amplifiers with Non-Powered Speakers

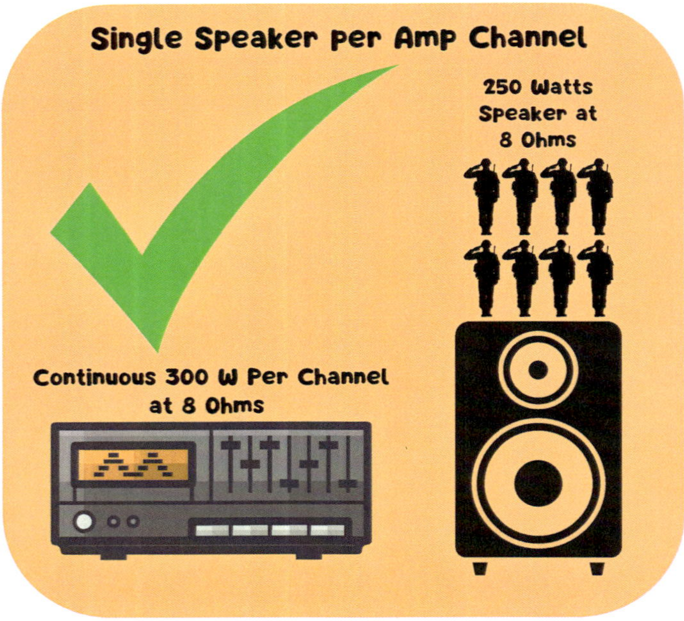

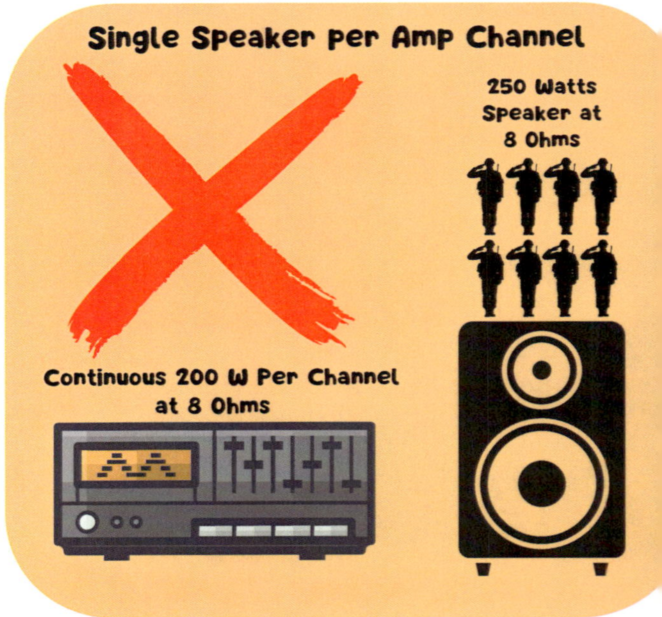

To prevent damaging the amplifier, make sure the amp can easily power the speaker while matching the impedance load. For best practice, it's better to have a little extra power (around 20% more wattage) than the manufacturer "continuous power" rating required to properly power the speaker. Furthermore, under powerin a speaker with a below spec "continuous" rating will create damage to an amplifier in time while hindering speaker performance. Lastly, if an amplifier is too powerful and far accedes the capacity of the speaker, damage to that speaker is likely to happen as well. However, adjusting or limiting the line input of the amplifier can help prevent damaging volume levels for the speaker components.

Tip*
Always evaluate an amplifier or speaker's performance based on its functionality rather than the name alone. An amplifier may have "2000W" or "3000W" in its title, but this typically represents "peak" wattage rather than "continuous" wattage. Essentially, while amplifiers often advertise their peak wattage in their names, the continuous performance is what truly counts in real-world applications.

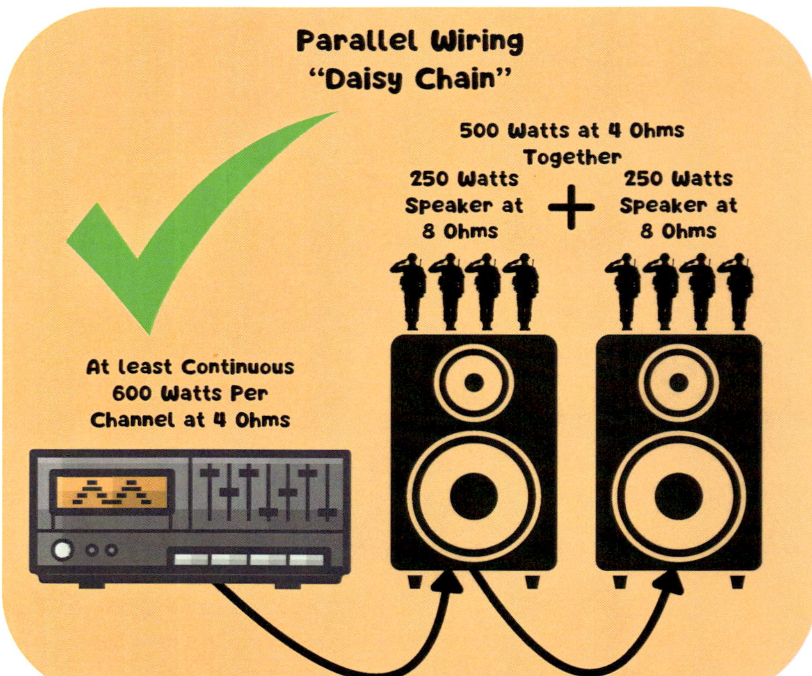

In situations where dais chaining non-powered speakers is necessary wiring them in paralle will divide the speaker ohm rating by the tota number of speakers while multiplying the wattage. In this exampl it would be crucial to verify the amplifier specifications for the new "4 Ohm" impedanc load. Keep in mind tha the more speakers yo daisy chain, the harde the amplifier has to work.

The Basic Pro Tools Shortcuts

Navigation Shortcuts

- Scroll Wheel = Scroll Up/Down
- shift + Scroll Wheel = Scroll Left/Right
- T = Zoom In
- R = Zoom Out
- ⌥ option + A = View All
- ^ control + Scroll Wheel = Track Height Up/Down
- ⌘ command + = = Toggle Edit/Mixer View

Track Shortcuts

- shift + ⌘ command + N = New Track
- shift + S = Solo
- shift + M = Mute
- shift + ⌥ option + D = Duplicate Track
- shift + ⌘ command + I = Import Audio

Editing Shortcuts

- A = Trim Right
- S = Trim Left
- D = Fade In
- F = Crossfade
- G = Fade Out
- B = Break Clip
- H = Heal Clip
- Z = Undo
- ⌘ command + D = Duplicate Clip
- shift + ^ control + ↑ ↓ = Clip Gain Up/Down

Record Shortcuts

- shift + R = Record Enable Track
- ⌘ command + (space bar) = Record
- ^ control + \ = New Playlist

*If you are using a different DAW from Pro Tools, discover the equivilant shortcut or workaround.

The Basic Shortcuts Challenge

Navigation Shortcuts

	=	Scroll Up/Down
	=	Scroll Left/Right
	=	Zoom In
	=	Zoom Out
	=	View All
	=	Track Height Up/Down
	=	Toggle Edit/Mixer View

Track Shortcuts

	=	New Track
	=	Solo
	=	Mute
	=	Duplicate Track
	=	Import Audio

Editing Shortcuts

	=	Trim Right
	=	Trim Left
	=	Fade In
	=	Crossfade
	=	Fade Out
	=	Break Clip
	=	Heal Clip
	=	Undo
	=	Duplicate Clip
	=	Clip Gain Up/Down

Record Shortcuts

	=	Record Enable Track
	=	Record
	=	New Playlist

Basic DAW Setup To Do List

- [ ] CREATE A NEW SESSION... NAME IT
- [ ] CHOOSE A 48KHZ SAMPLE RATE
- [ ] CHOOSE A 24BIT BIT DEPTH
- [ ] SAVE SESSION TO YOUR "EXTERNAL DRIVE"
- [ ] BE SURE PLAYBACK IS SET TO YOUR INTERFACE
- [ ] SET THE BUFFER TO 124 SAMPLES FOR RECORDING
- [ ] SET SESSION GRID VALUE TO 1 BAR
- [ ] VIEW "BARS AND BEATS" IN THE TRANSPORT
- [ ] SET SESSION TEMPO
- [ ] CREATE CLICK TRACK
- [ ] IMPORT TRACKS
- [ ] CREATE AUDIO TRACKS
- [ ] CREATE 1 STEREO "MASTER FADER"
- [ ] DOUBLE CHECK INPUTS AND OUTPUTS
- [ ] DOUBLE CHECK BUSSES
- [ ] ROUTE ALL AUDIO TRACKS TO MAIN OUTPUT
- [ ] SAVE

The Acoustic Environment and Monitoring

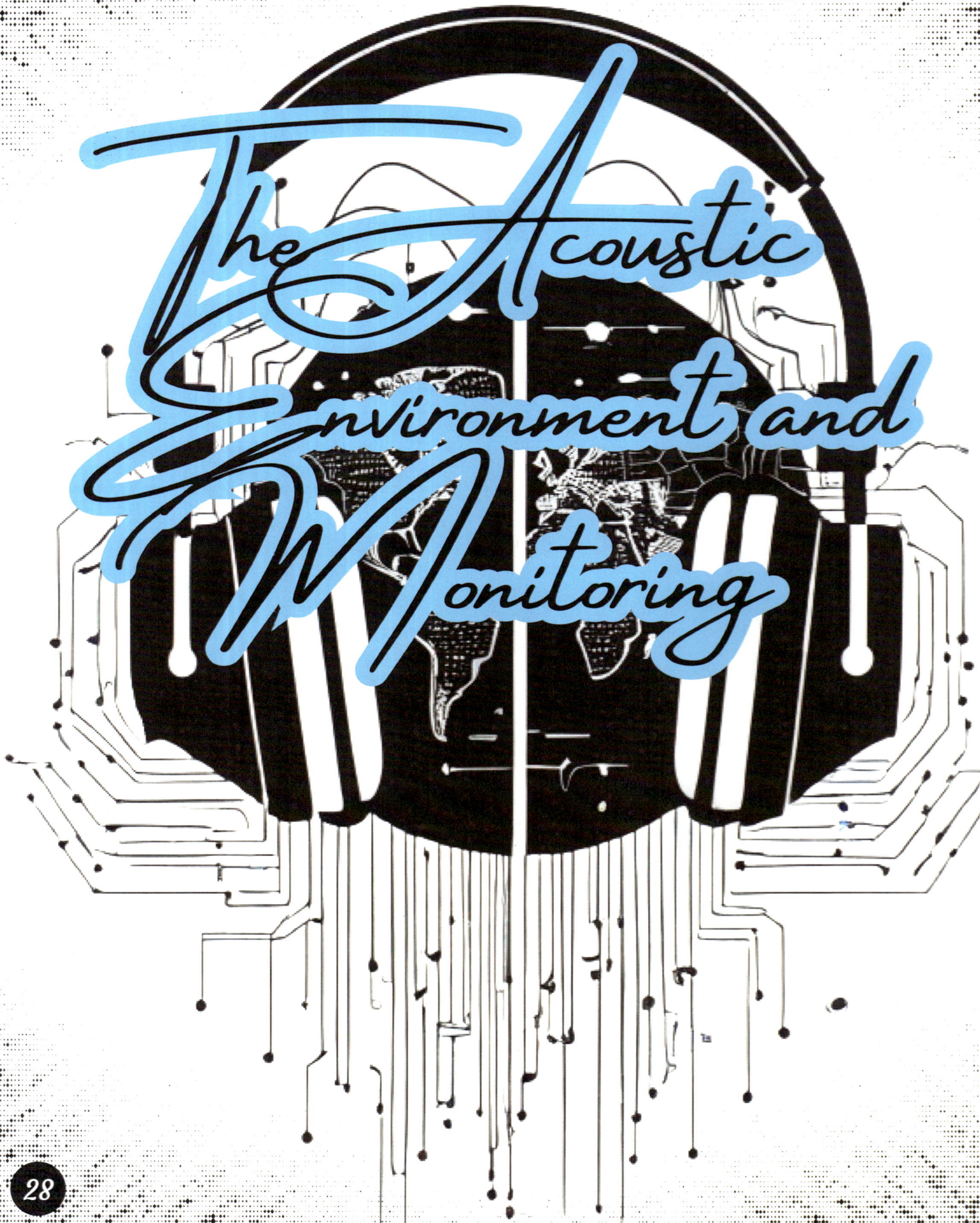

Types of Acoustic Surfaces

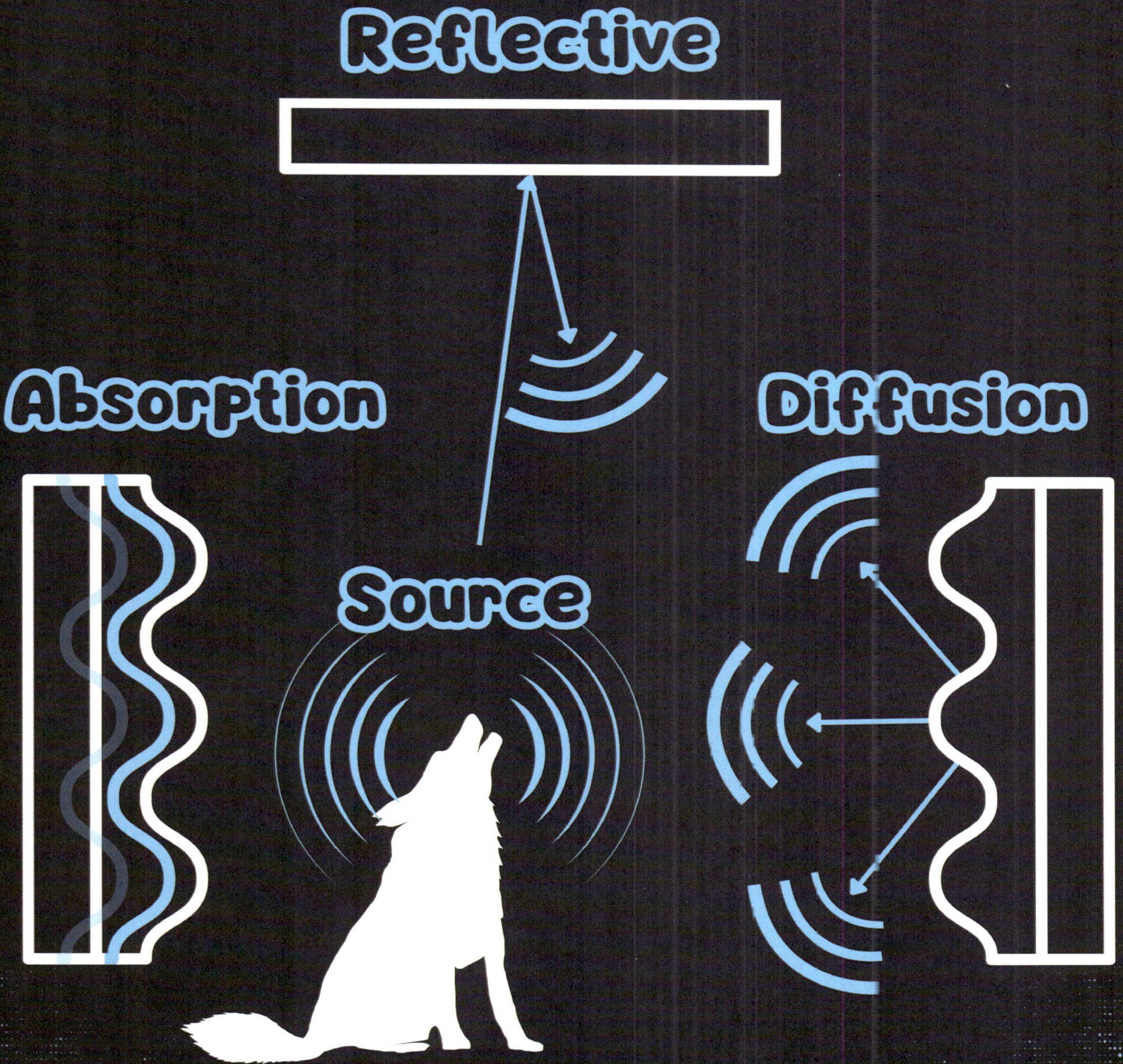

Monitor Position

For a proper mixing position, place your studio monitor tweeters at ear height, while forming an equilateral triangle between both speakers and the back of your head.

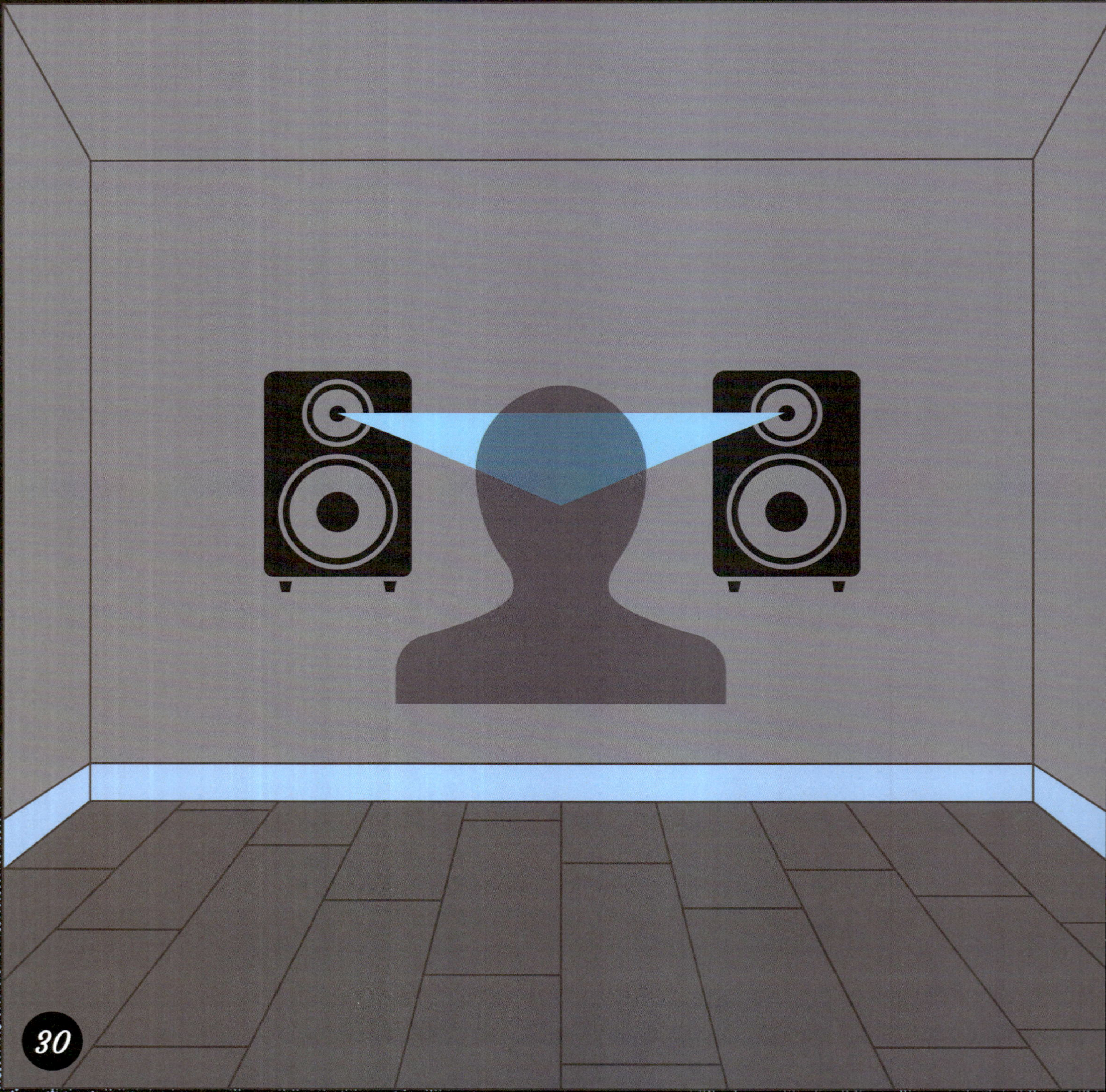

Direct Sound

The term refers to the sound emitted directly from the speakers. Critical listening heavily depends on proper monitoring with direct sound.

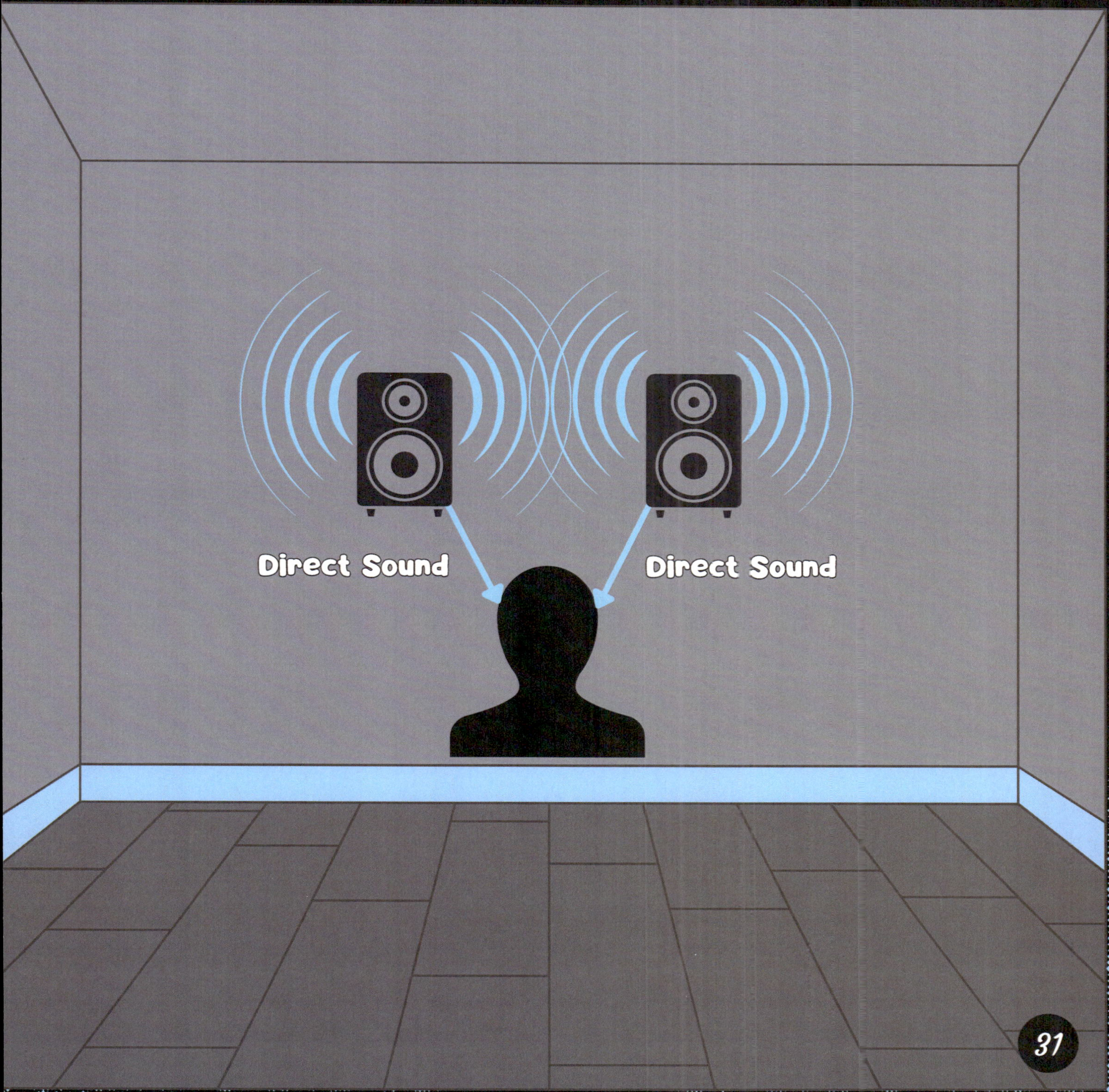

1st Reflections

These points in a room refer to specific locations where sound reflects off a single surface and returns to the listener's ear. Untreated first reflection points can be problematic, creating interference in a critical listening environment. Placing absorption and diffusion in first reflection points can dramatically improve the sound of a room.

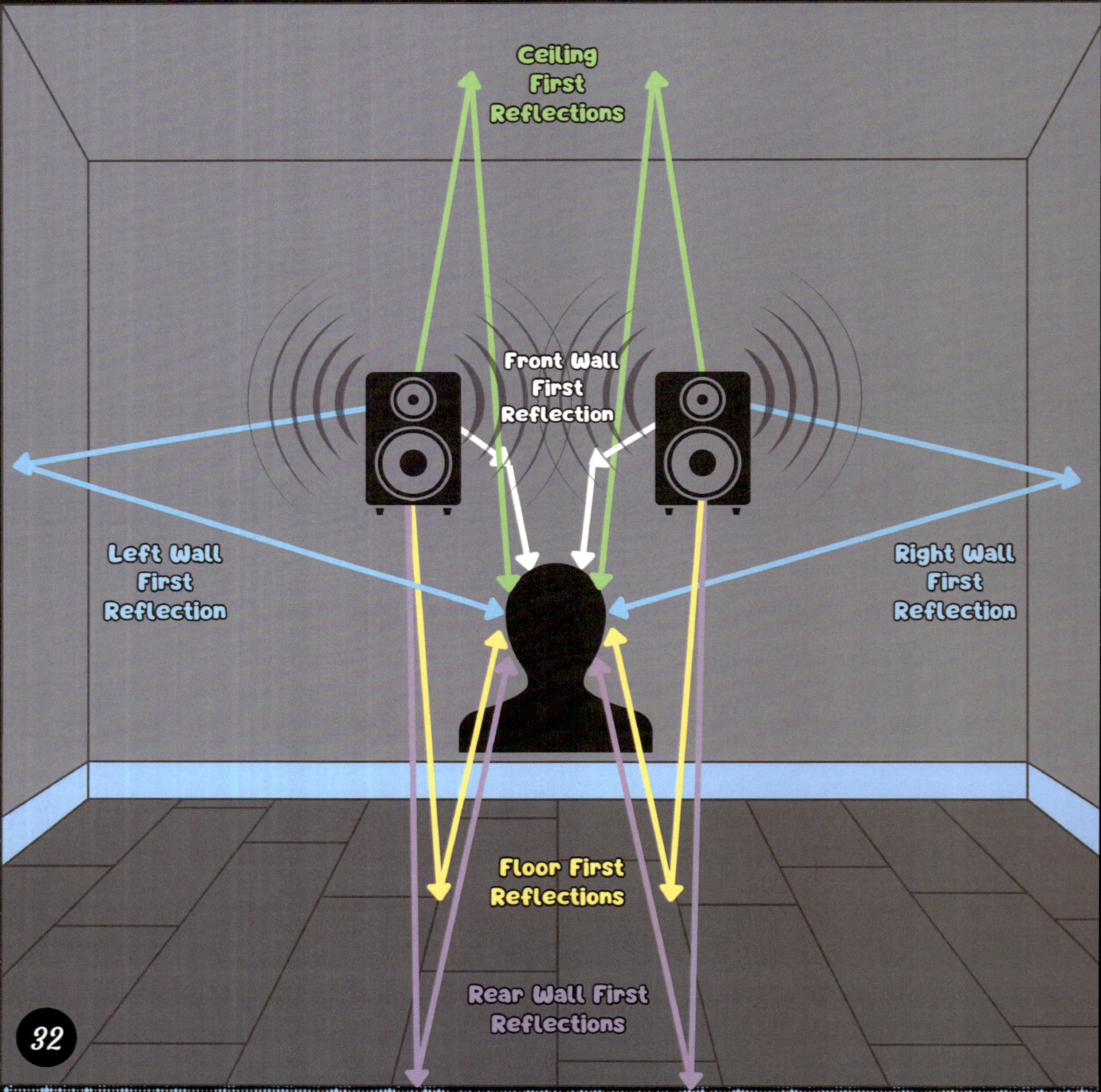

Room Nodes

Reflections in a room will eventually lead to cancelation at particular frequencies in specific areas of a room. These points of cancelation are known as "Nodes".

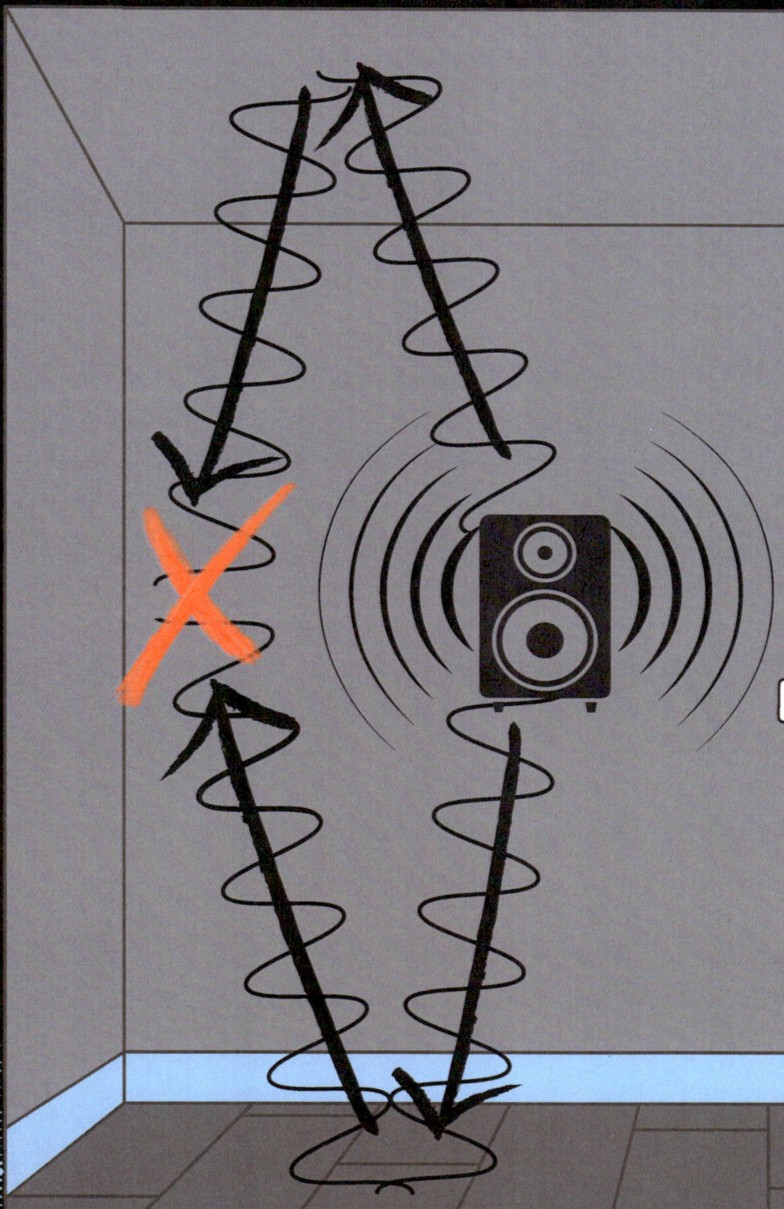

Avoid critical listening or placing clients in these "Node" areas, particularly in spots of a room that lack fullness in the lower frequencies.

Monitoring

There are 3 major obstacles to consider....

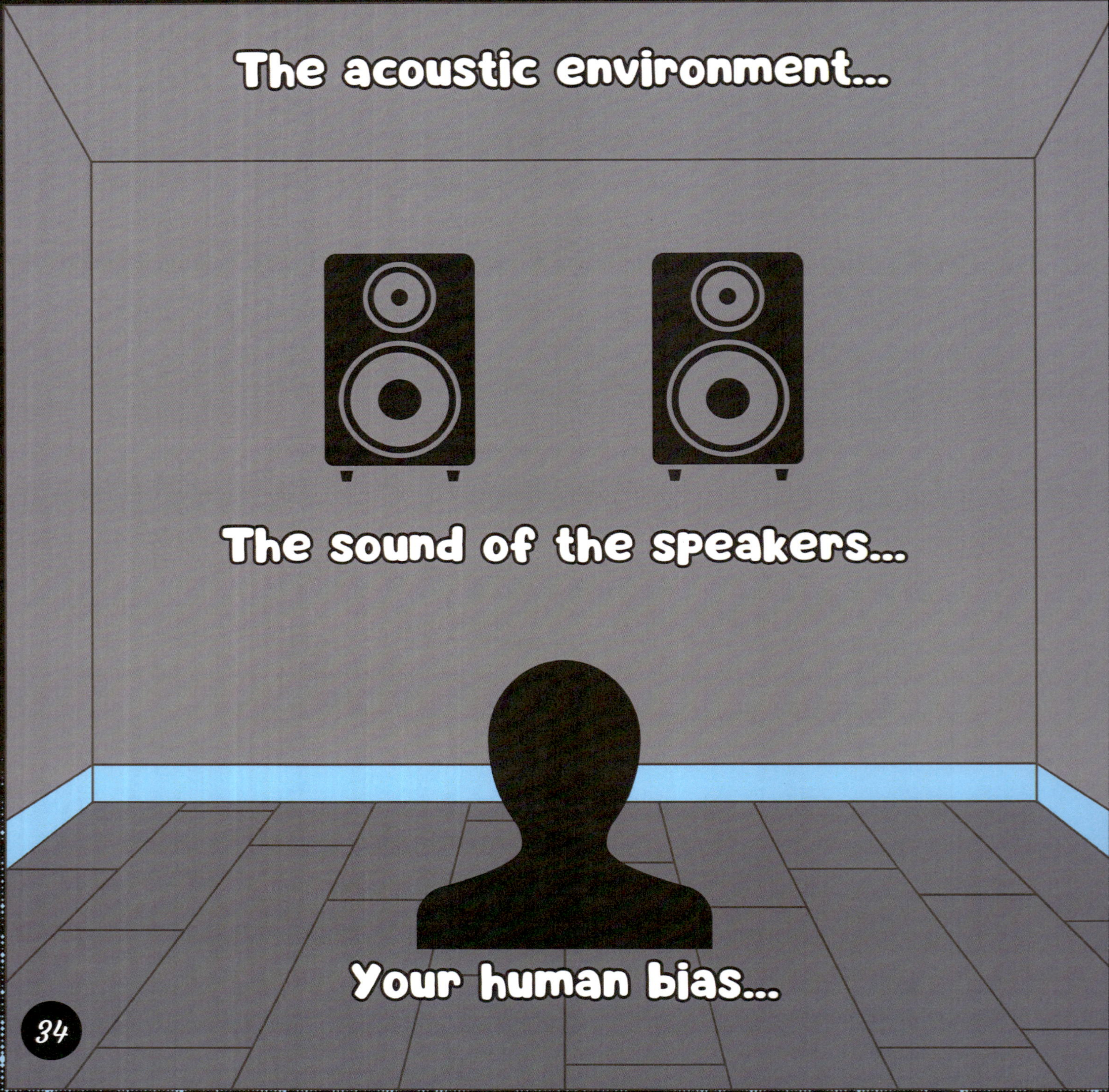

Your Reference Mix

Select three songs that you feel best showcase the frequency spectrum according to your preference. These songs can be personal favorites or tracks that exhibit a frequency balance you admire. For instance, you might say, "this song has just the right balance of bass." Listening to these songs on a sound system can assist you in acclimating your ears to the listening environment. The objective is to adjust and calibrate the system so that your reference songs sound the way you know they sound....

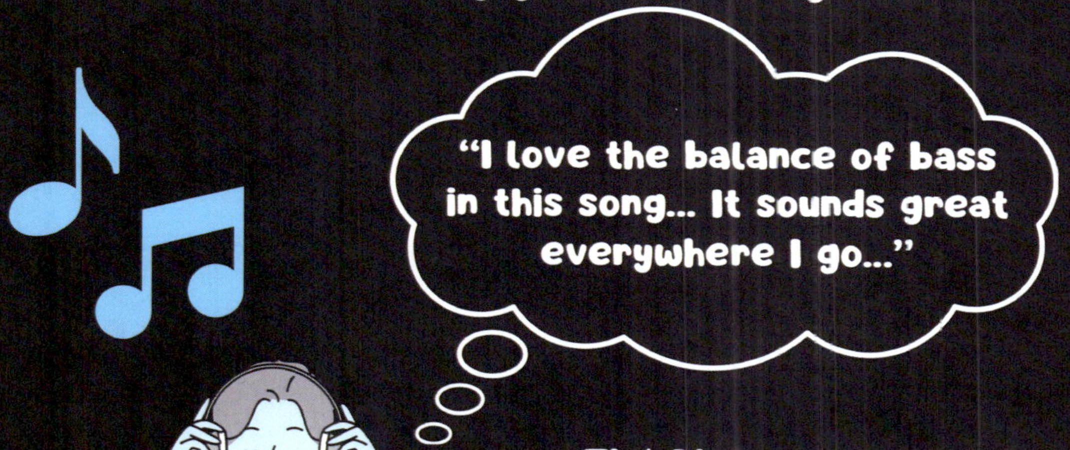

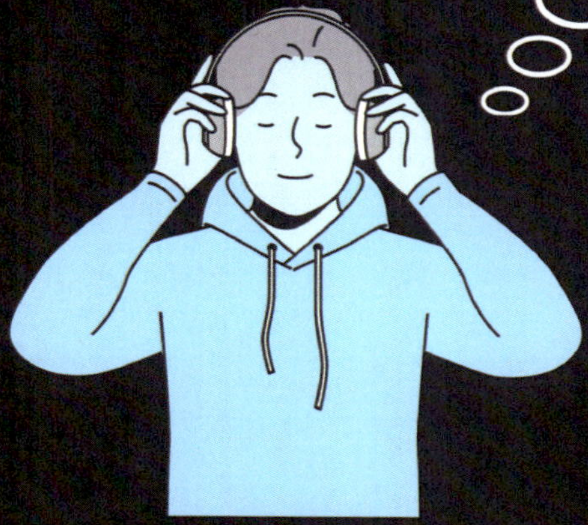

Tip* Pick songs that challenge the extremes of certain sonic qualities.. These songs make it easier to expose potential problems in speakers or rooms...

Tip: Several well-known reference songs are often glamorized by engineers. Yet, if you are not familiar with these songs, they may not be effective as references.

Applying your Reference Mix

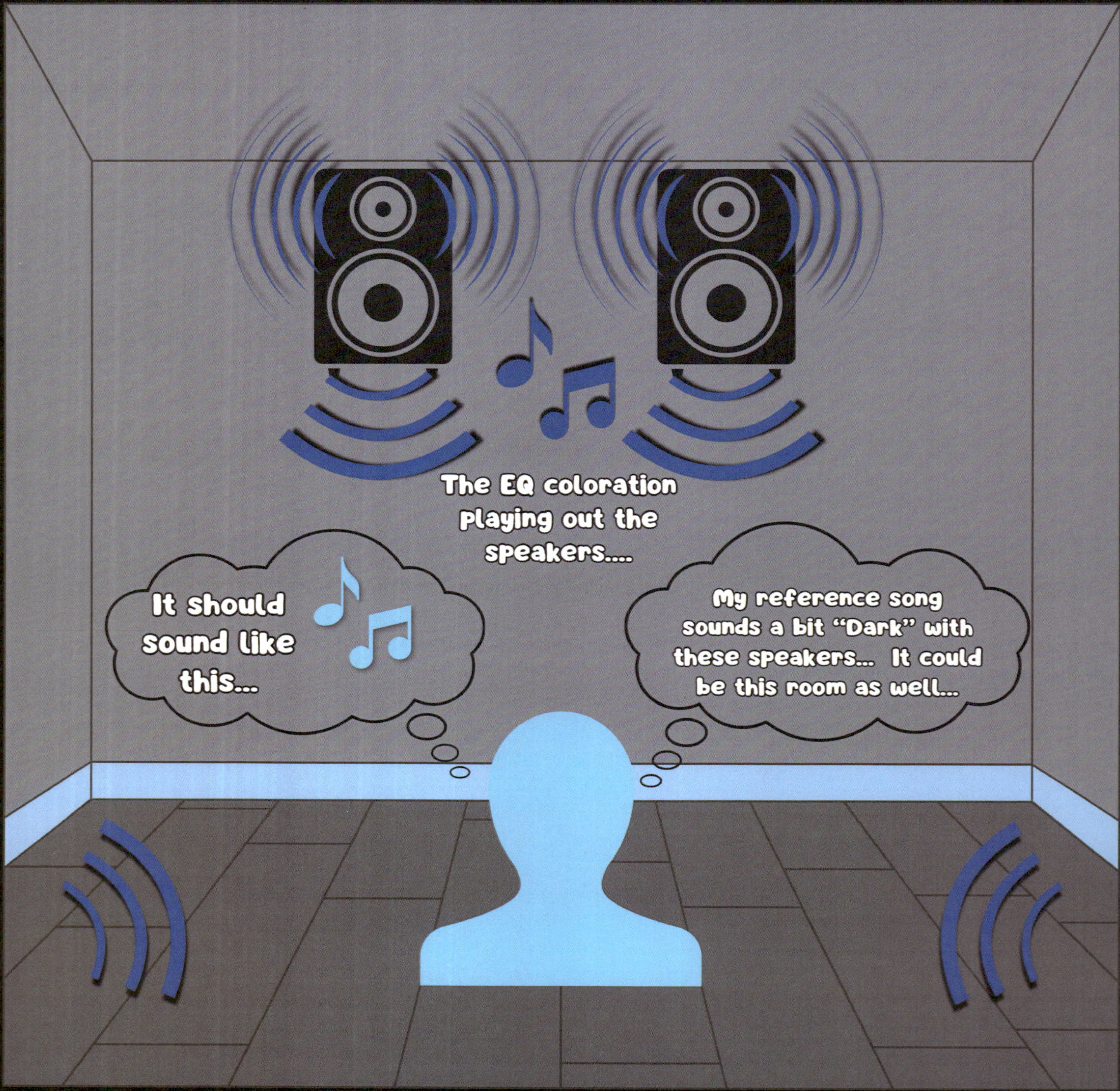

Note: The engineer is noticing a lack of brightness in the room with the reference mix. Mixing in this environment may cause this engineer to overcompensate high frequencies in the mix.

Headphone Monitoring

Listening with headphones is an excellent method to eliminate room interference issues. Moreover, it adds flexibility to your workflow and reduces the need for a flawless environment when making important mixing choices.

Taking that into consideration, it's important to note a few things when using headphones. When using speakers, both ears hear sound from both speakers. However, when using headphones, the left ear only hears the left speaker, and the right ear only hears the right speaker. This distinction can affect how you perceive the depth and imaging in your mix. Lastly, headphones are prone for fatigue due to the physical pressure on the head.

Headphones

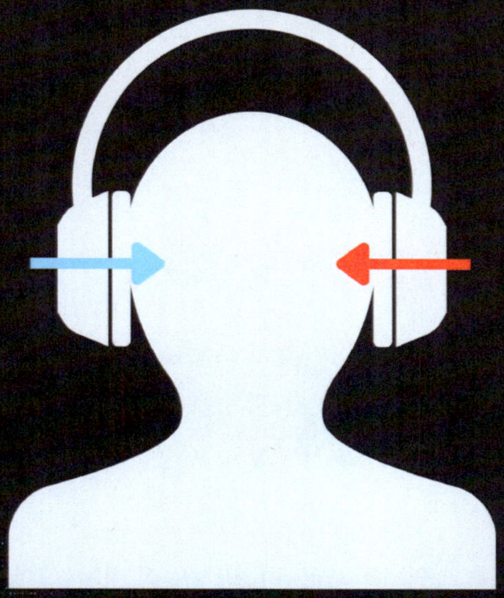

Speakers

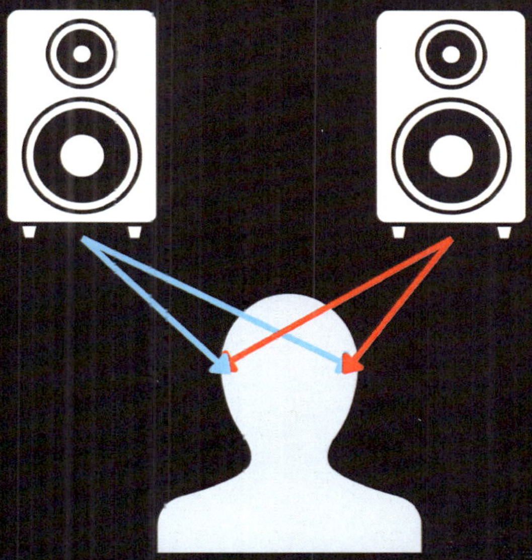

Microphones

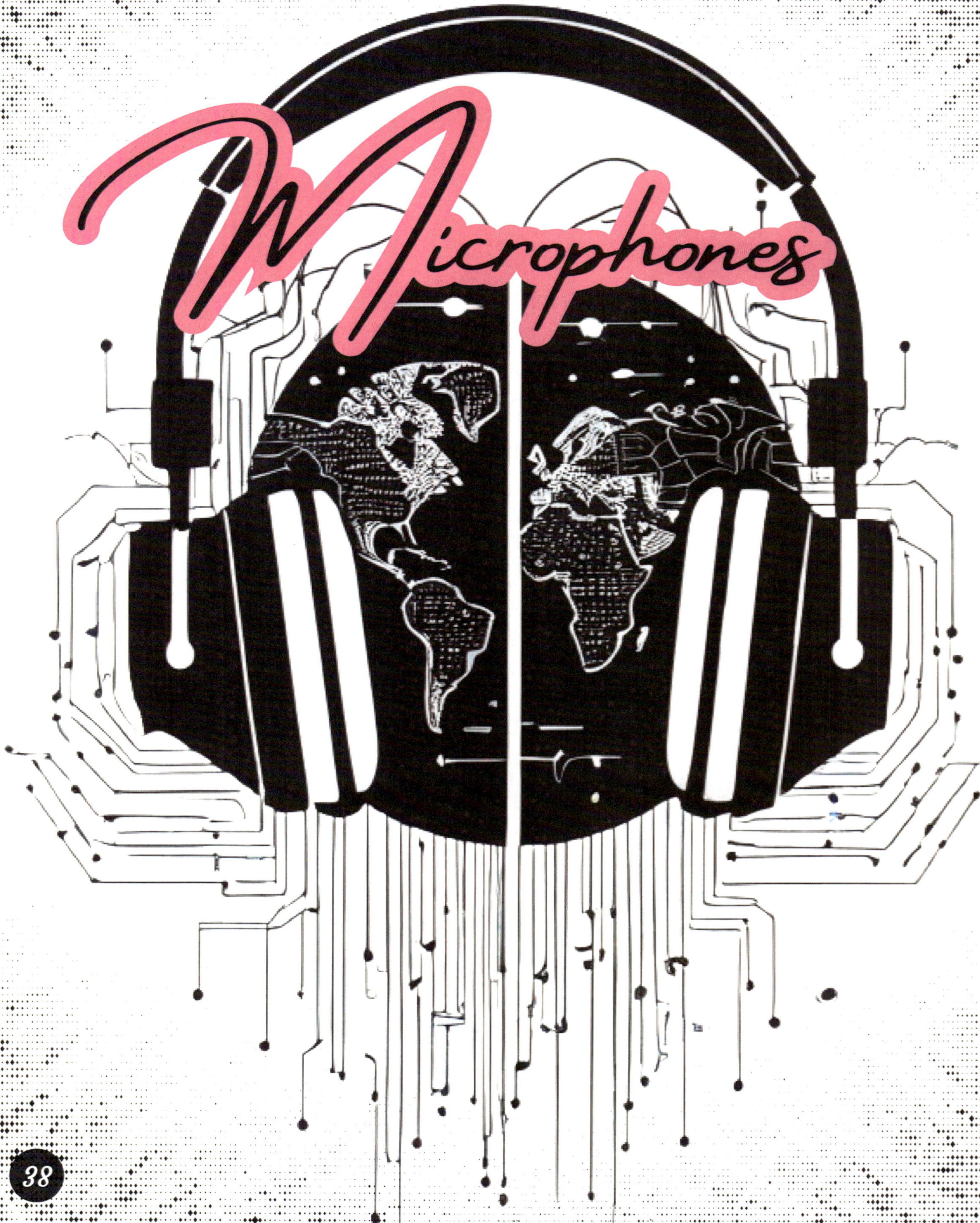

3 Types of Microphones

The primary types of microphones are...

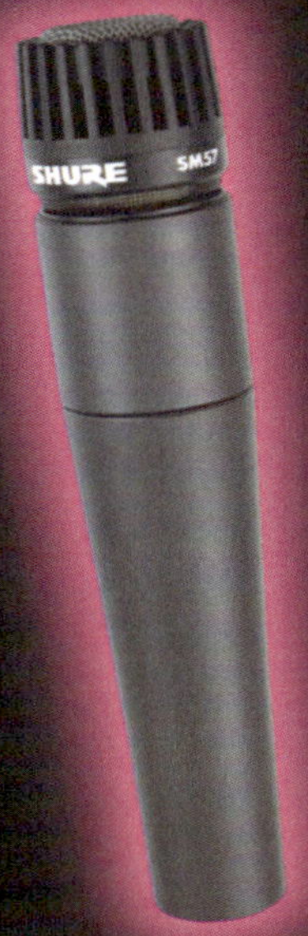

Dynamic

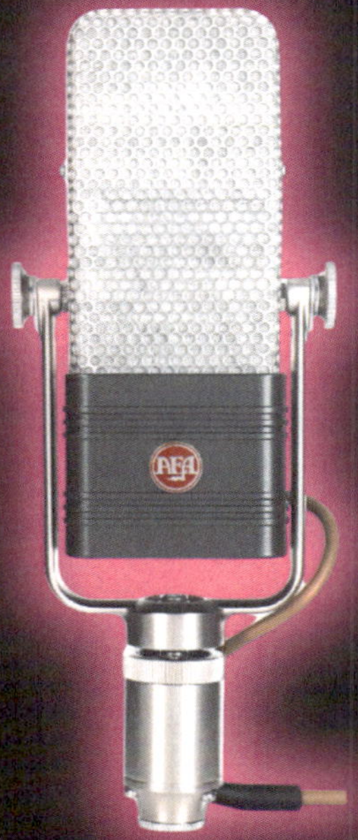

Ribbon

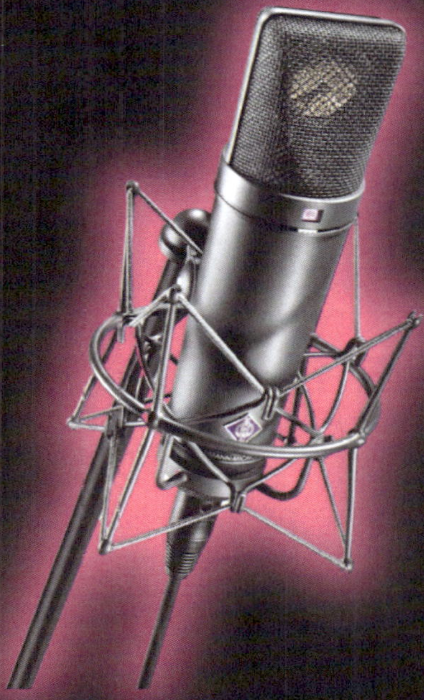

Condenser

3 Types of Polar Patterns

The primary types of polar patterns are...

Cardioid

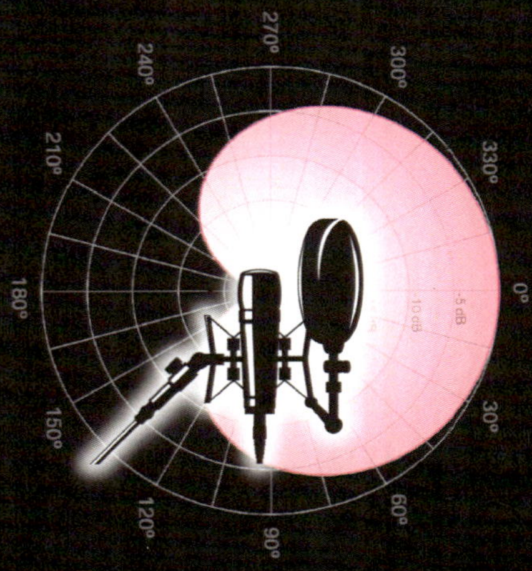

A cardioid pattern enables a microphone to capture sound primarily from the front position, while effectively minimizing sound from the sides and rear.

Bi-Directional

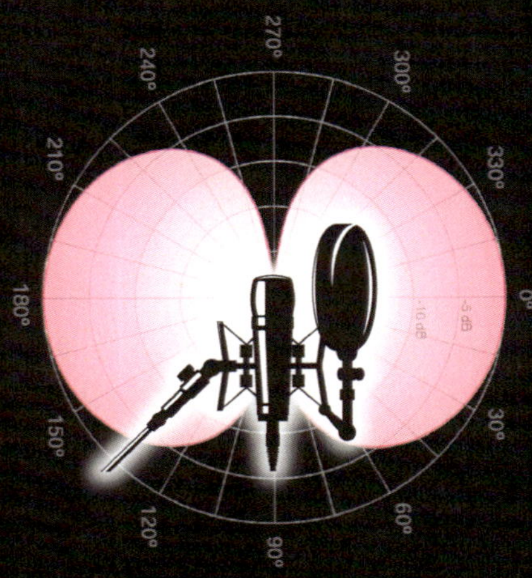

A Bi-Directional or "Figure-8" pattern allows a microphone to capture sound primarily from the front and rear position, rejecting sound from the side.

Omni-Directional

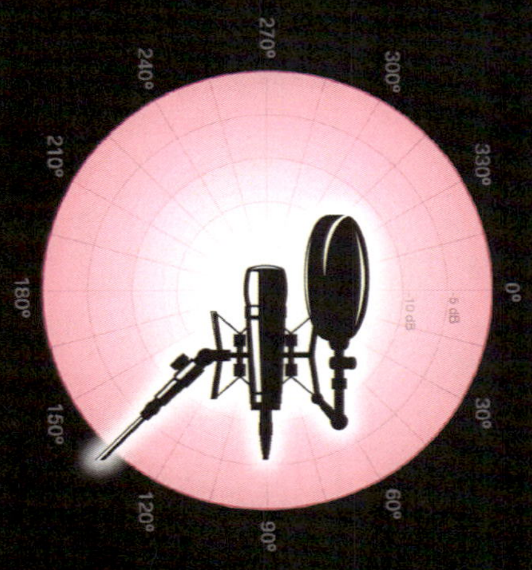

A microphone featuring an omnidirectional pattern will equally capture sound from all directions, front, sides, and rear.

Tip: Considering the level of ambiance or isolation you wish to achieve can help inform your decision when selecting a polar pattern.

Dynamic Mirophone

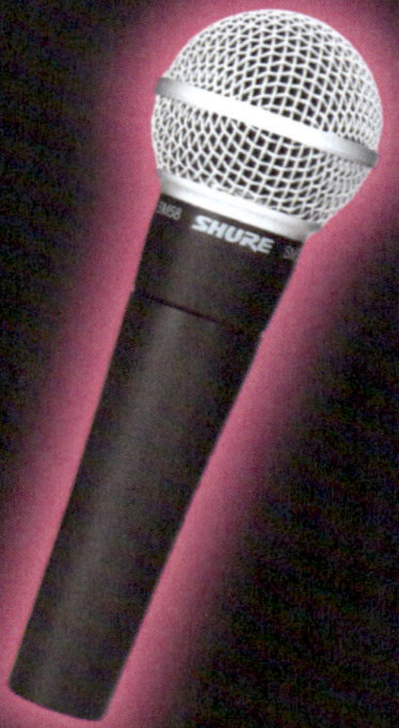

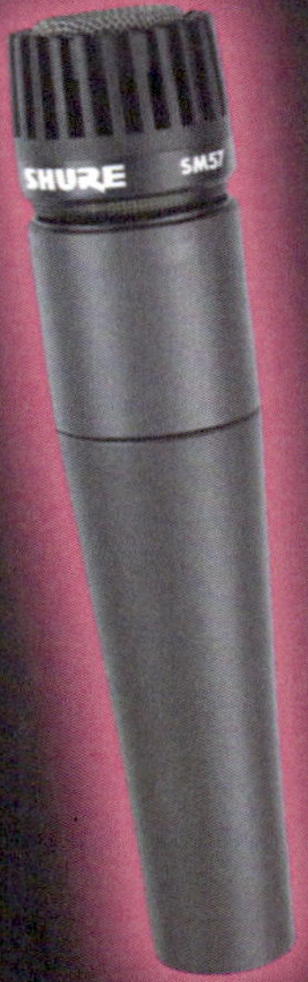

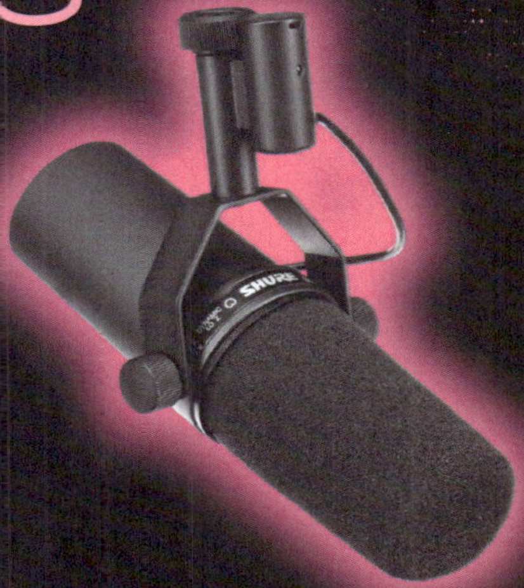

- Great for Isolation
- Dependent on Close Proximity

- Mid Range Focused
- Rugged
- Relatively Inexpensive

Dynamic microphones are known for their excellent sound isolation, which requires close proximity to achieve optimal results. Their frequency response is primarily focused on the midrange of the spectrum, resulting in reduced effectiveness at the lower and higher ends. With a robust design and affordability, these microphones are widely accessible in both studio and live applications..

Ribbon Microphone

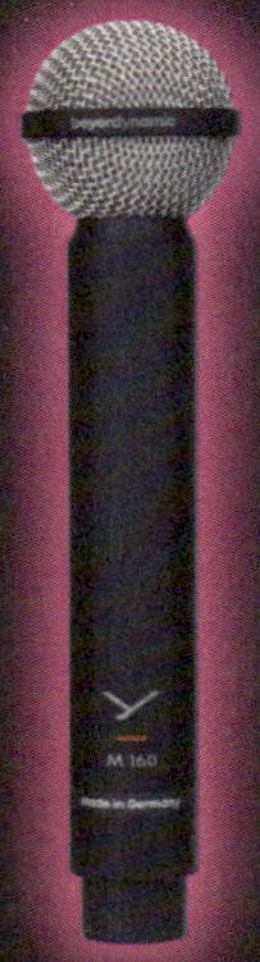

- Smooth / Flat Frequency Response
- Great for Softening Harsh Sound
- Bi-Directional

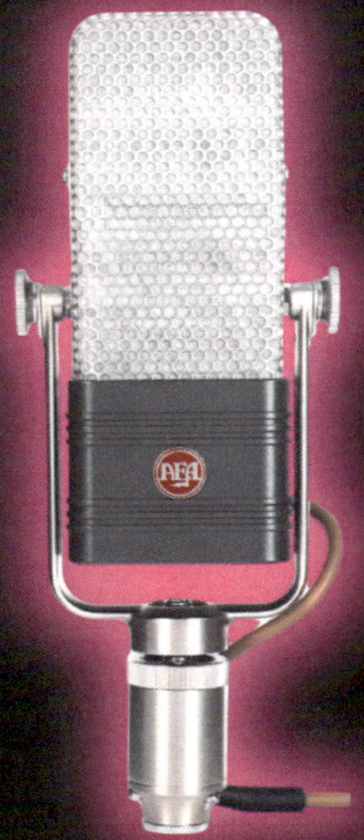

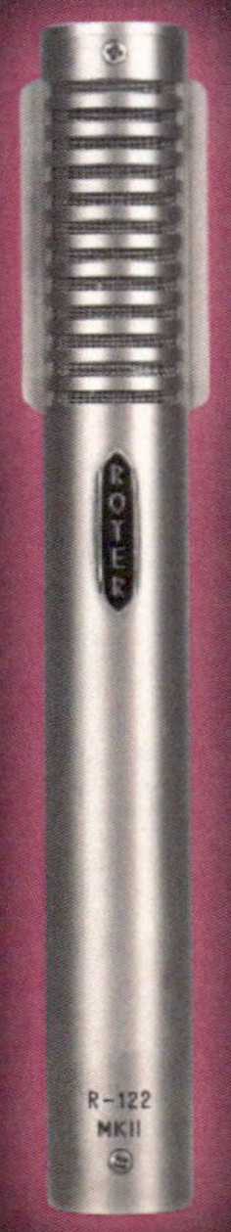

- Delicate / Fragile
- Low Output
- Require Lots of Gain

Ribbon microphones are renowned for their "smooth and flat" frequency response. They excel at softening the harshness, yet capturing the details of instruments such as cymbals and piercing guitars. These mics are inherently bi-directional, which means they will capture a touch of room ambience. Due to the delicate ribbon, they require careful handling and should not receive +48V phantom power (unless specified). Not only are they delicate, but the ribbon produces minimal energy, creating a low-level output signal. This means it's essential to amplify the signal using a clean, high-gain pre-amp.

Condenser

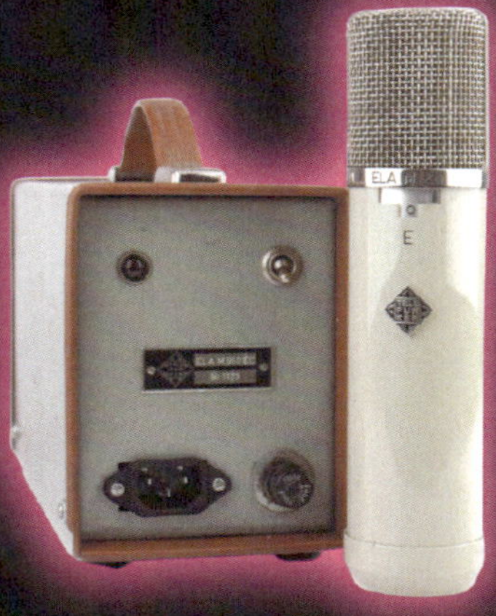

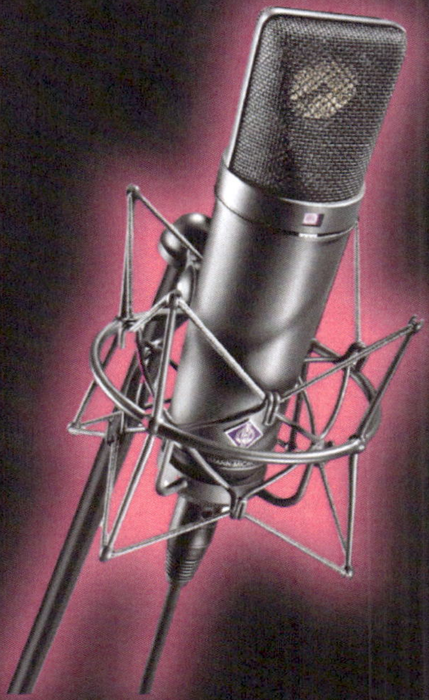

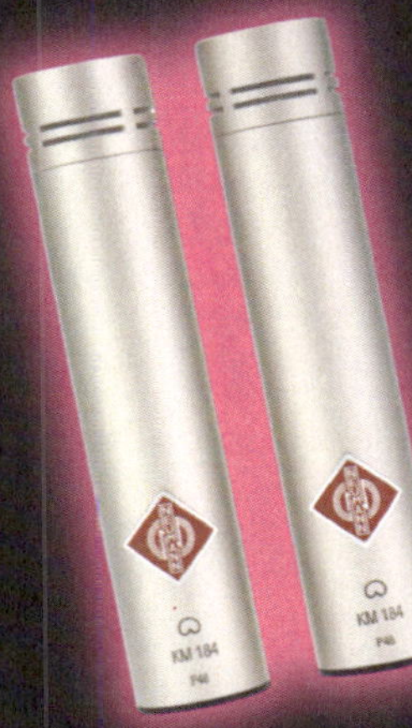

- Widest Frequency Response of All Mics
- High Output
- Requires +48v Phantom Power or Power Supply

- Large or Small Diaphragm
- Variable polar patterns.

Condenser microphones are renowned for their high output, broad frequency response, adjustable polar patterns, and overall versatility. These microphones utilize active electronics and need either +48V phantom power (for solid state models) or an external power supply (for tube models) to function. They come in small or large diaphragm designs, offering exceptional versatility for a variety of applications. Lastly, be sure to pay close attention to the sonic properties of the source and acoustic environment, as a condenser microphone captures "everything you want" and "everything you don't want".

Ambisonic Microphone

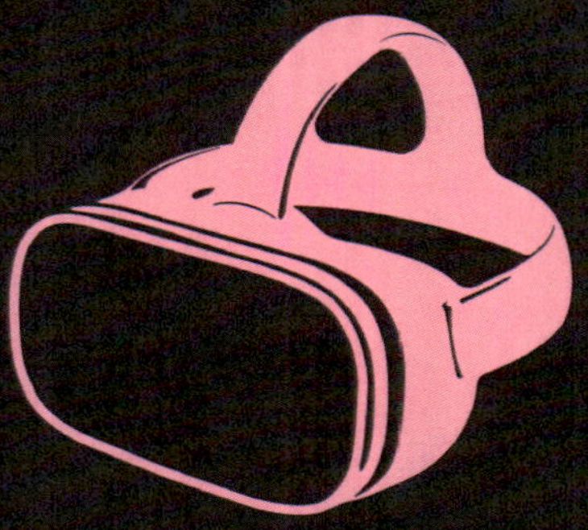

Commonly used for VR, Video Games, and other types of immersive entertainment...

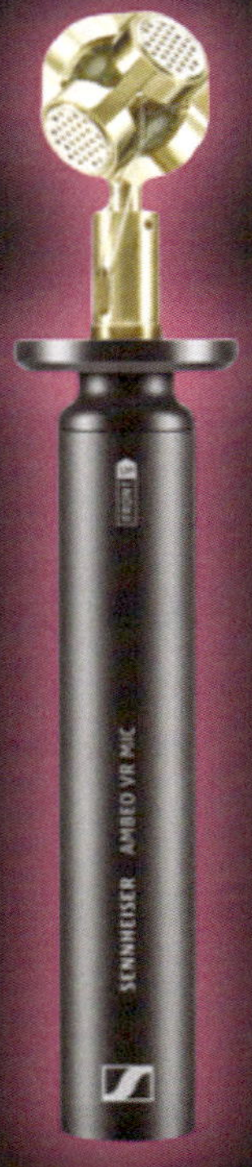

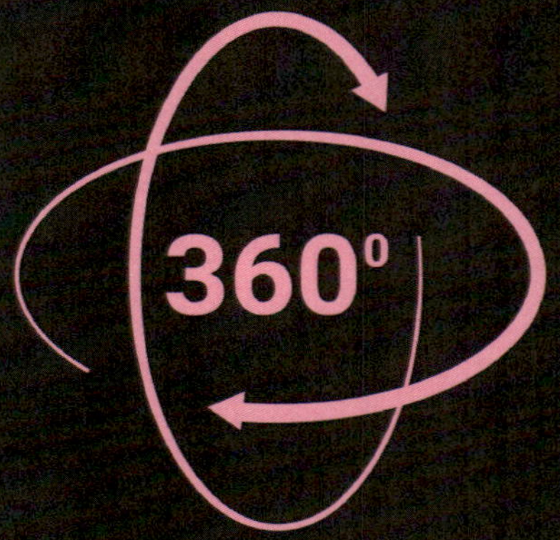

These microphones utilize a minimum of 4 separate audio channels to record Ambisonics content...

The idea behind an Ambisonic Microphone is simple; to capture a 360-degree sound image of an environment. These microphones are equipped with a minimum of four cardioid condenser capsules pointing in a surround "ball like" configuration. By consolidating multiple capsules into a single enclosure, these mics are not only convenient for setup, but offer excellent phase coherency and lots of flexibility for post production.

Mic Positioning

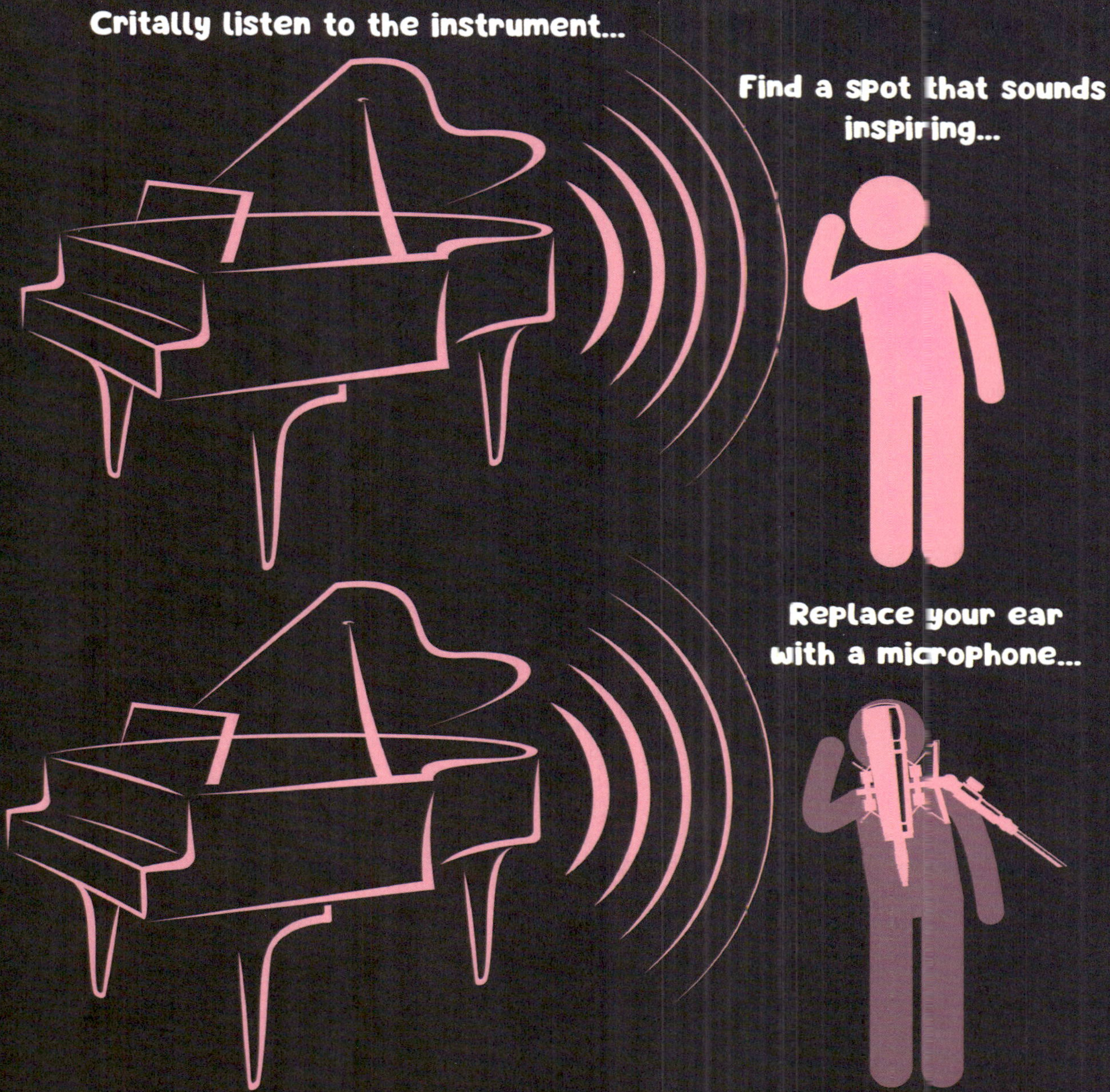

Merging your understanding of microphones with your awareness of the environment can be an excellent starting point...

XY

Stereo Mic'ing Technique

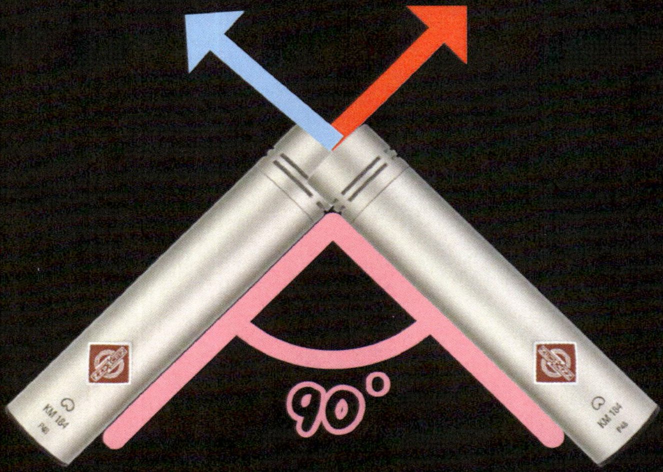

The XY stereo mic'ing technique stands out as one of the easiest and most efficient ways to capture a stereo image. By placing both capsules directly on top of one another equidistant from the source, the XY method guarantees remarkable phase coherence. Although the XY is reliable, sometimes the information from left to right is too similar resulting in a lack of "width" or "imaging". The XY works great in almost any situation, whether mic'ing at close or far distances.

✓ Exelent Phase Coherency

✓ Fail Safe

ORTF

Stereo Mic'ing Technique

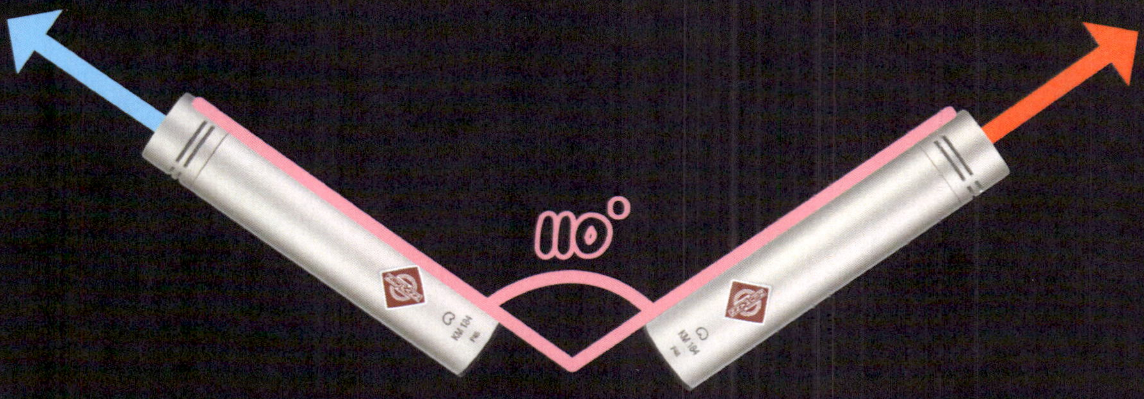

The ORTF technique, much like the XY method, is an excellent stereo mic'ing approach that provides good phase coherence and a wider sound image than the XY. To implement this technique, position two cardioid condenser microphones equidistant from the sound source while forming an angle of about 110 degrees. While this method may occasionally lack a strong center focus, it excels at capturing stereo ambience. It is particularly effective for drum overheads, acoustic guitars, pianos, and room ambience.

- ✓ Good Phase Coherency
- ✓ Wide Image

Blumlein

Stereo Mic'ing Technique

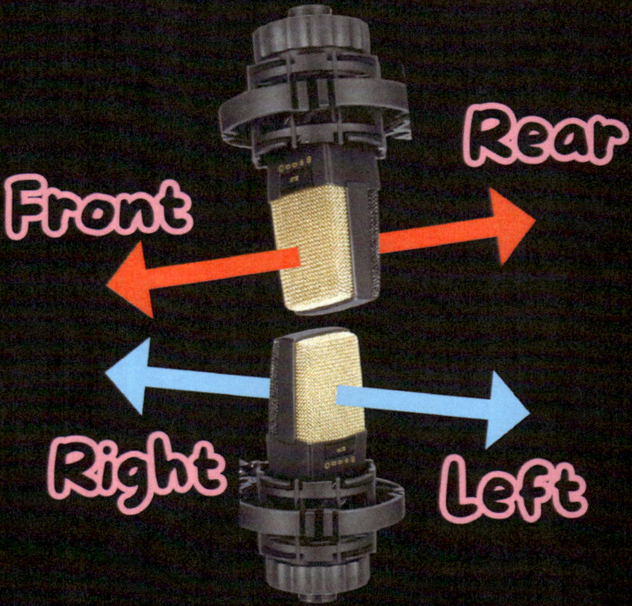

The Blumlein stereo mic'ing technique offers a distinctive omnidirectional method by utilizing two figure-8 pattern condenser microphones to achieve a full 360 degrees of sound capture. This approach ensures excellent phase coherence, as both microphones are positioned directly on top of each other. Due to it's omni directional nature, the Blumlein pair excels at capturing detailed room ambience.

✓ Exelent Phase Coherency

✓ Omni Directional

Mid-Side

Stereo Mic'ing Technique

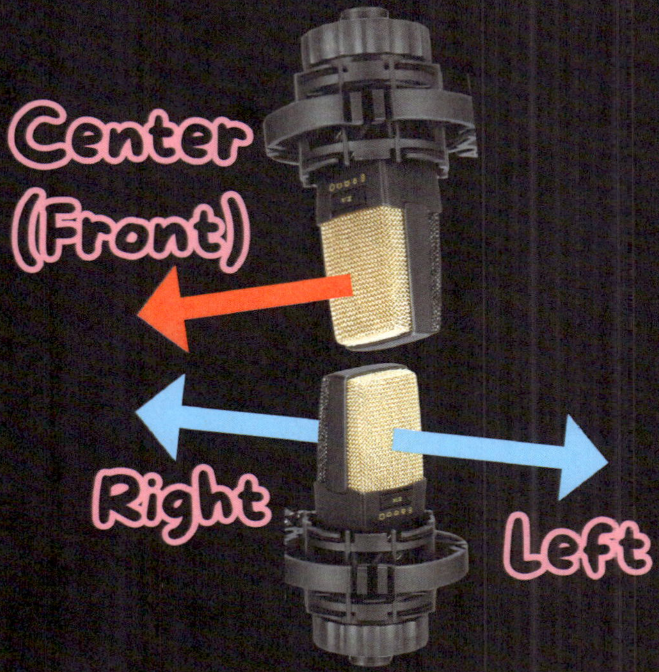

The Mid-Side stereo microphone technique creates a pronounced frontal focus while also capturing the sound from the left and right. This method involves placing one cardioid microphone above a figure-8 pattern microphone. You can visualize this technique as a "Left, Center, Right" strategy, effectively rejecting sound from the rear.

- ✓ Exelent Phase Coherency
- ✓ Exelent Width and Center Focus

Analog vs Digital

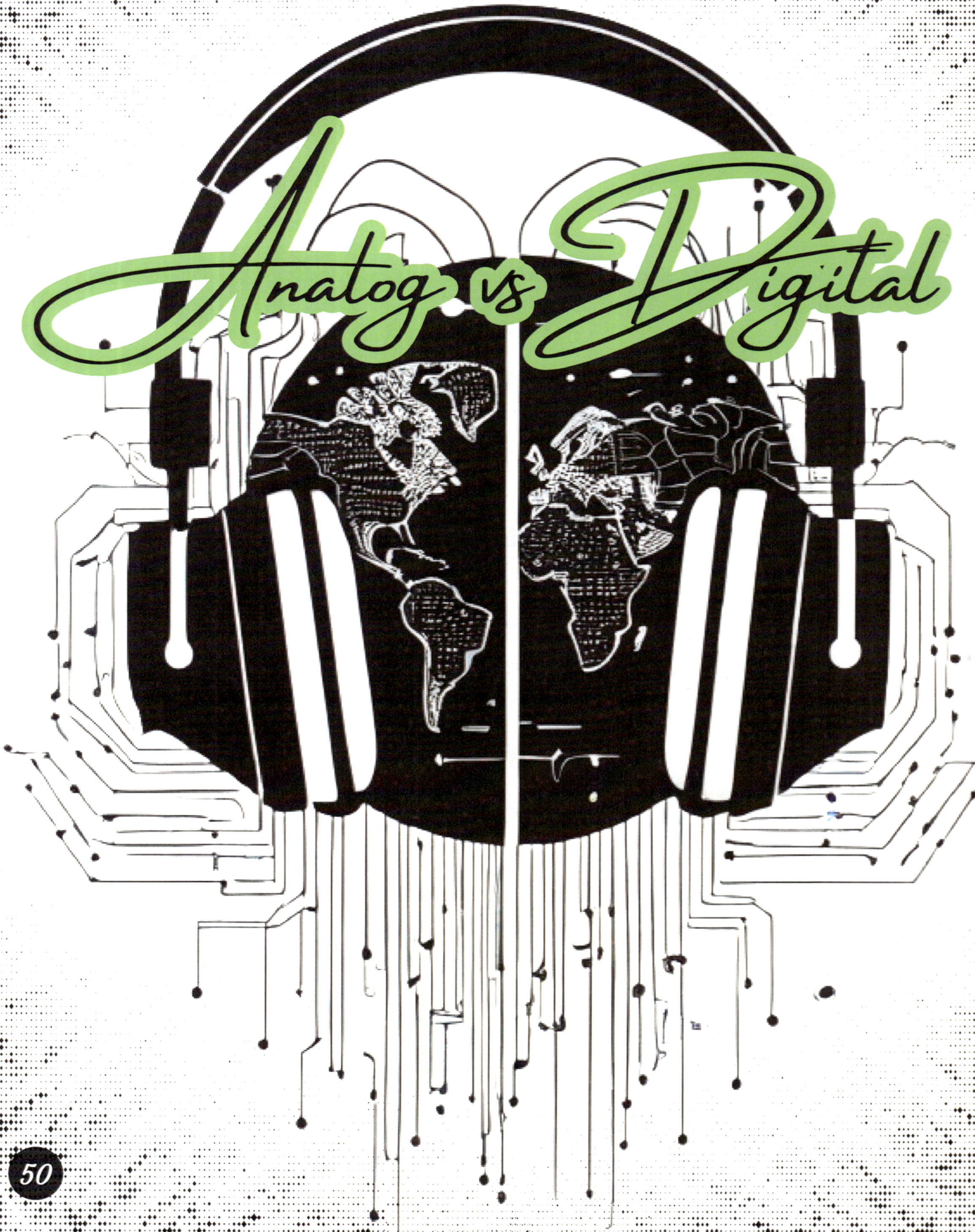

Analog vs Digital

Analog	Digital
1 Component = 1 Function	1 Component = Multiple Functions
Requires More Physical Space	Saves Physical Space
More Expensive	Less Expensive
Manual Recall Settings	Automated Recall Settings
Requires Maintenance	Requires Updates
Less Dynamic Range	More Dynamic Range
Noisy (Can Sound Good)	Clean (Can Sound Good)
Tactile Interface	Computer Interface
Linear Workflow	Nonlinear Workflow
More Cables	Less Cables

Analog Concept

1 Physical Component = 1 Function

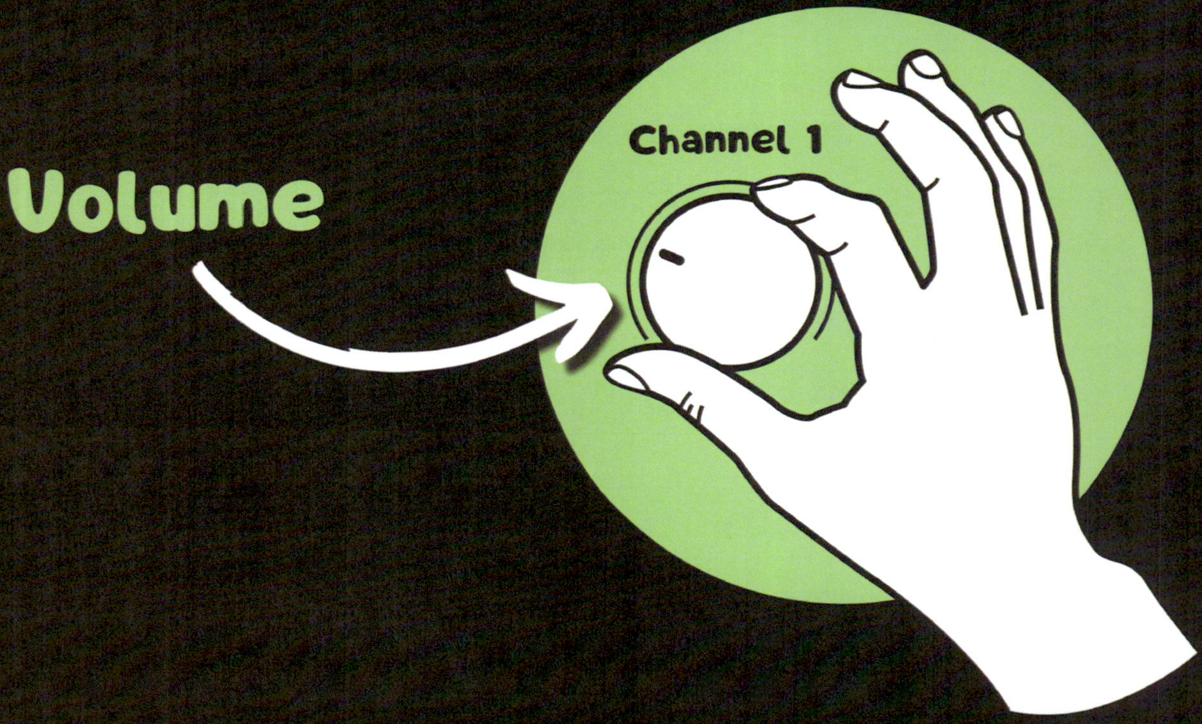

This knob only functions as the volume for channel 1...

Analog Examples

You

Cables

(the wire itself)

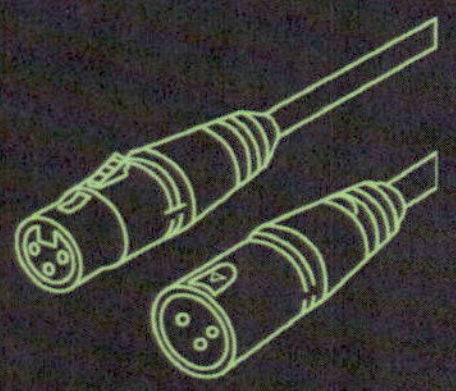

Power

Microphones

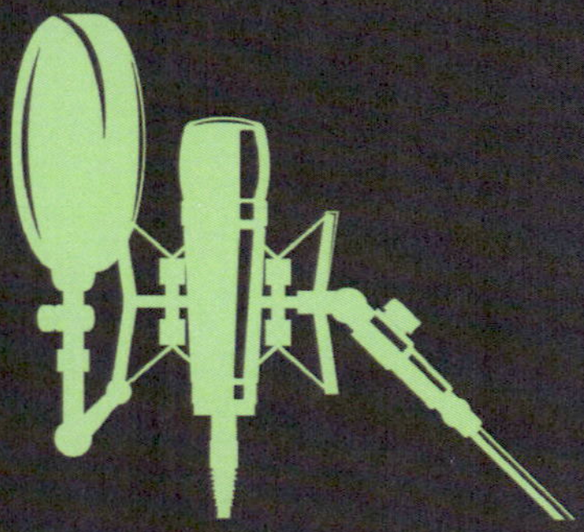

Amplifiers

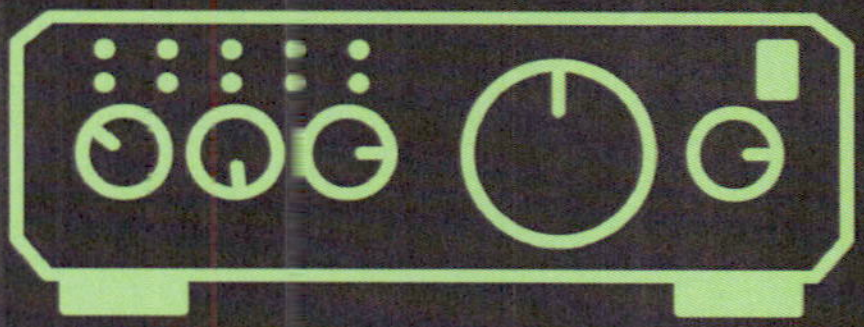

Reel to Reel Tape Recorder

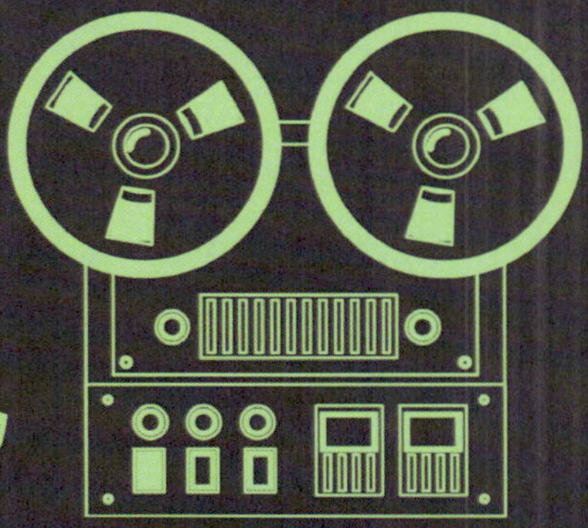

Analog Mixer

(All knobs and faders have 1 function)

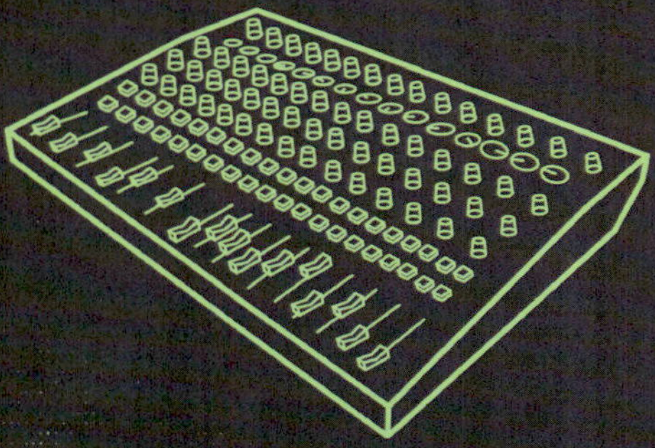

Speakers

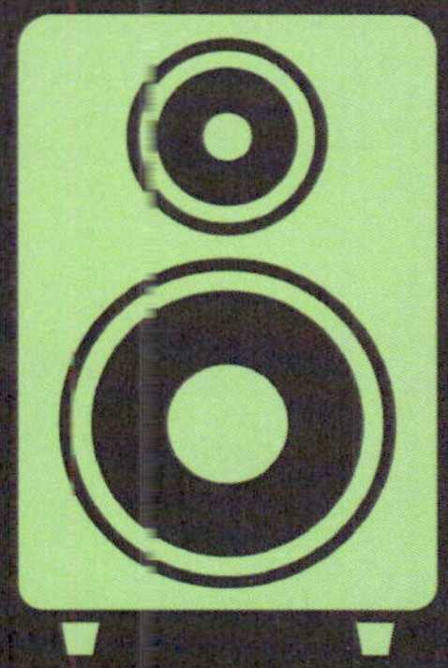

Digital Concept

1 Physical Component = Multiple Functions

Functions as Volume for any selected channel

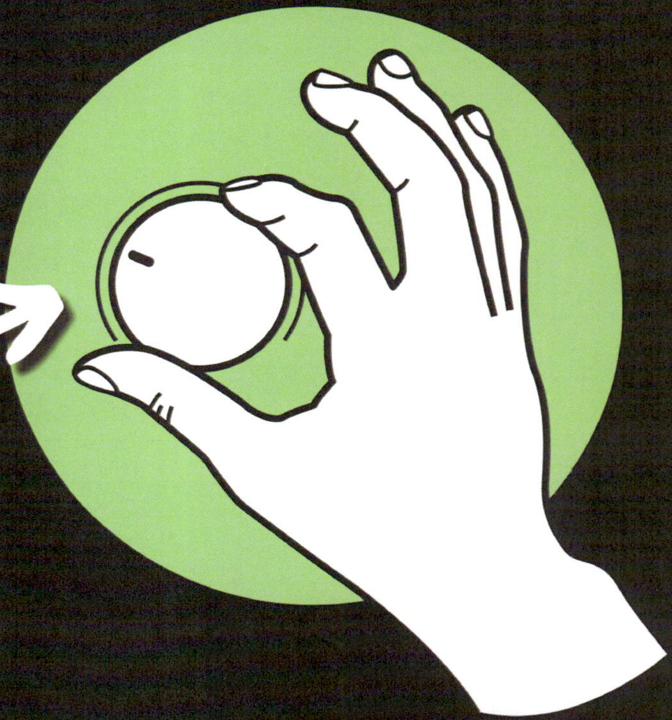

Channel 1 SELECT

Channel 2 SELECT

This knob can funcation as the volume for any channel selected on the mixer...

AD/DA Conversion

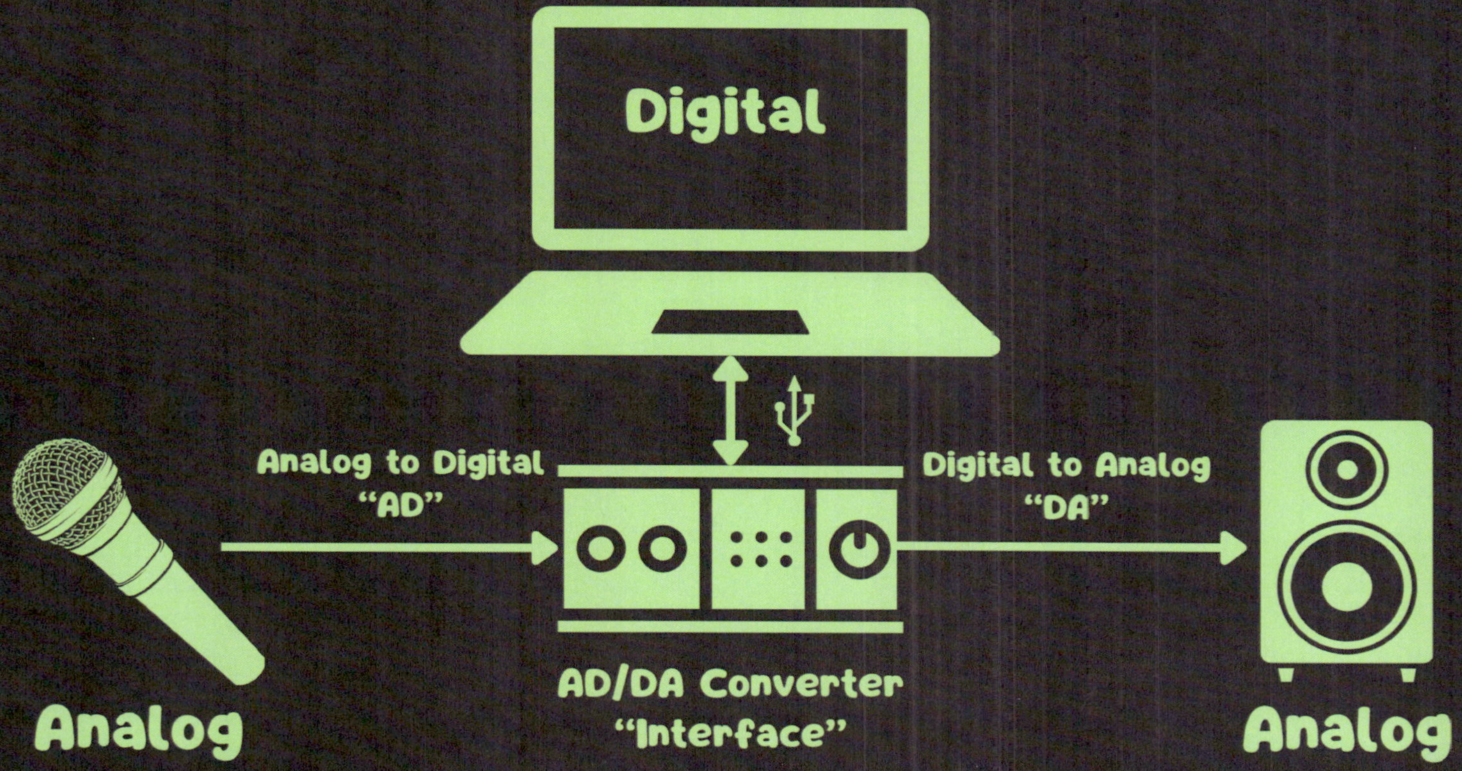

Pulse Code Modulation

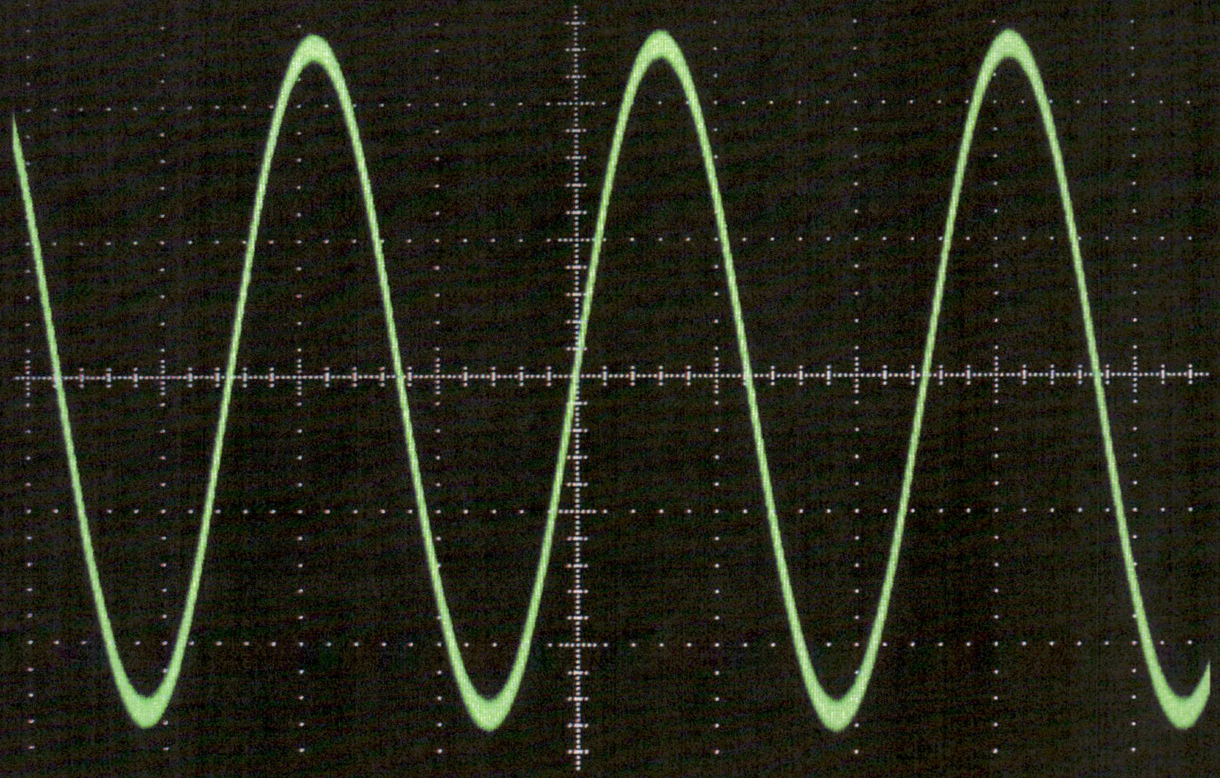

"PCM" is a technique for converting an analog waveform into digital data, quantified by samples and bits.

Sample Rate

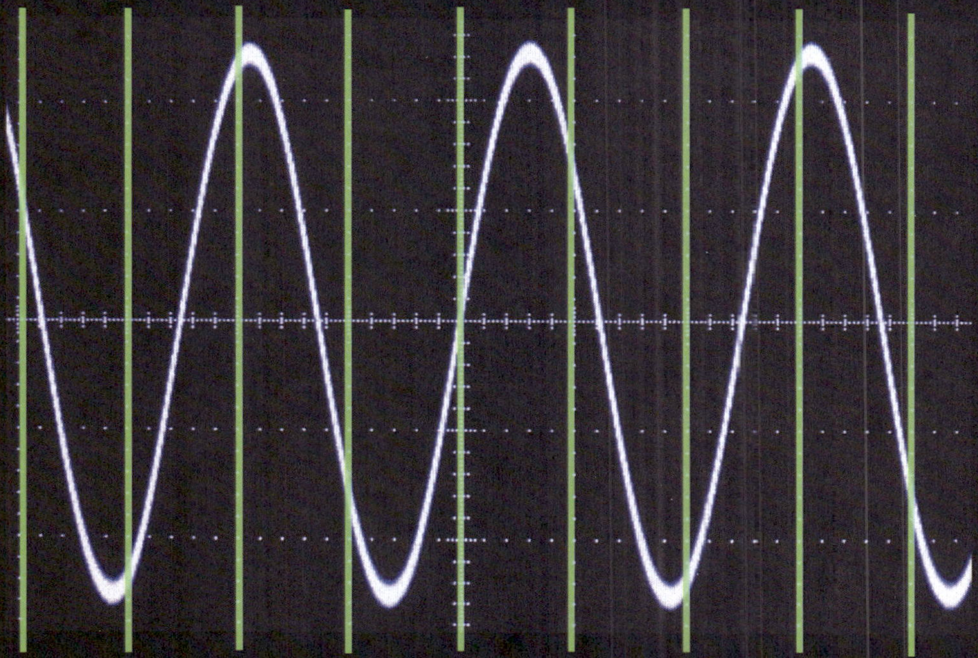

Measured in samples per second, a sample rate refers to the amount of samples captured over time. This method helps a device calculate the frequency (or pitch) of a waveform. As stated by the "Nyquist Theorem", to accurately digitize an analog waveform, the sample rate must be at least twice the rate as the highest frequency in the human hearing spectrum. This is where the standard sample rate 44.1 kHz (more than 2x 20 kHz) came to fruition.

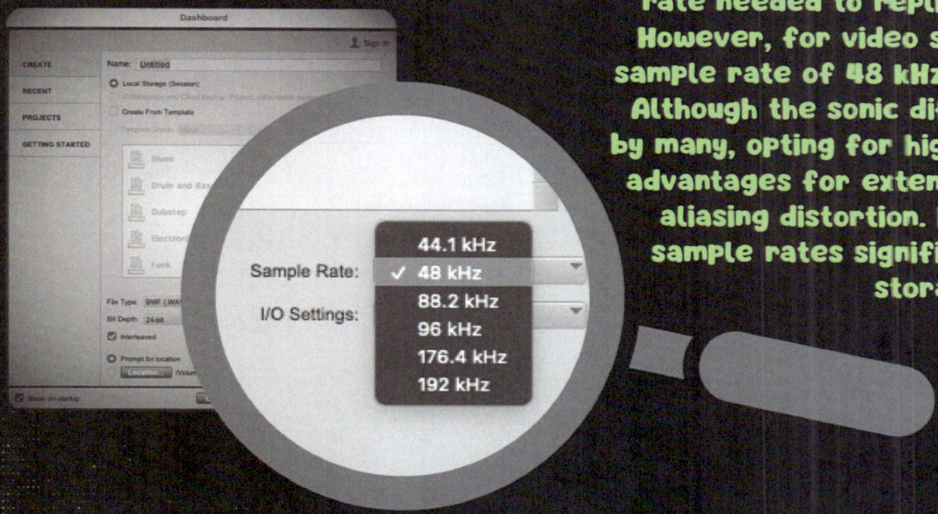

44.1 kHz is the minimum non-compressed sample rate needed to replicate an analog waveform. However, for video synchronization purposes, a sample rate of 48 kHz or 96 kHz is recommended. Although the sonic difference may be unnoticed by many, opting for higher sample rates can offer advantages for extensive editing and preventing aliasing distortion. However, working at high sample rates significantly increases CPU and storage usage.

Bit Depth

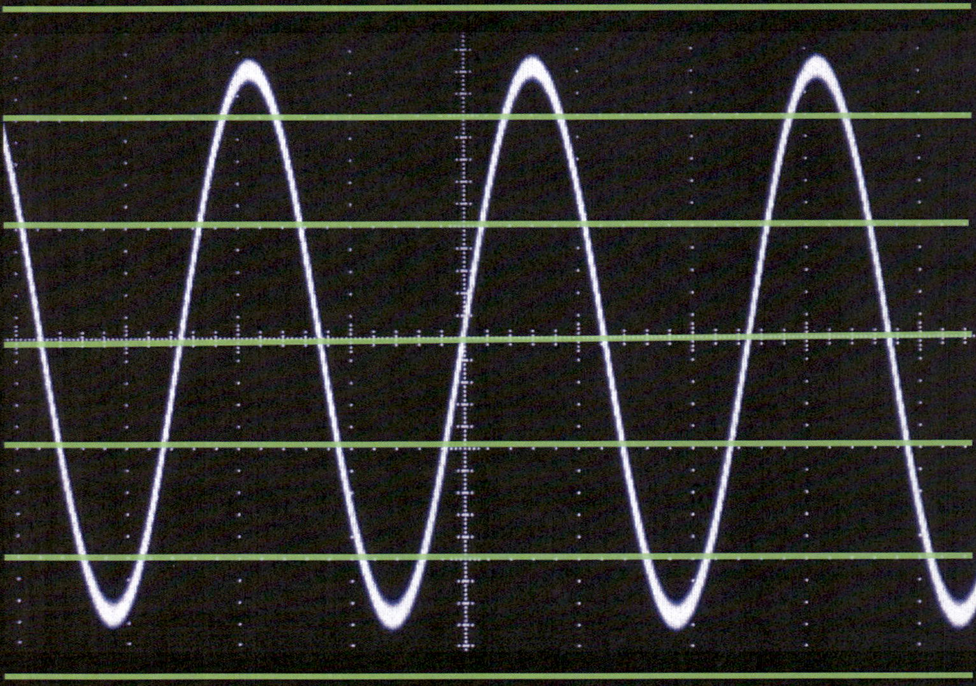

The Bit Depth quantifies the dynamic range of a digital waveform. As the number of bits increase, so does the amount of information from loud to soft.

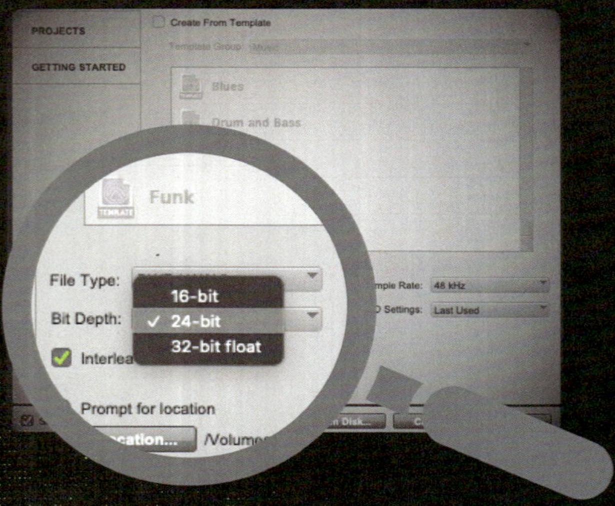

The ideal Bit Depth for a recording session is 24-bit. However, if storage space and processing power are not limitations, 32-bit float can offer an increased dynamic range and headroom for your recordings.

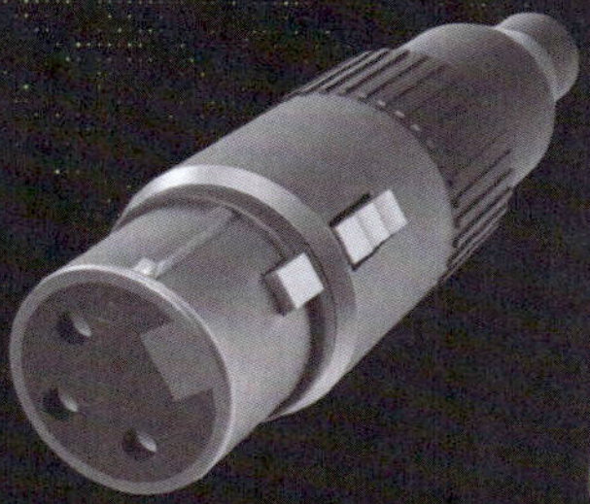

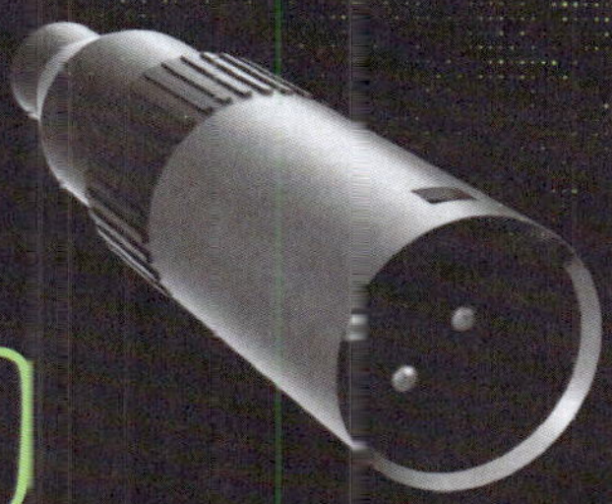

AES/EBU

"Audio Engineering Society / European Broadcast Union"

The AES/EBU digital audio protocol allows for the transmission of two audio channels up to 100 meters using a single balanced XLR cable (25-gauge, 110-ohm). It's essential that the AES/EBU outputs of one device connect to the AES/EBU inputs of another. This protocol is frequently utilized for distributing a "main stereo mix," which can include stereo outputs from a console, a media player feed, or the primary output from a digital DJ deck to a digital speaker processor's input.

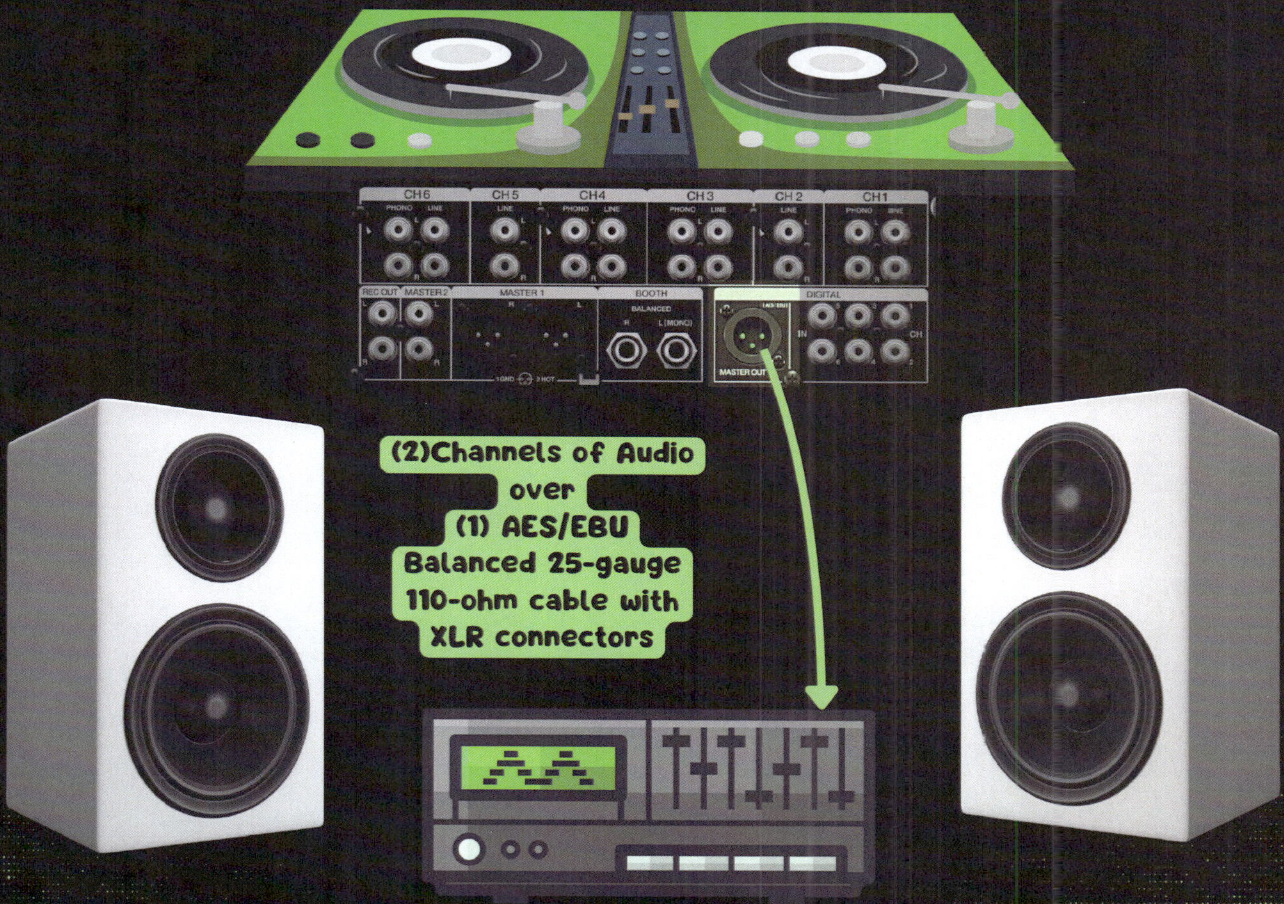

S/PDIF

“Sony / Philips Digital Interface

S/PDIF is a digital audio protocol designed to transmit two channels of audio over short distances. It is widely utilized in home theater setups and consumer audio systems. S/PDIF typically employs coaxial cables with either RCA or BNC connectors.

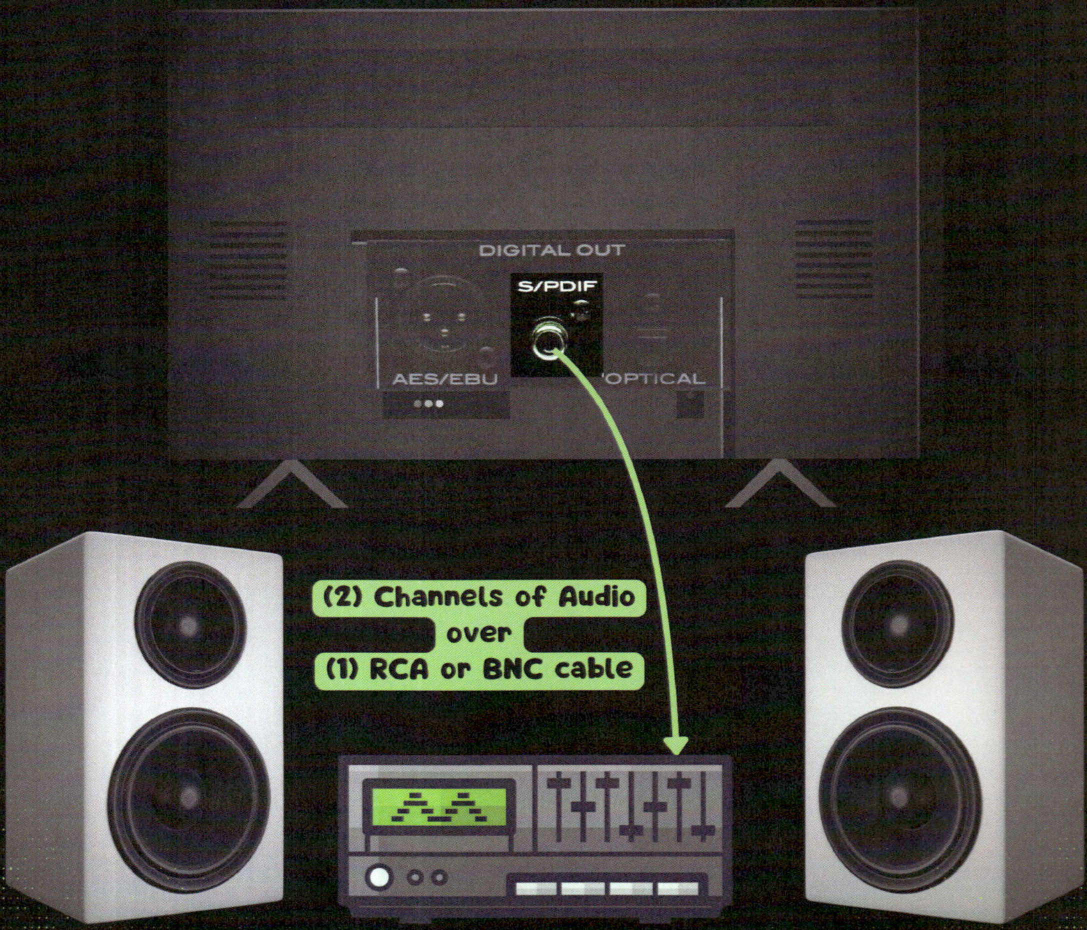

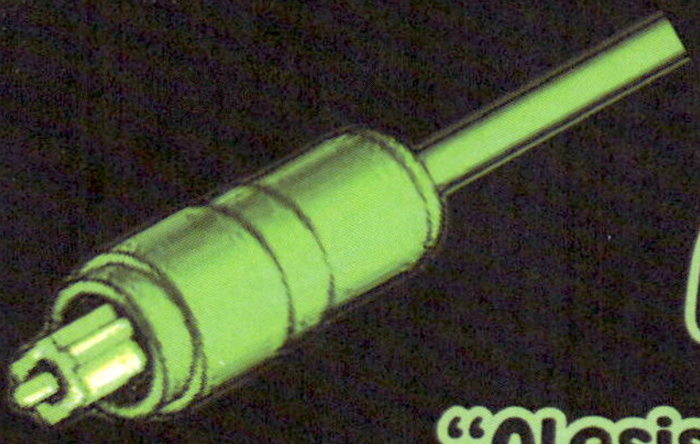

ADAT

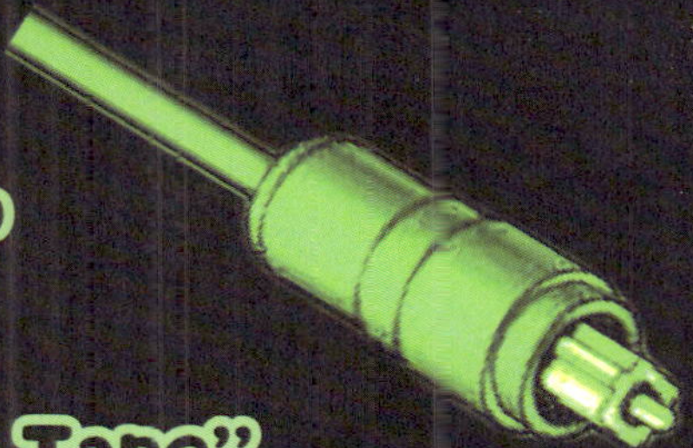

"Alesis Digital Audio Tape"

ADAT is a digital audio protocol that can transmit up to 8 channels of audio in one direction. ADAT is commonly used to expand the I/O of audio interfaces with a 8 channel mic pre unit connected via 1 lightpipe cable.

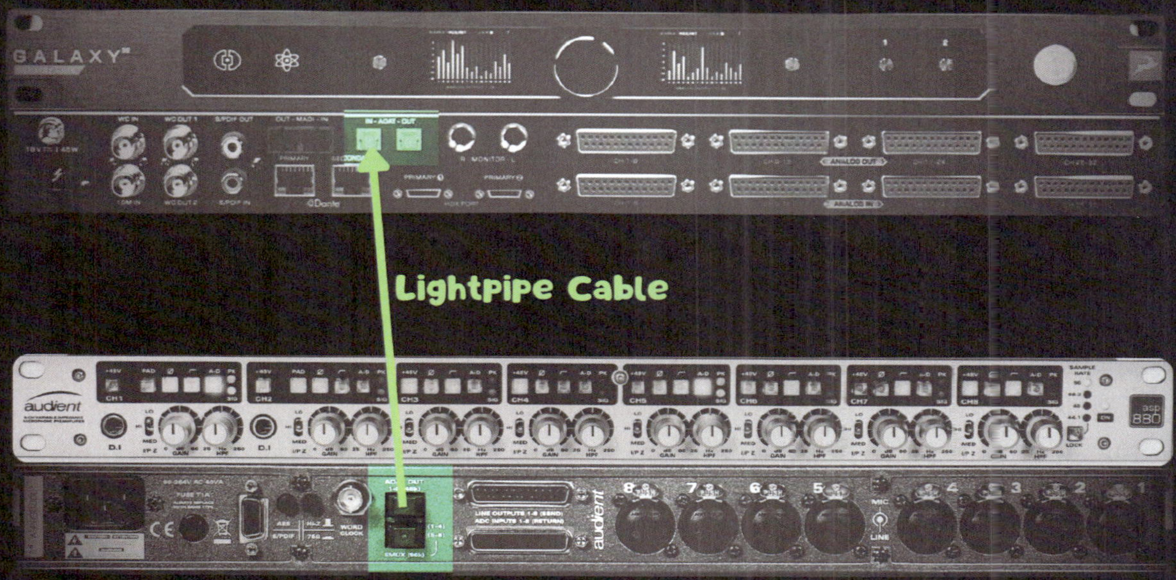

MADI

"Multichannel Audio Digital Interface"

MADI is a digital audio protocol capable of transmitting 64 channels of audio in a single direction. Thanks to its high channel capacity, MADI is frequently utilized for primary digital transmission between mixers and other MADI-compatible audio devices. A MADI signal can be carried using either BNC coaxial or fiber optic cables. One significant advantage of MADI is its reliability in transmitting audio at long distances (100s of Meters over Fiber).

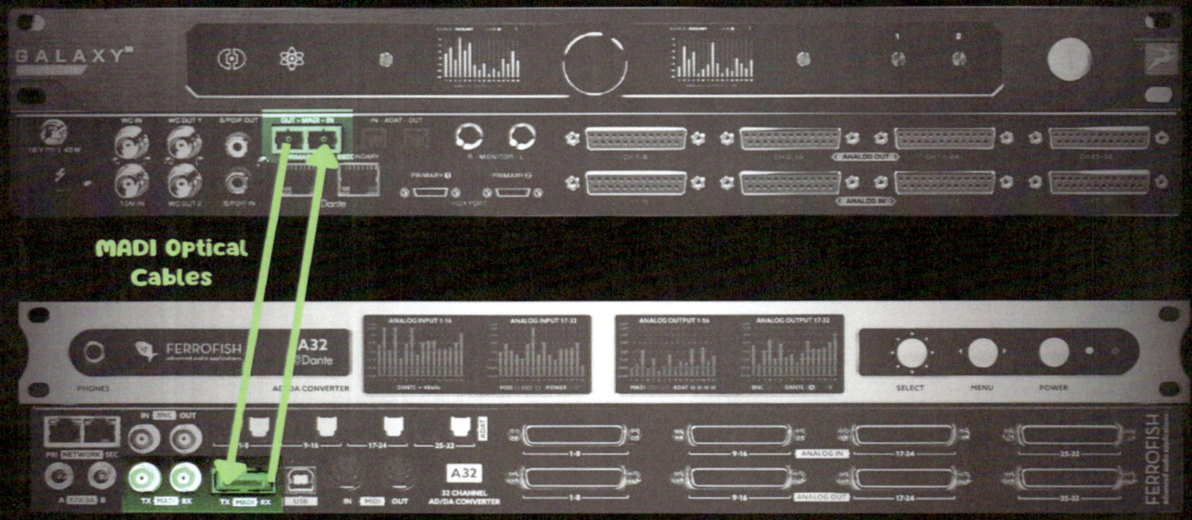

DANTE

Network Based Digital Audio

DANTE provides a networking solution that facilitates the routing and connection of both audio and video devices. By using a gigabit Ethernet switch, numerous Dante-compatible devices can link together through standard Cat5e or Cat6 cables. With a sample rate of 48kHz, each gigabit ethernet link can handle up to 512 channels each direction. Additionally, DANTE features an application called Dante Virtual Soundcheck, which functions as a soundcard and transforms any computer into a DANTE device.

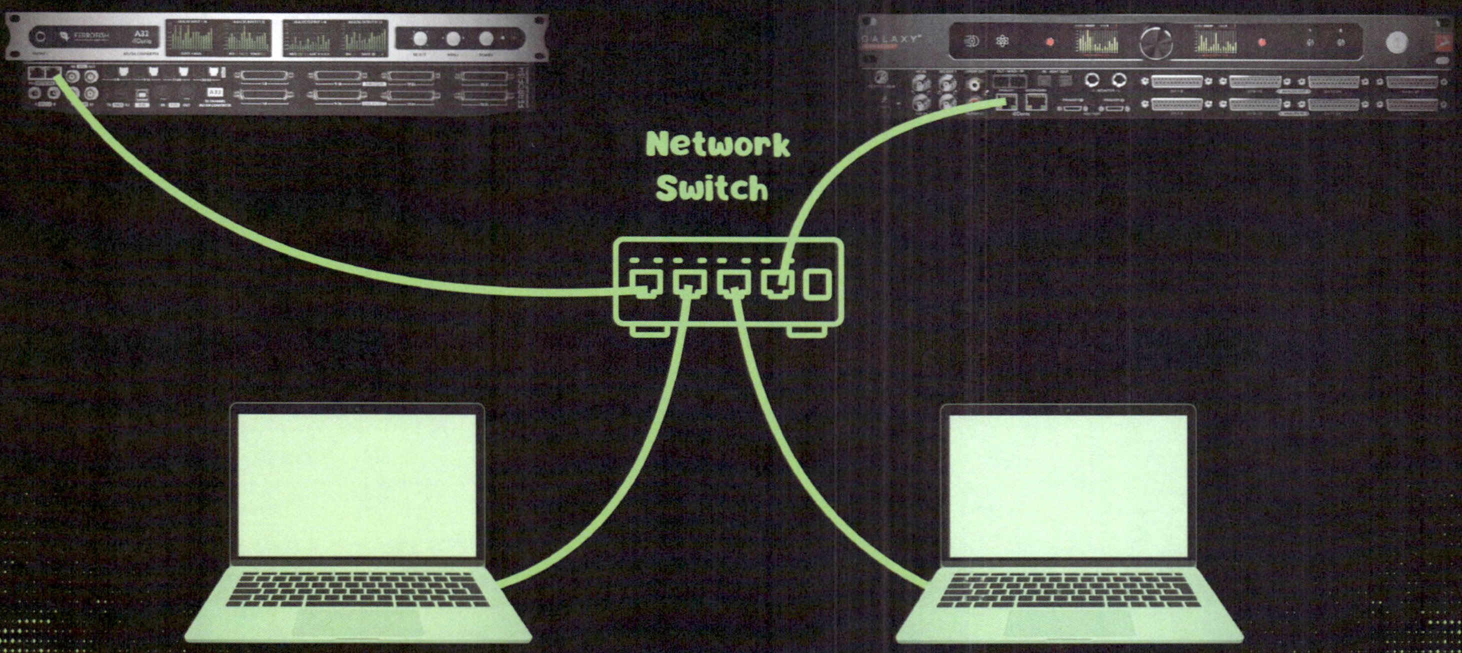

Word Clock "Sync"

When linking devices using digital transmission methods like ADAT, AES-EBU, MADI, or DANTE, it's essential to determine a "master" device. This "master" device will establish the common rate of sampling for all connected or "slave" equipment. To achieve this relationship, a BNC cable can send sync data "out" of the master device, while the BNC "in" of a slave device receives that sync data. Lastly, be sure that the master device is set to its internal clock source, while the slave device's clock source is set to "word clock". This will create a proper "digital sync" where each device will clock the sample rate at the exact same rate.

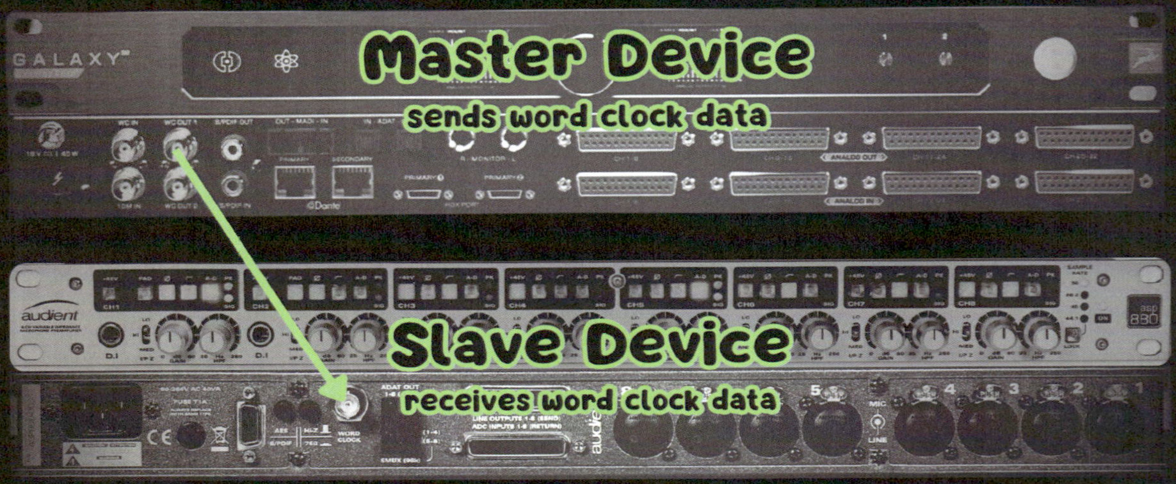

Comprehending the idea of a word clock is much like understanding the connection between these three different clocks. There's no assurance that each clock will maintain the same time as time passes. Even if they all start together, it's likely that over time, they will gradually drift out of sync with each other. This is the reasoning for a master device within a digital connection.

MIDI

Musical Instrument Digital Interface

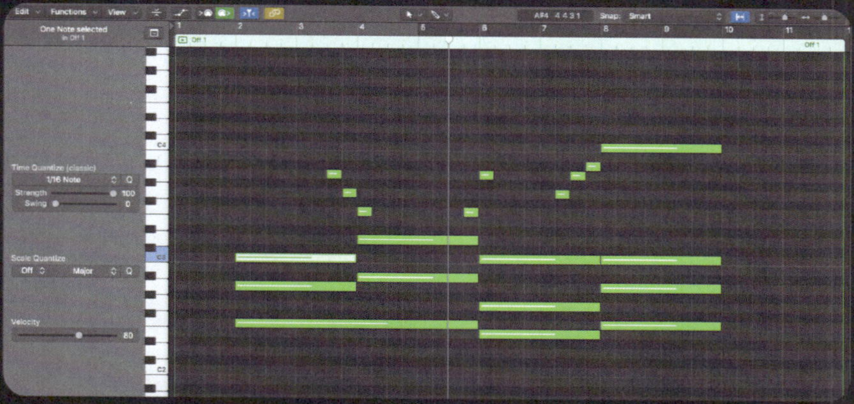

&

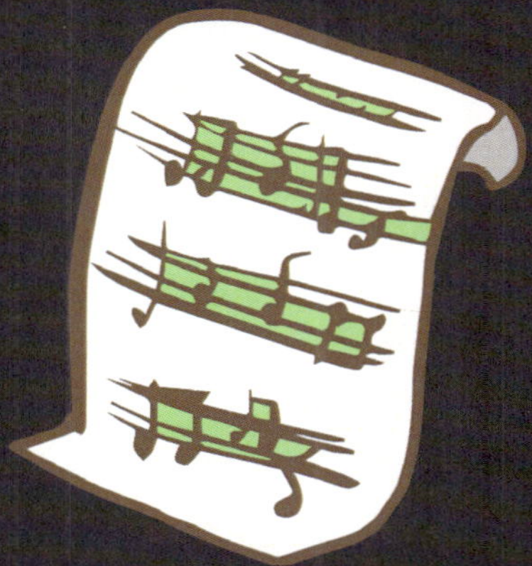

MIDI is a communication protocol that enables digital devices to transmit and receive performance data. Similar to sheet music for a musician, MIDI does not create sound, rather it depends on a sampler or virtual instrument to interpret the message. In other words, the output of one MIDI device generates a message, while another MIDI device receives that message and executes a specific sound or function.

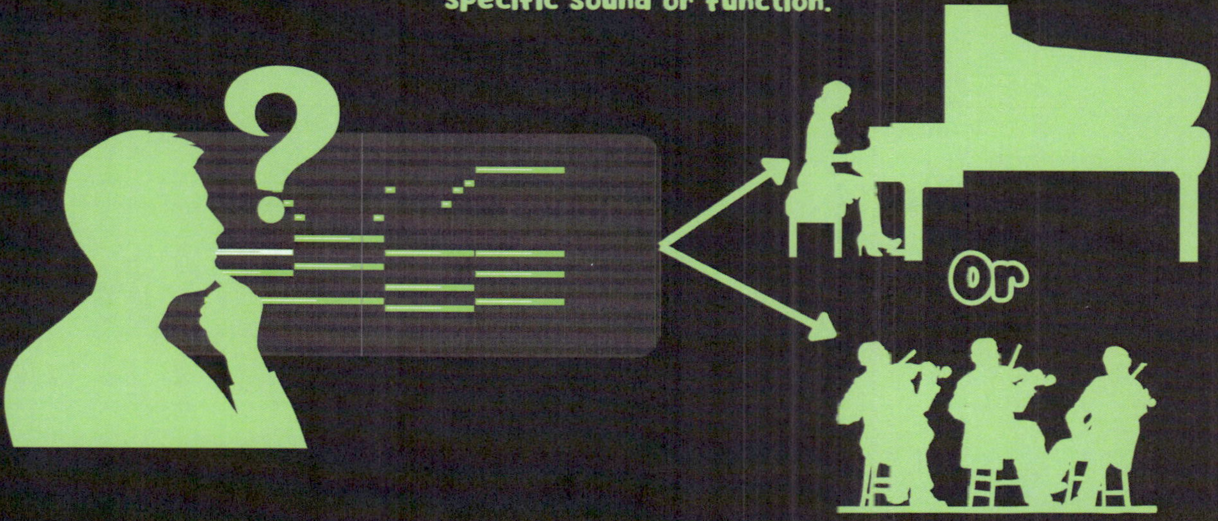

In many situations, MIDI serves as a creative tool for notating compositions within a DAW. It enables us to capture a performance without being tied to a specific sound or particular interpretation. Additionally, MIDI provides composers the flexibility to manually create or adjust notes when needed. A fun workflow involves initially recording a performance and then enhancing or modifying the instrument afterward. Remember, the sound of the instrument or environment significantly influences a performance. Therefore, aim to discover the ideal sound from the start.

MIDI

The Message and Routing

Status Byte		Data Byte	Data Byte
1001	0000	00111100	01111111
Note On	Channel 1	C4 on Piano Note Number 60	Note Velocity 127

Take a closer look at the MIDI message shown above. Notice how the middle Data Byte aligns with the notes on the piano. Keep in mind there are 128 notes on a MIDI piano roll, given that there are 128 combinations within an 8-bit code. Furthermore, the Data Byte on the right hand side offers 128 (0-127) velocity values for each note, representing the dynamic range of the performance. The message also includes a "Status Byte," which specifies the channel routing and the current status of the note (e.g., Note On, Note Off). This knowledge becomes vital when creating your own sample-based instruments. Like audio, it's important to grasp the concept of MIDI routing. Each connection via 5 pin DIN or USB cable can transfer up to 16 different MIDI messages.

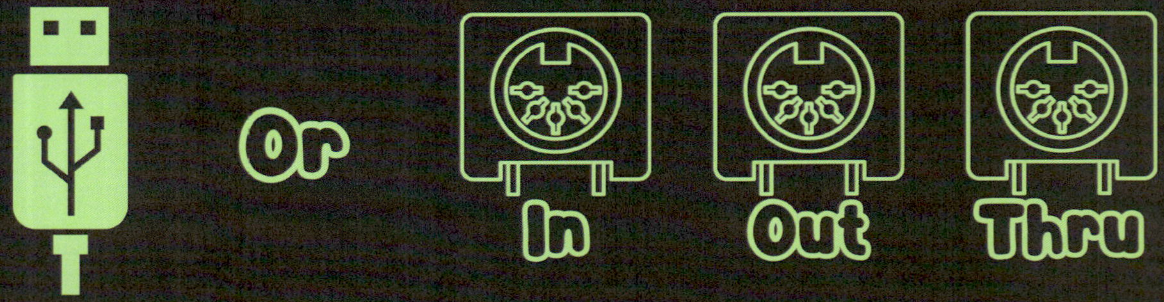

Regardless of whether a MIDI message is transmitted through DIN or USB, the principles of MIDI In, MIDI Out, and MIDI Thru remain the same. The MIDI Out port produces and transmits the message to an external device. Then, the MIDI In port on a device receives the message and executes a specific task. Meanwhile, the MIDI Thru port duplicates the MIDI In data and forwards the message to another device without altering the original MIDI In information. Ensure that the MIDI channels match on both ends of the message, and avoid using MIDI cables longer than 30 feet to prevent drop outs or "stuck notes".

Virtual Instruments and Synthesizers

All software-based instruments within a computer can be classified in one of two categories. The first category is a software synthesizer, which involves sound sources generated by the plugin's internal oscillators or pre-programmed sound. The second approach consists of loading recorded samples into a sampler and playing them with a MIDI device. Let's compare these two instruments a little more below.

Software Synthesizer

- Commonly used for non-acoustic, synthesized instruments.
- Relatively small file size
- Less CPU intensive
- Highly adjustable parameters
- Recommended to store instrument on local hard drive
- No need for a hosting sampler plugin

ARTURIA
_The sound explorers

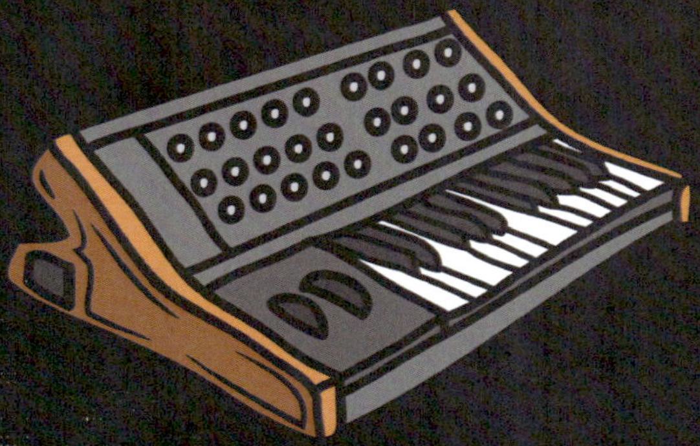

Sample Based Virtual Instruments

- Commonly used for sampling Organic Instrument for Orchestral Scoring
- Large file size
- High CPU usage
- Less modifiable parameters
- Recommended to store sample data on external SSD drive
- Host sampler plugin may be produced by a different company from the sample library company.

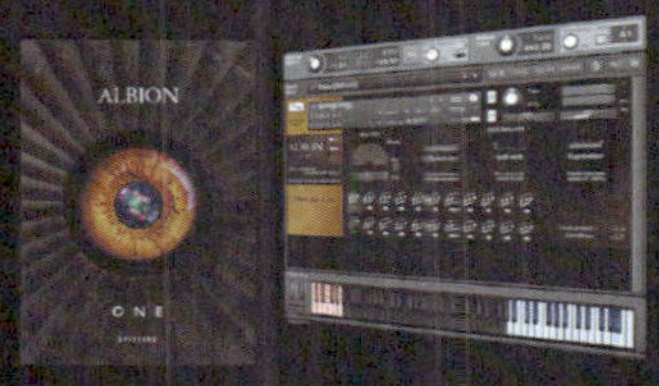

Fundamental Audio Concepts & Terms

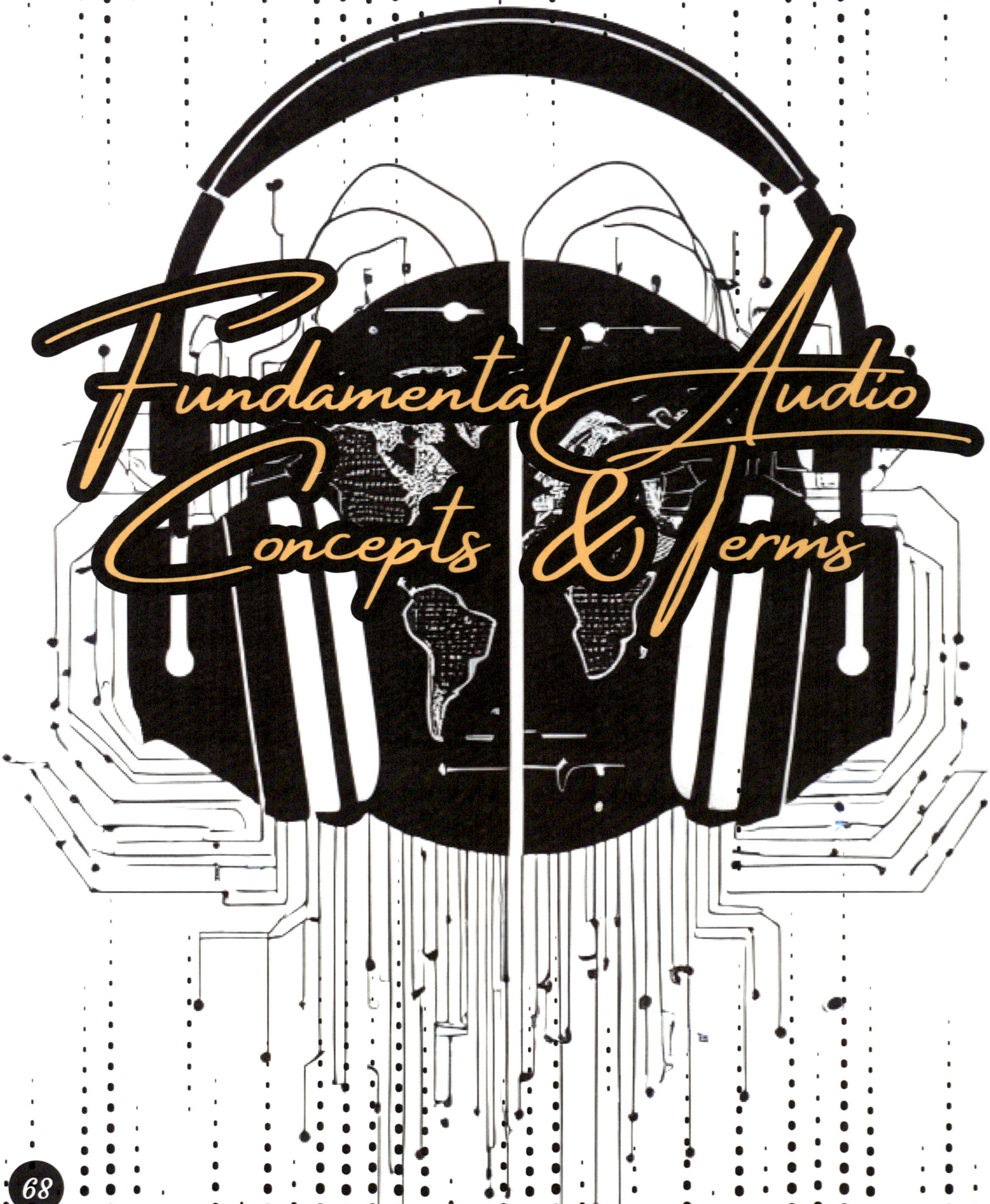

Nominal Signal

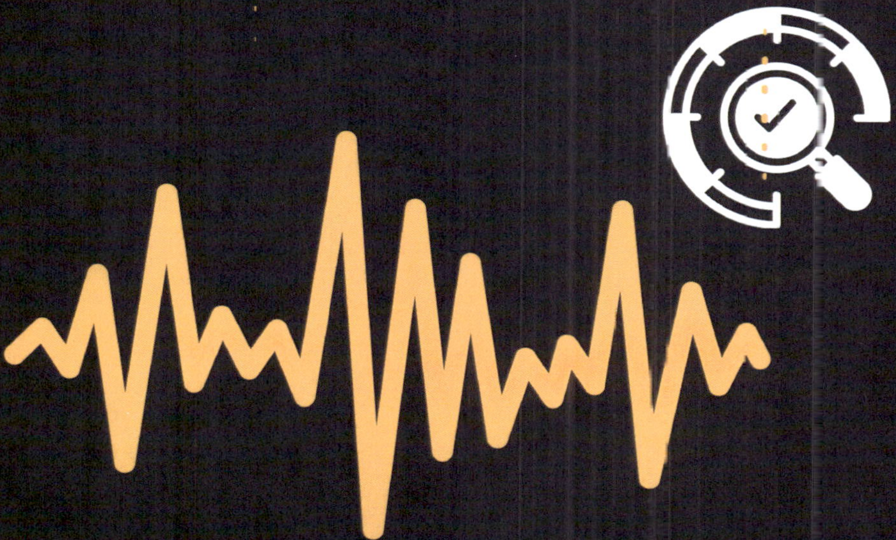

A Nominal Signal represents the ideal signal level at which electronic devices operate most efficiently.

Note: Each piece of equipment will include a manufacturer-recommended optimal signal strength referred to as the "Nominal Signal."

Noise Floor

"The sum of undesirable systematic and ambient noise...."

Note* A lower noise floor will enhance the clarity of your audio.

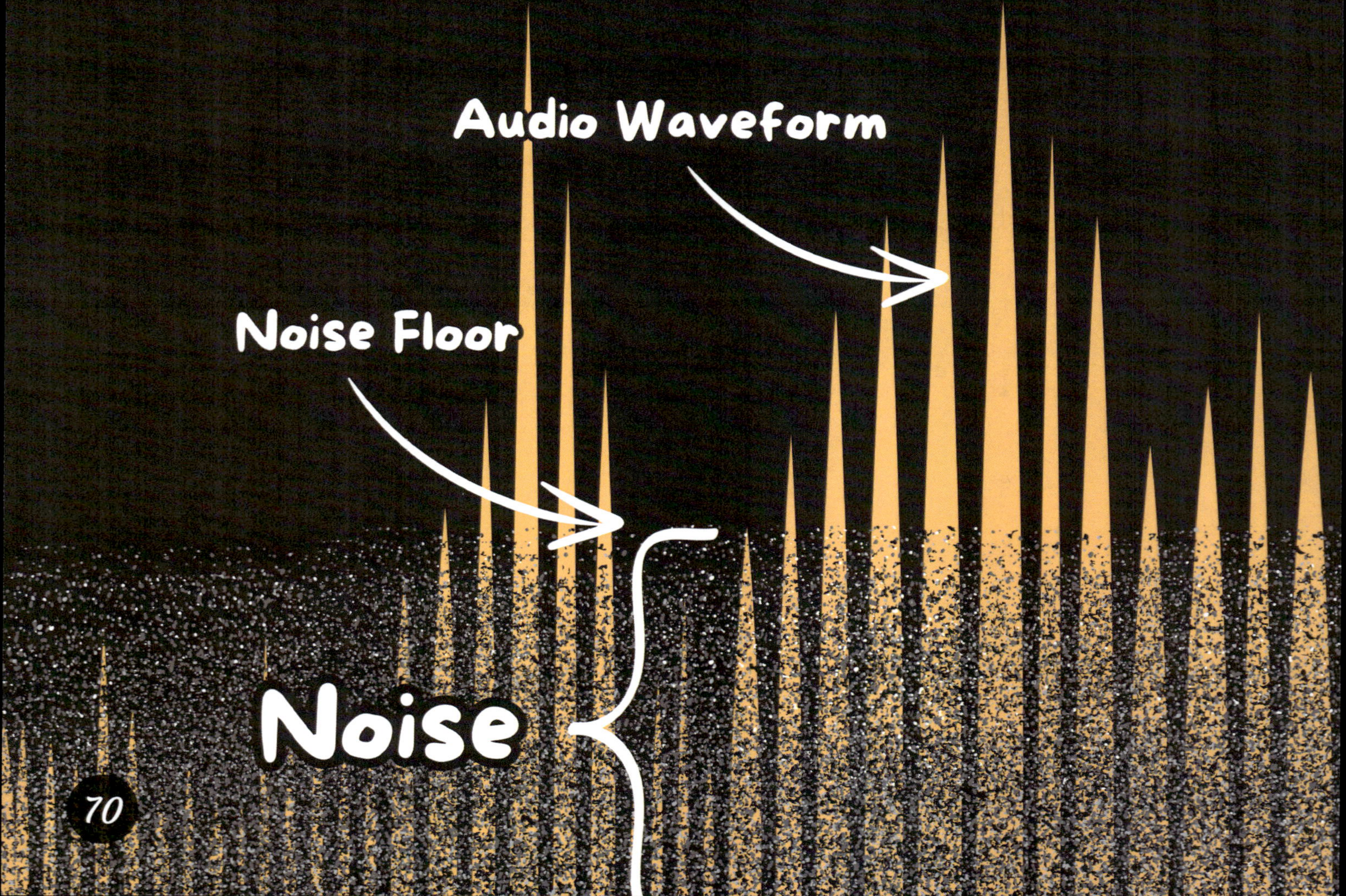

Signal to Noise Ratio

"The range between a nominal signal and the noise floor..."

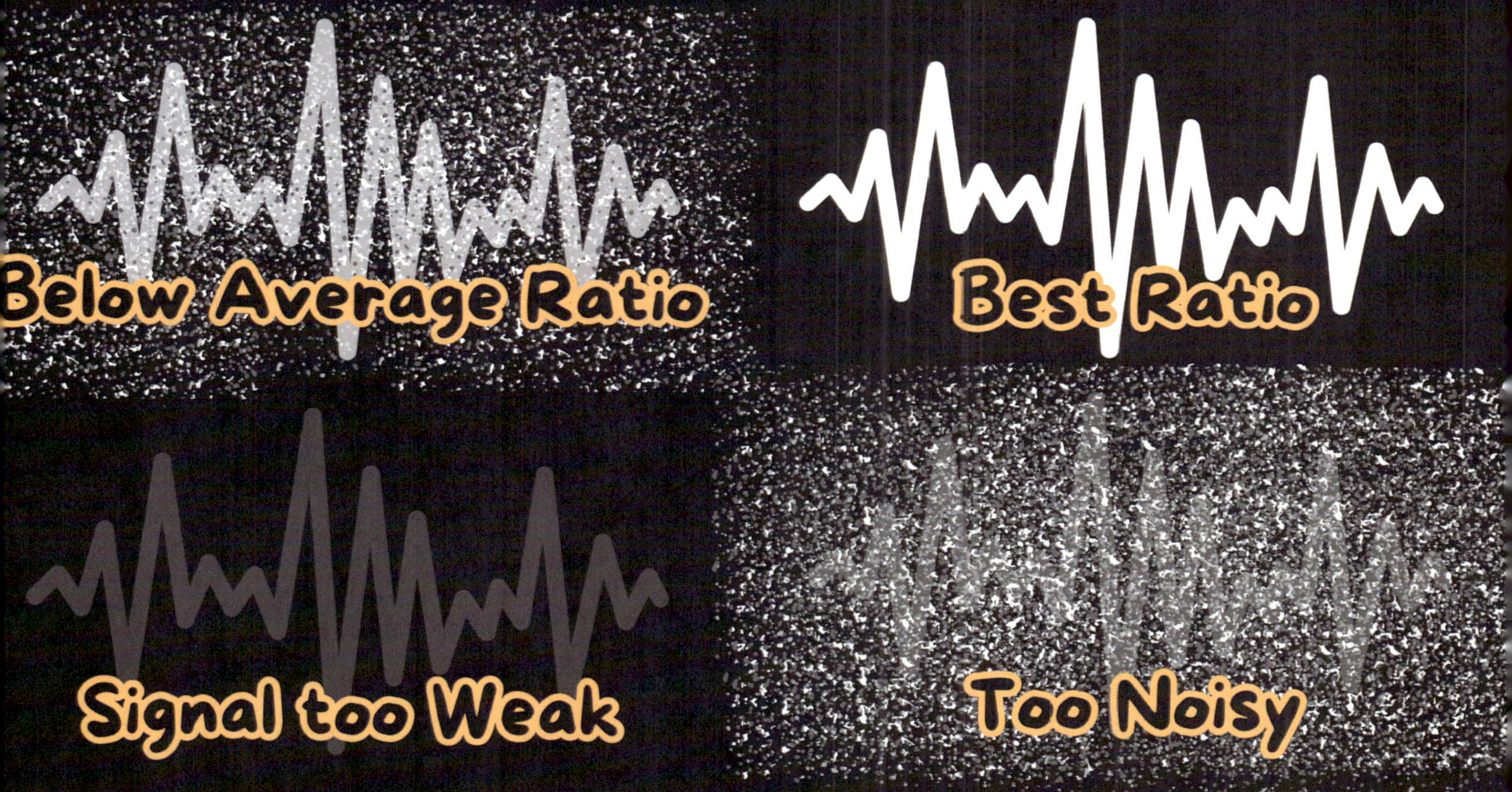

Tip: A great signal-to-noise ratio results in higher audio fidelity.

Dynamic Range

"The range of volume spanning from softest to loudest in a given context..."

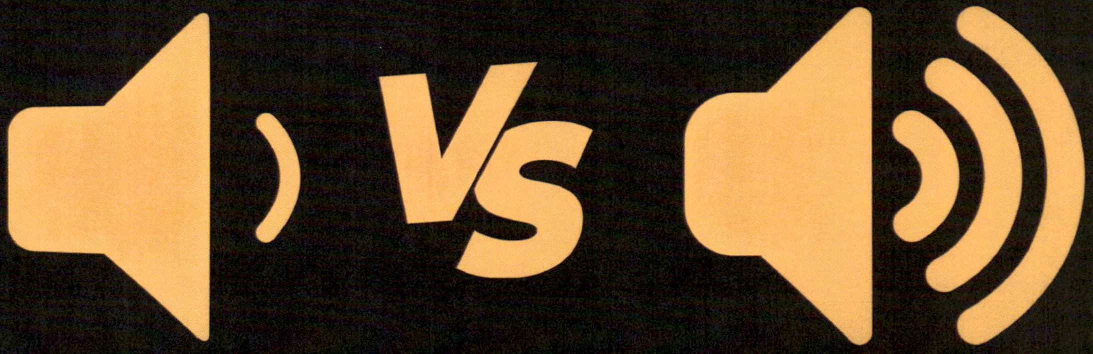

Note* A song that is consistently loud has a limited dynamic range. Conversely, a song that alternates between very quiet and very loud moments has a broad dynamic range.

Headroom

"The remaining signal amount between a nominal signal and unwanted distortion..."

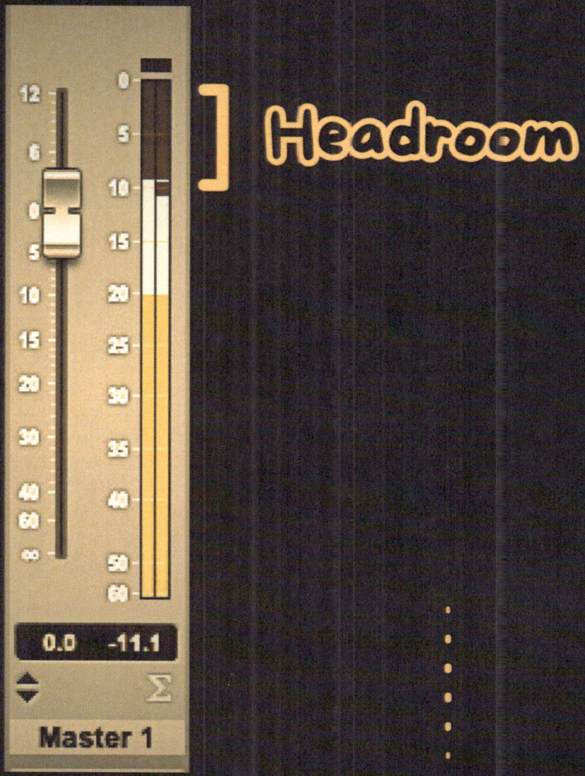

Tip* Leaving extra headroom can help avoid distortion in case of a signal spike during a high-energy performance.

Unity

"The fader position (usually represented as 0dB) that symbolizes no level change to the incoming signal as it passes through an audio channel..."

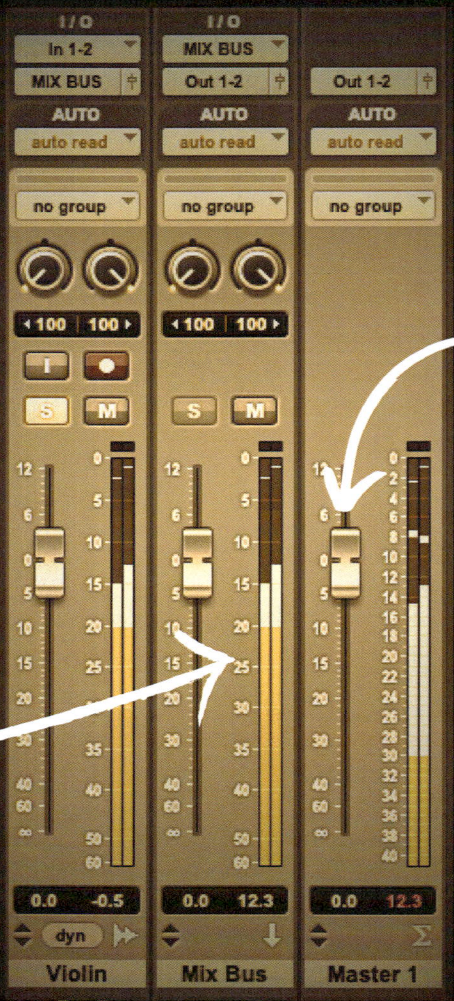

Note* All the faders are at Unity, which assures no alteration t the Violin as it passes through the Mix Bus and Master.

Note* See how the signal level of the "Mix Bus" is the same as the "Violin" at the unity position.

Panning

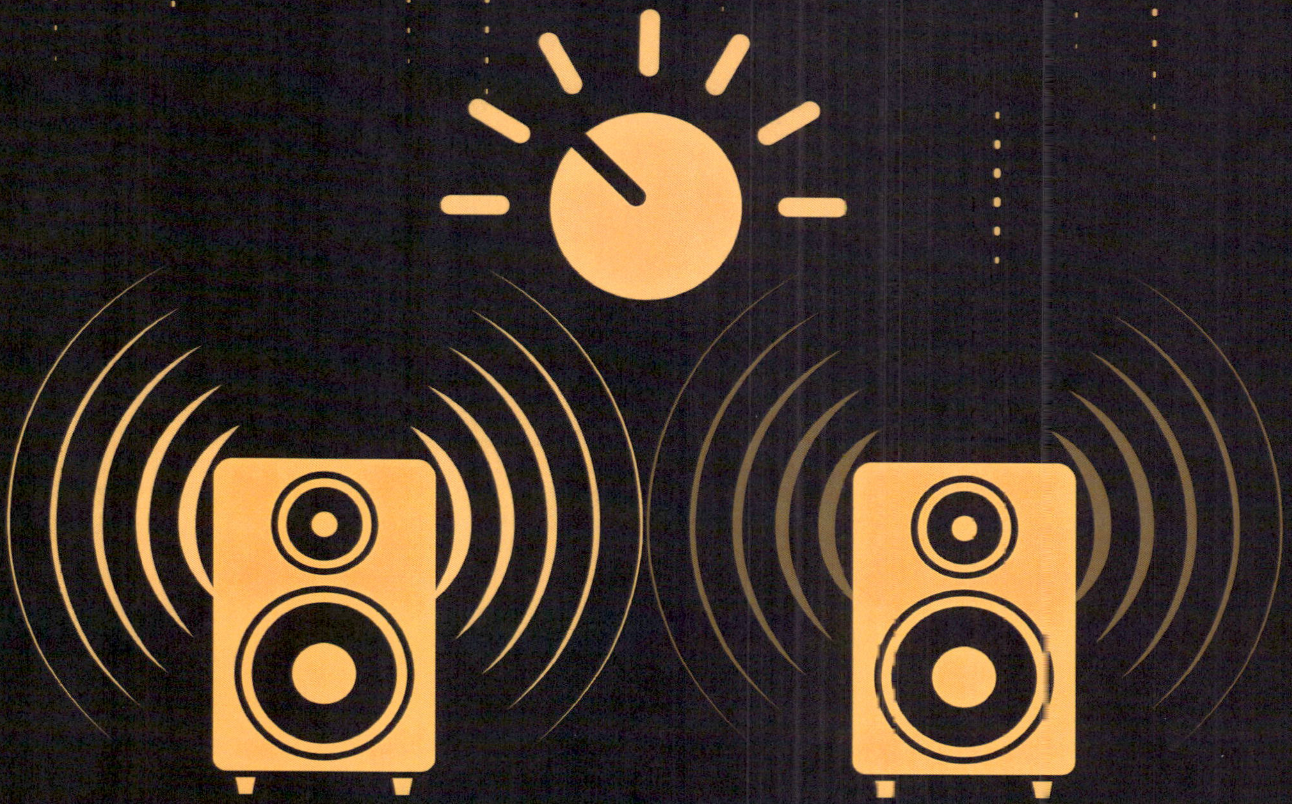

Panning is the technique used to position a sound source within a listening field. This is typically accomplished by reducing the volume sent to a specific speaker. For instance, a pan knob placed in the left position would lower the volume sent to the right speaker, creating the perception of sound coming from the left side of the stereo system.

ADSR

Attack, Decay, Sustain, and Release (ADSR) refers to the envelope or shape of a particular sound in relation with time.

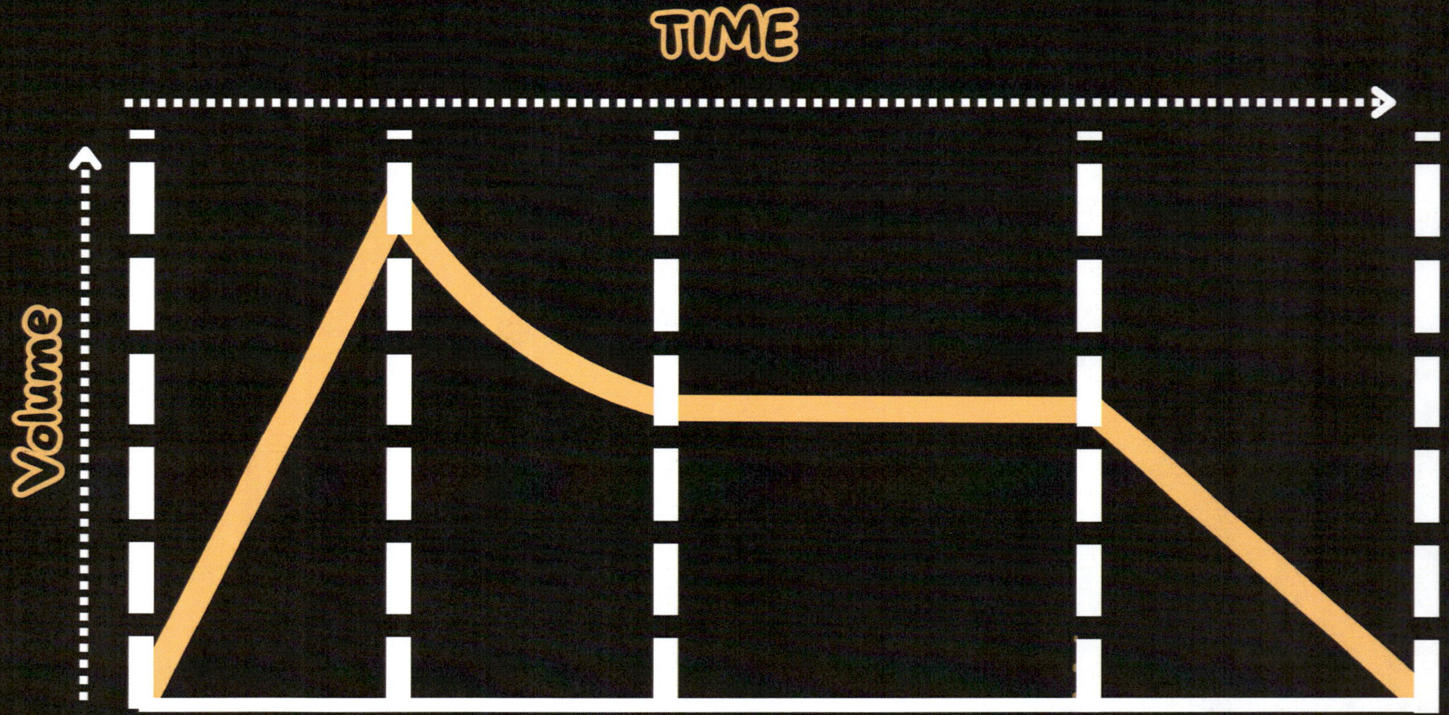

The **Attack** refers to the time it takes to reach the maximum amplitude of the waveform.

The **Decay** refers to the time it takes for the maximum amplitude to decend into the sustained information.

The **Sustain** pertains to the consistent amplitude produced when a sound is held out.

The **Release** represents the duration it takes for the sound to fade from a sustained level to complete silence.

Polarity

Refers to the positive or negative pressure of a given sound wave.

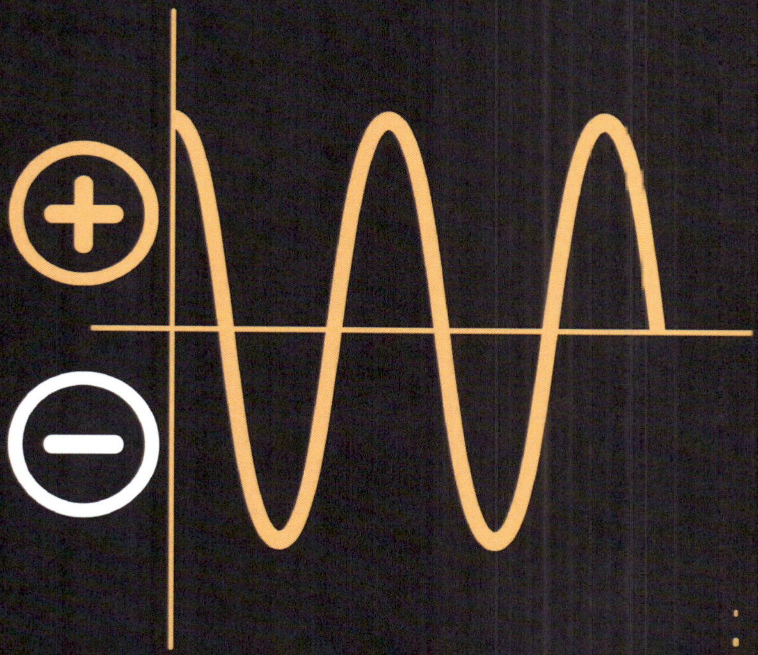

Note* The polarity of a wave form will have a significant effect on how it interacts with other wave forms.

Polarity Flip

The action of inverting the polarity of a given waveform.

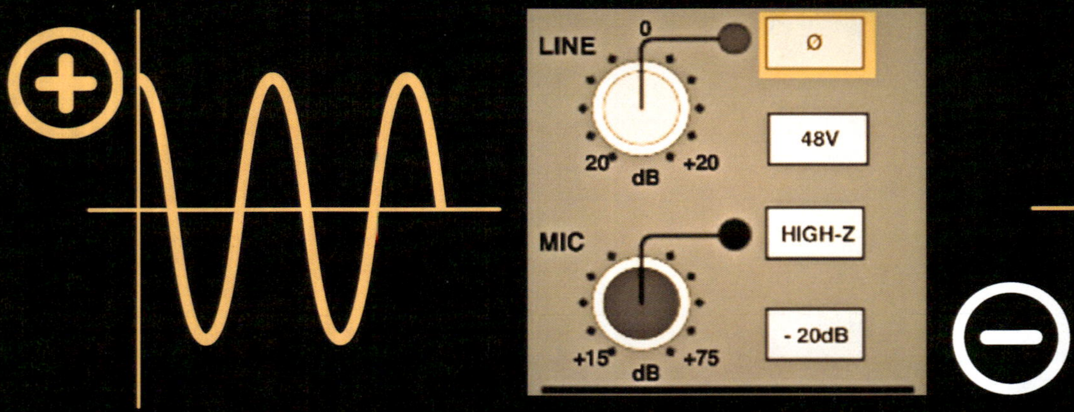

Note* To flip the polarity on a mixer, engage the polarity flip button usually located near the pre-amp section of the channel... It looks like a circle with a slash through it...

In Phase

"When a single sound source reaches multiple destinations with the same polarity in the wave cycle."

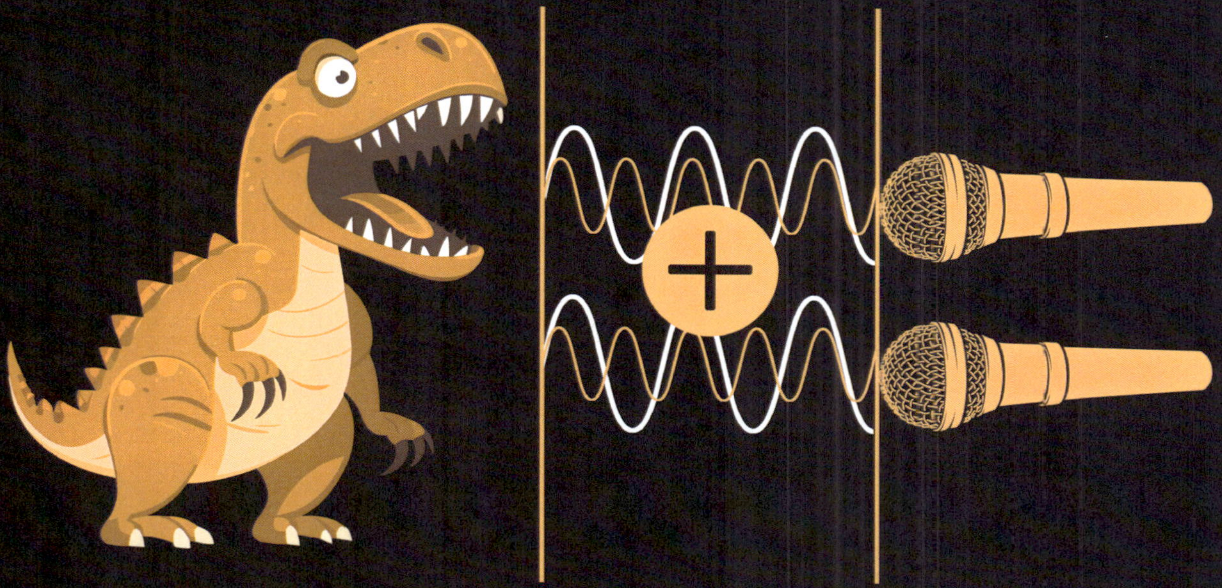

Note* These two microphones are positioned at an equal distance from the dinosaur, enabling them to capture the roar at the same point in the wave cycle.... When combined, they will compliment or "Sum" with each other.

Out of Phase

"When a single sound source reaches multiple destinations at different times within the wave cycle."

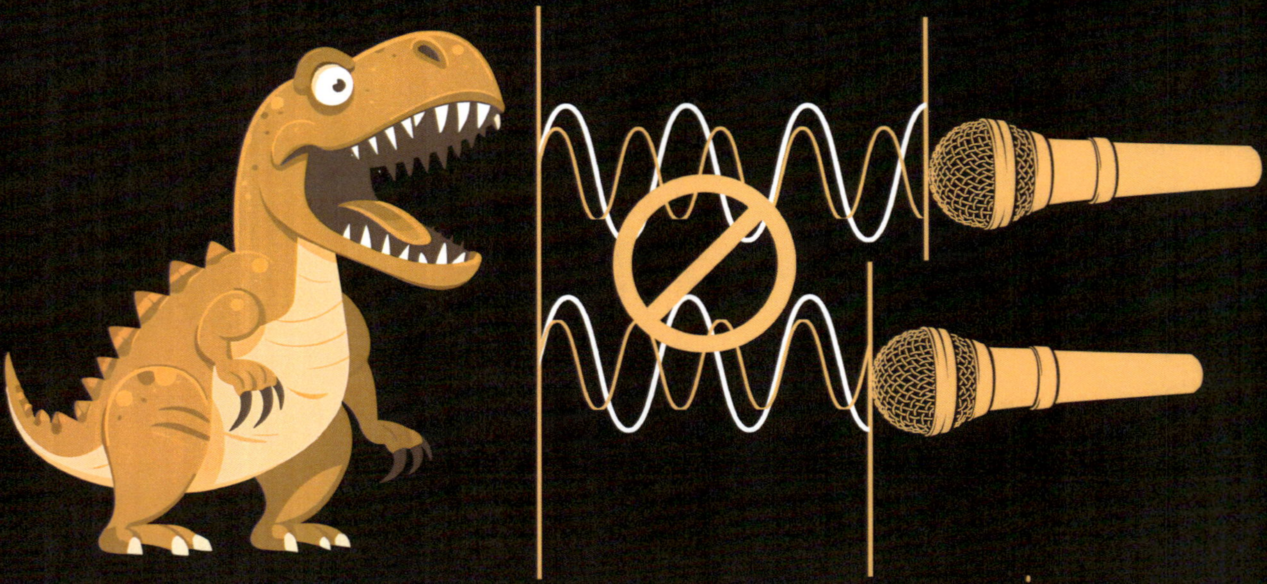

Note* These two microphones are placed at slightly varying distances from the dinosaur, capturing sound at different points in the wave cycle. When merged, they will interfere with each other.

Speaker Alignment

In Alignment

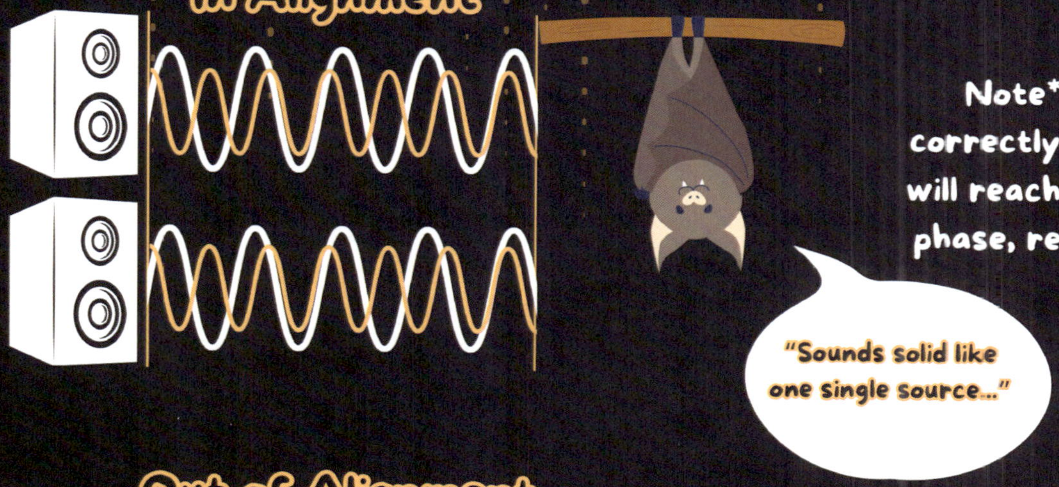

Note* When speakers are correctly aligned, all frequencies will reach the listening position in phase, resulting in "summation".

Out of Alignment

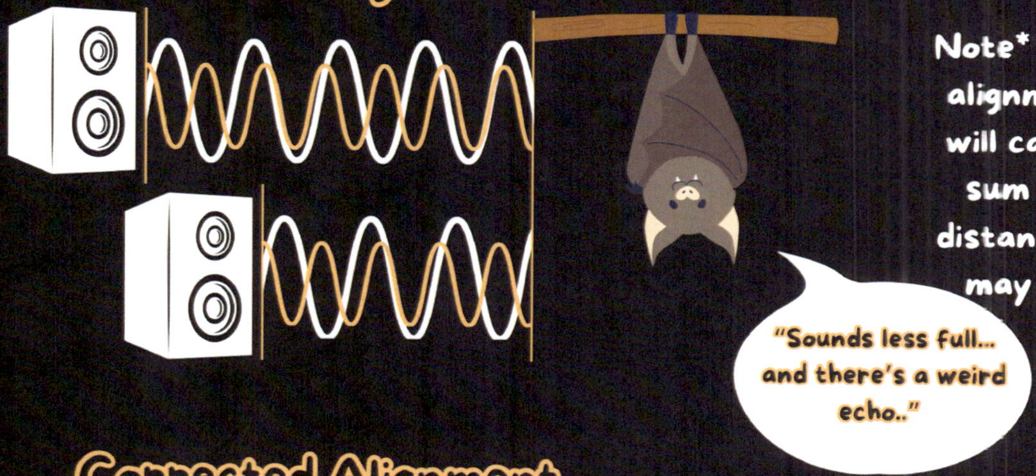

Note* When speakers are not in alignment, certain frequencies will cancel out while others will sum together.... With enough distance between speakers, you may start to hear an echo....

Corrected Alignment

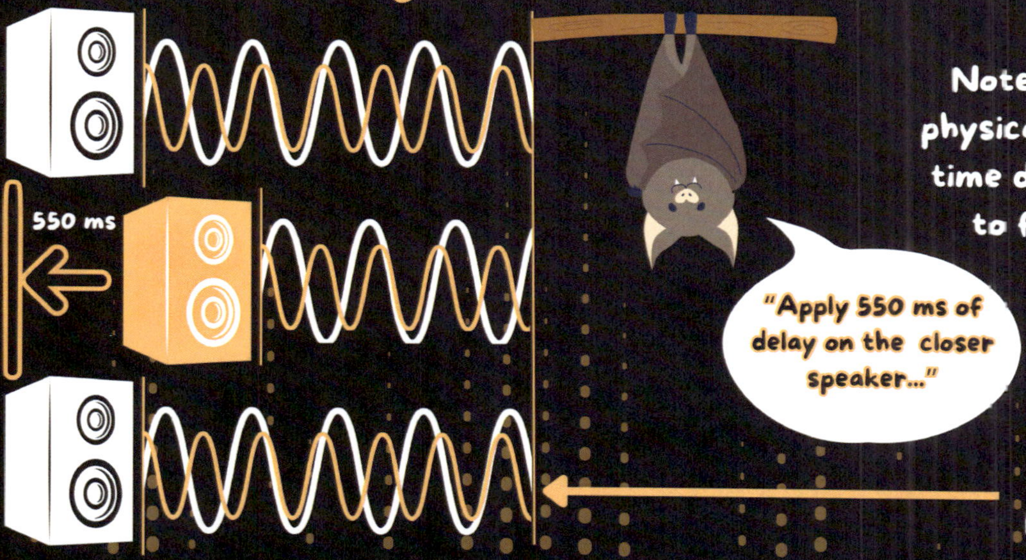

Note* If the speakers are not physically aligned, you can apply a time delay on the closer speaker to fix the time discrepancy.

Corrected speaker placement using delay...

Decorrelation

Is the act of changing a duplicate waveform while maintaining a similar sound to the ear. This can combat many of the summation and interference challenges we face combining multiples of the same waveform in a given listening environment.

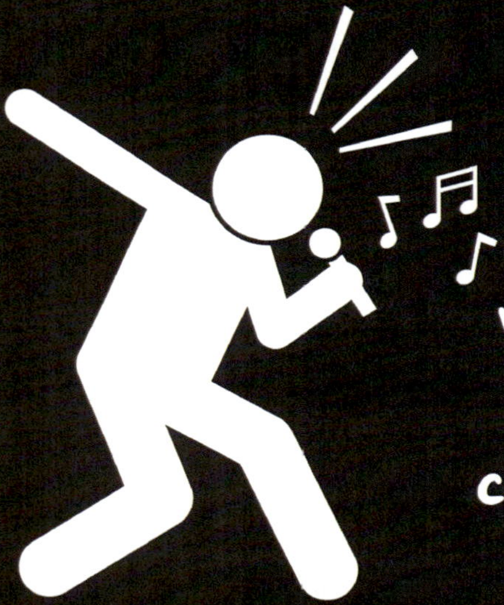

Original Vocal Sound

"These 2 versions of the Vocal will compliment each other."

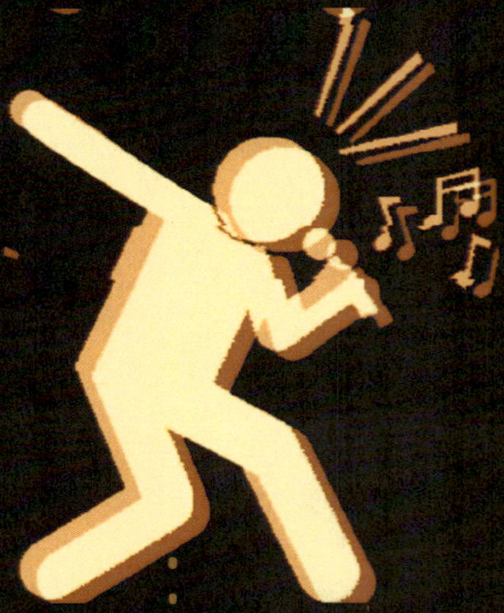

Decorrelated Vocal Sound

Tip* This concept will play a significant role in crafting effective immersive content.

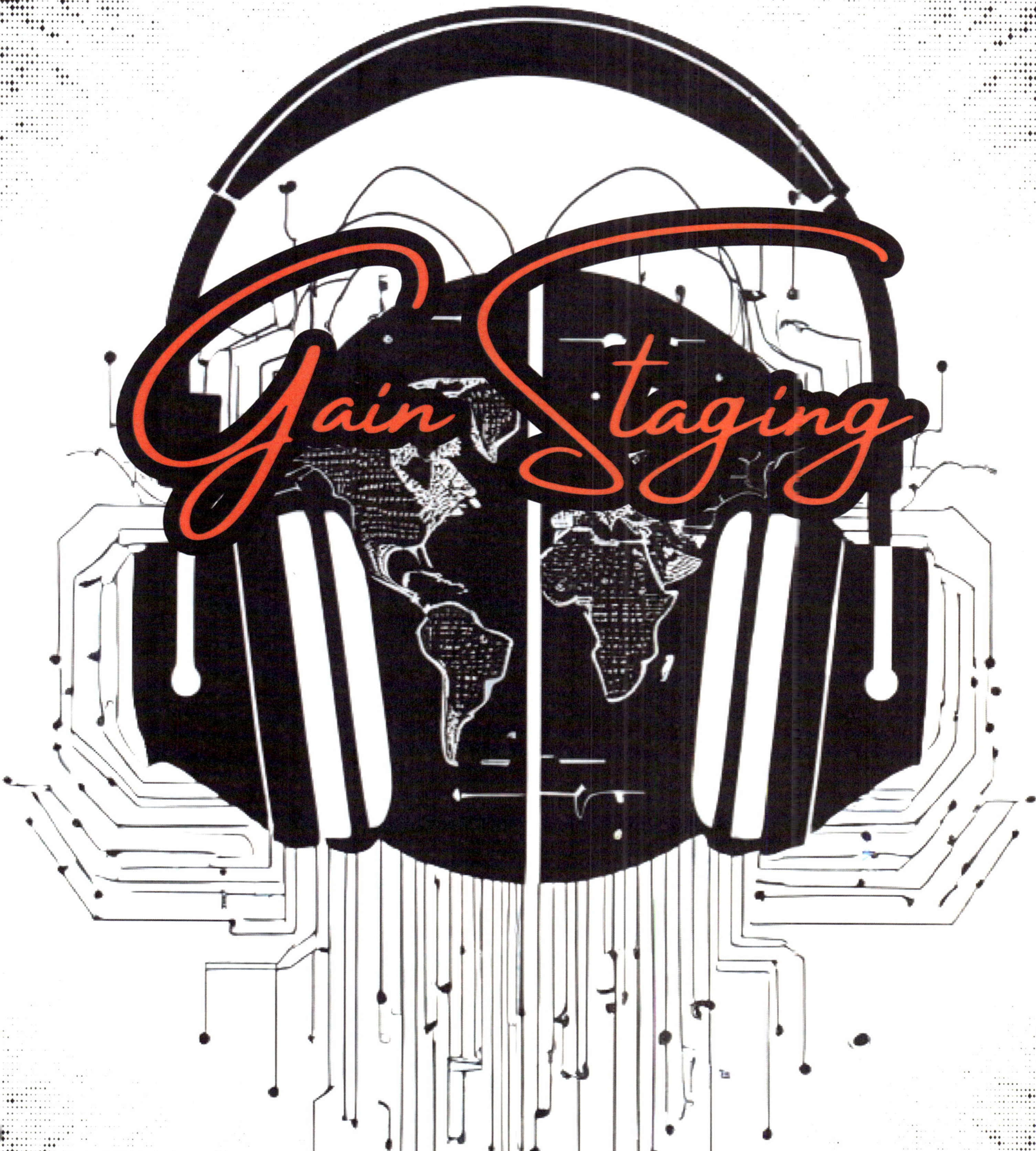
Gain Staging

What is Gain Staging?

Gain Staging is the act of optimizing audio signal levels by achieving the best signal to noise ratio, while demonstrating relative fader position to what you hear...

Much like a doctor practices medicine....
An audio engineer practices gain staging...

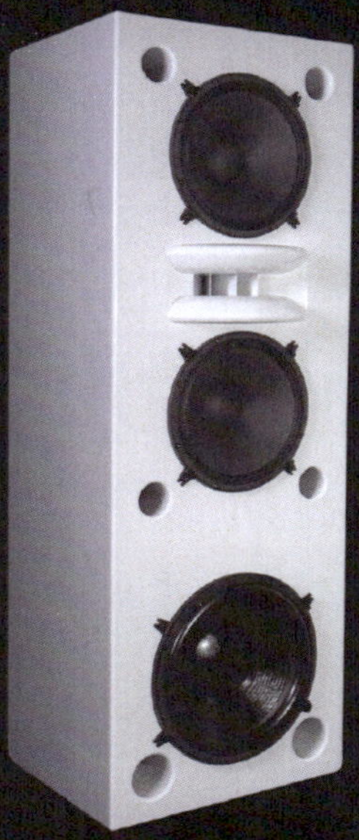

Speaker Level: The output level of the speaker amplifier

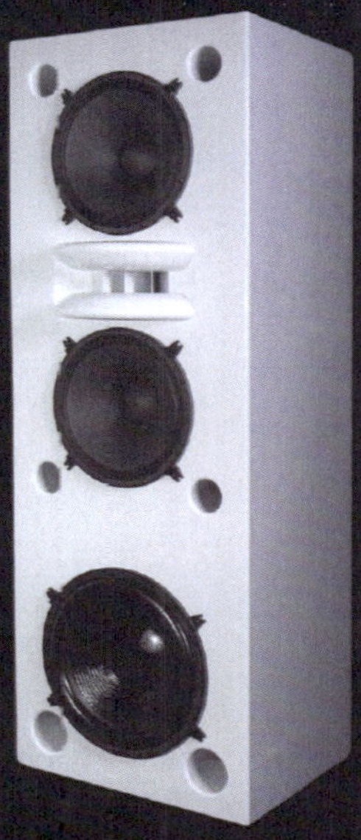

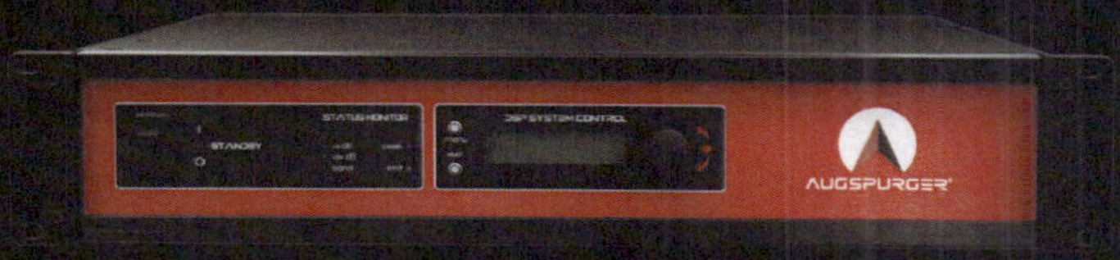

Speaker Amplifier

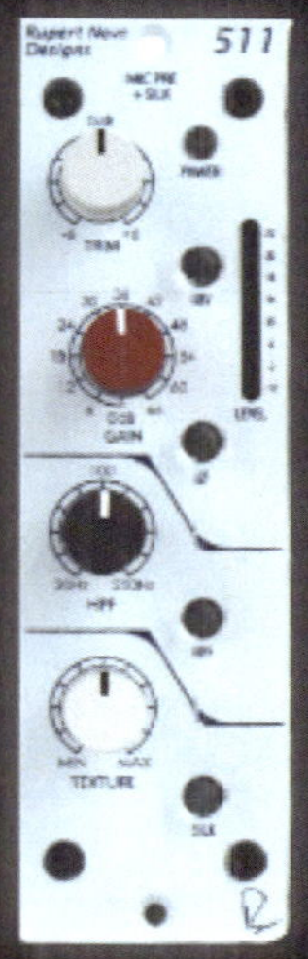

Microphone Pre-Amplifier

Line Level: The output level of a pre-amp and mixing console

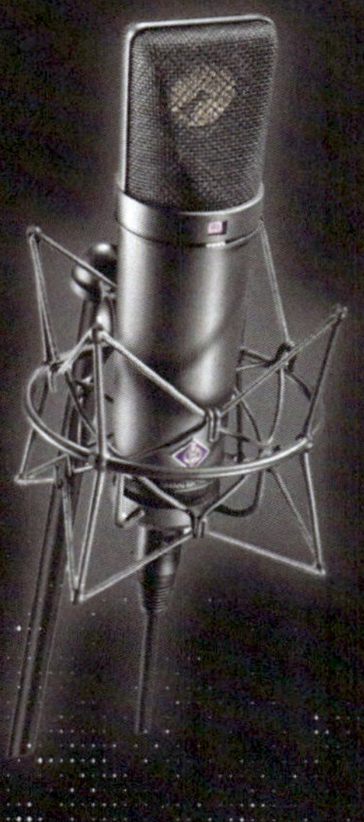

Mic Level: The output level of a microphone

Preamp Technique

Start with the "Trim" or "Output" at Unity (0dB)...

Raise the "Gain" until the signal is reading between -18dB and -12dB on your DAW input...

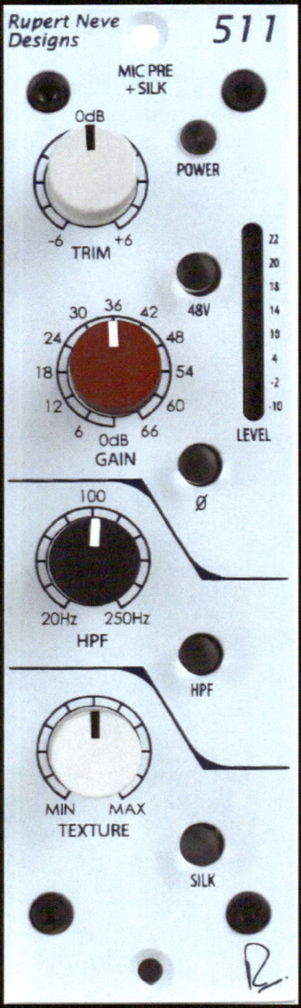

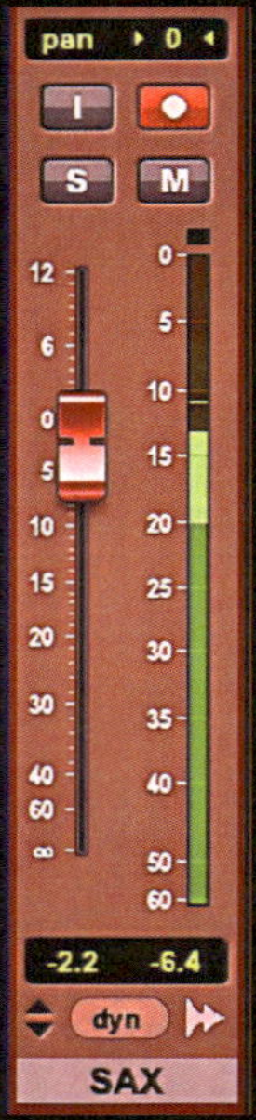

Try some different flavors...

Boost the "Gain and reduce the "Trim" to enhance saturation.

Reduce the "Gain and boost the "Trim" for a cleaner sound.

Note: While "Gain" and "Trim" are frequently used interchangeably, it's essential to understand their distinctions. Gain refers specifically to the amplification of a signal as it travels through the circuit. In contrast, Trim indicates the attenuation or slight adjustment to the gain level. You can observe the difference in the knob values: Gain begins at 0dB and increases with positive numeric values, whereas Trim shows 0dB in a neutral position, permitting both positive and negative dB adjustments.

Recording Levels

Turning up a weak signal will amplify and expose the noise floor.

Turning down a distorted signal won't get rid of the distrotion.

This balance is ideal! You can tweak this signal up or down without any negative consequences.

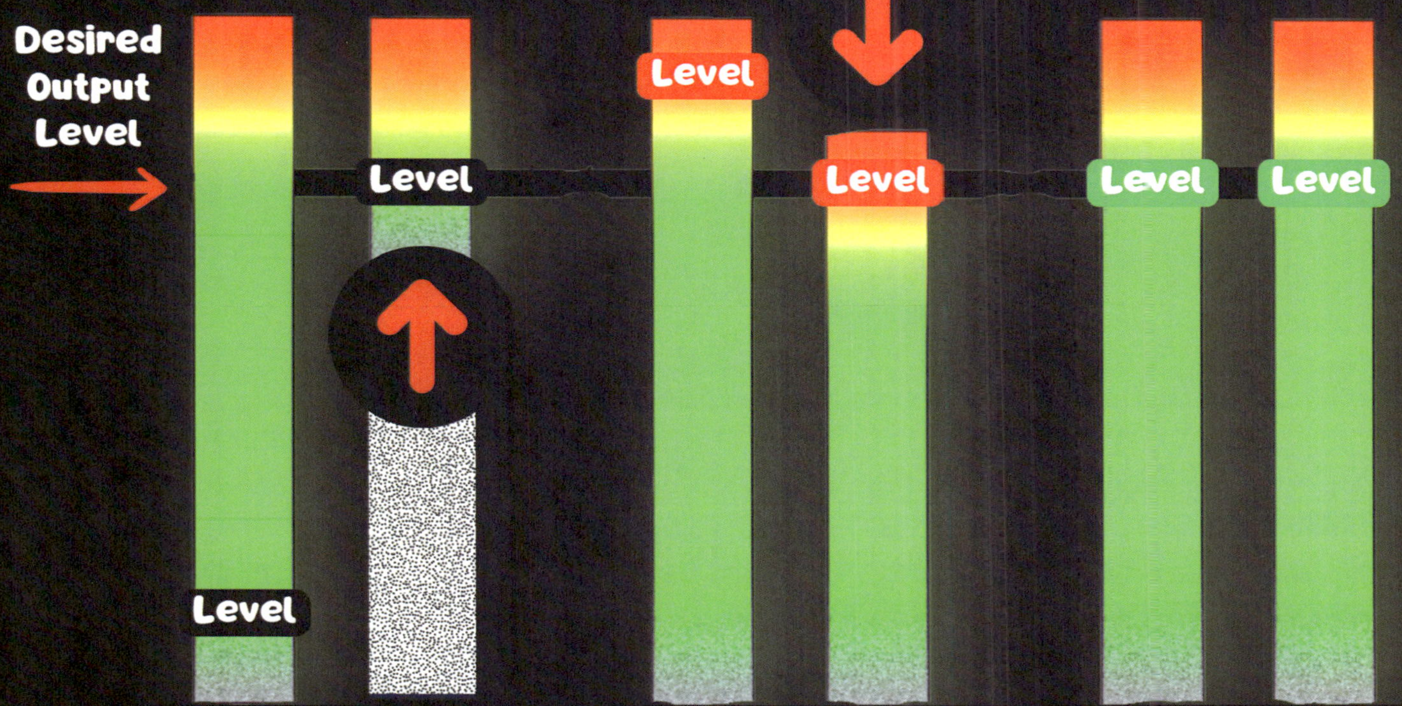

Recording a signal with too weak of a signal will result in a noisy signal.

Recording a signal too strong can result in a distorted signal.

Recording a signal at the same level as your desired output will result in the best signal to noise ratio.

Relative Fader Position

At the end of the day, the "sound" is what matter most. However, it's important to prepare your "gain structure" in a way that creates an intuitive workflow. This will make balancing and distributing mixes a much easier and efficient process. The relationship between how signal strength looks and the volume that is perceived by a listener can be quite misleading. This is do to the speed in which various types of meters react to transients, as well as how our ears perceive volume at particular frequencies.

By first setting the faders in a position "relative" to the balance you want to hear, you can then balance the mix at the "pre-amp level" (setting the gain structure for everything in the mixer). This does not mean "Don't touch your faders!". Remember this is a technique for calibrating your gain structure, not the mix itself. Avoid losing sleep over perfect gain structure, as sound and music is a moving target. After all, you can always "re-gain structure" if you find yourself fighting your faders, overdriving a channel, or capturing a noisy weak signal. Remember, this is a practice, much like how a doctor practices medicine, an engineer practices gain structure.

1

Turn both the pre-amp and speaker levels and all the way down..

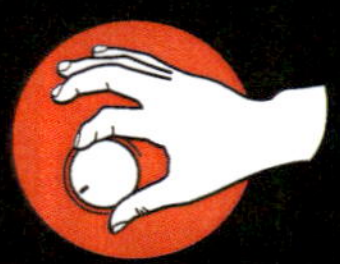

2

Raise your master fader to unity and your input faders to a uniform level (either unity or -5dB)

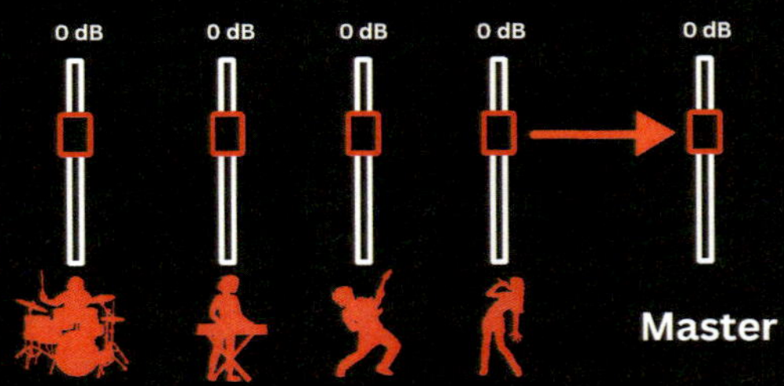

3

Increase the pre-amp gain for either the vocal or kick drum until your input signal reaches around -12dB to -8dB at its loudest points. This will leave you headroom in case the artist gets excited during their performance....

These tracks tend to meter a little higher than other instruments, making them a great signal reference for setting your speaker volume in the next step...

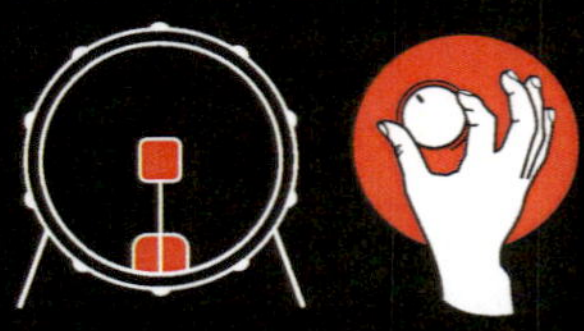

4

Determine your listening level by turn up the speakers until your "kick drum" feels appropriately loud for the situation...

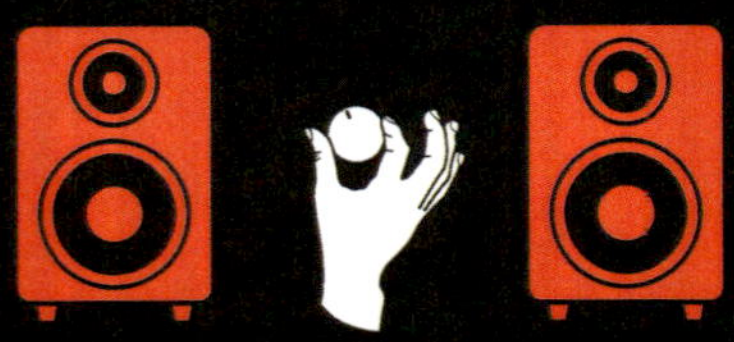

5

Now since you have established the appropriate volume for your speakers, balance the remaining inputs using only pre-amps to establish a basic mix...

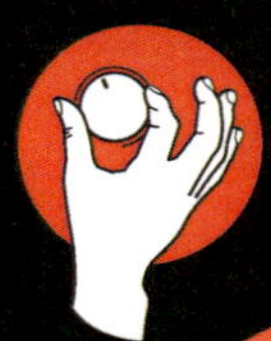

6

If you have done all 5 steps, you will realize the mix "sounds" balanced with faders in a uniform position... Now your fader movements will show "relativity" to the mix you are hearing. Notice the "relative fader position" demonstrated with these mixes below.

Vocal is most prominent

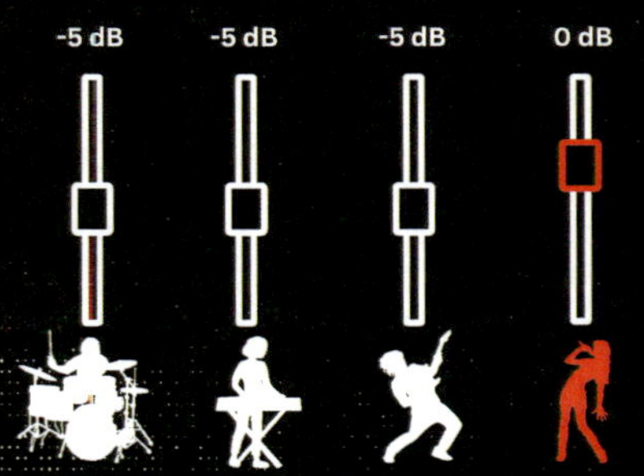

Guitar is most prominent

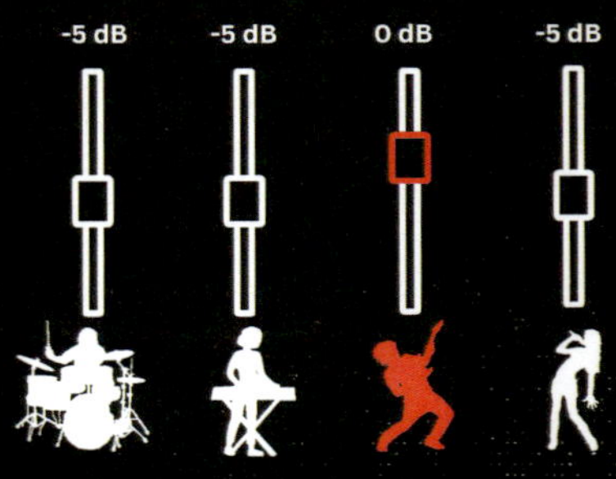

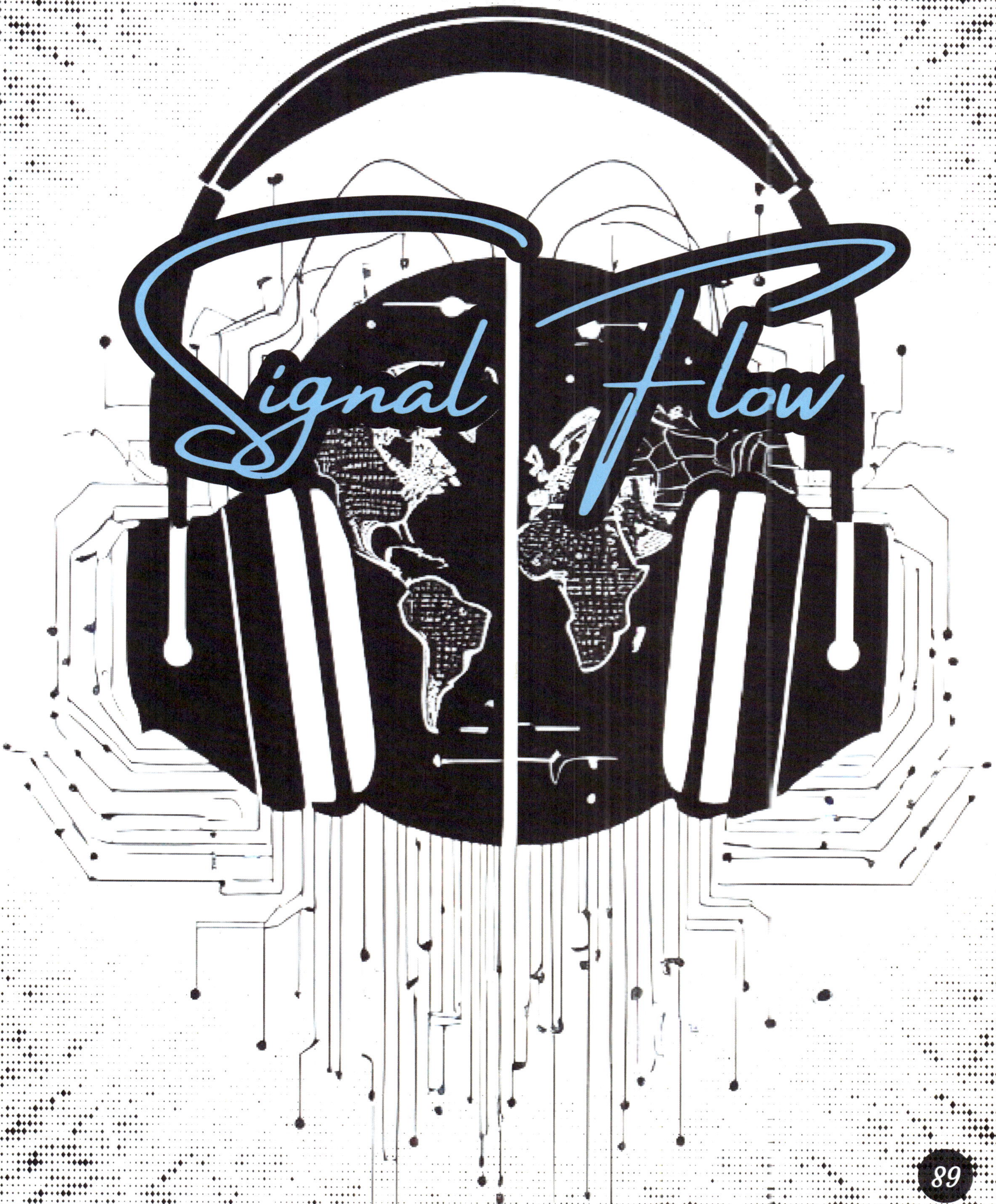
Signal Flow

What is Signal Flow?

Signal flow is the journey in which an audio signal travels...

The technical proficiency of an audio engineer relies significantly on their ability to manage audio routing and follow signal flow.

Mixing Console Basic Concept

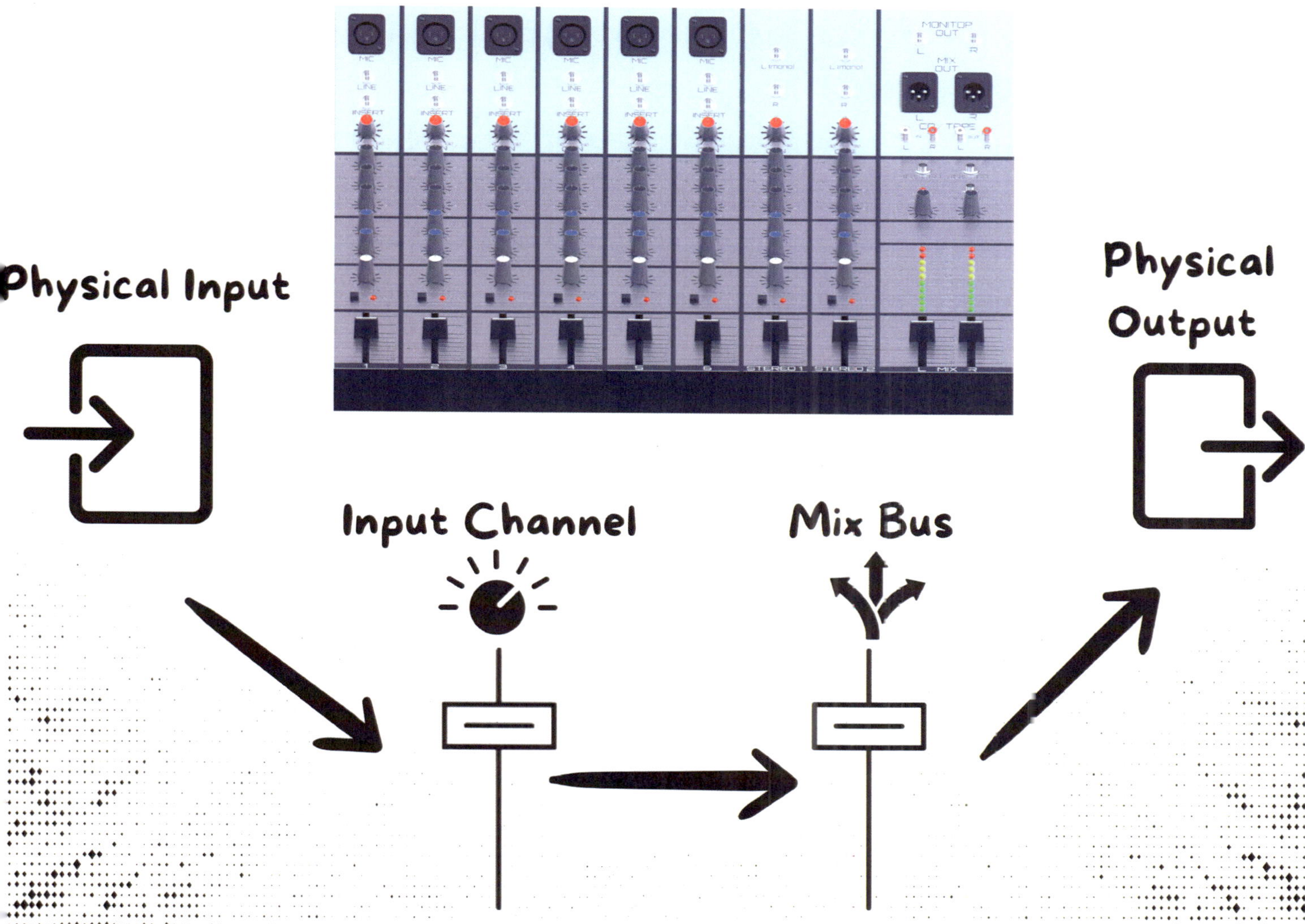

Physical Input

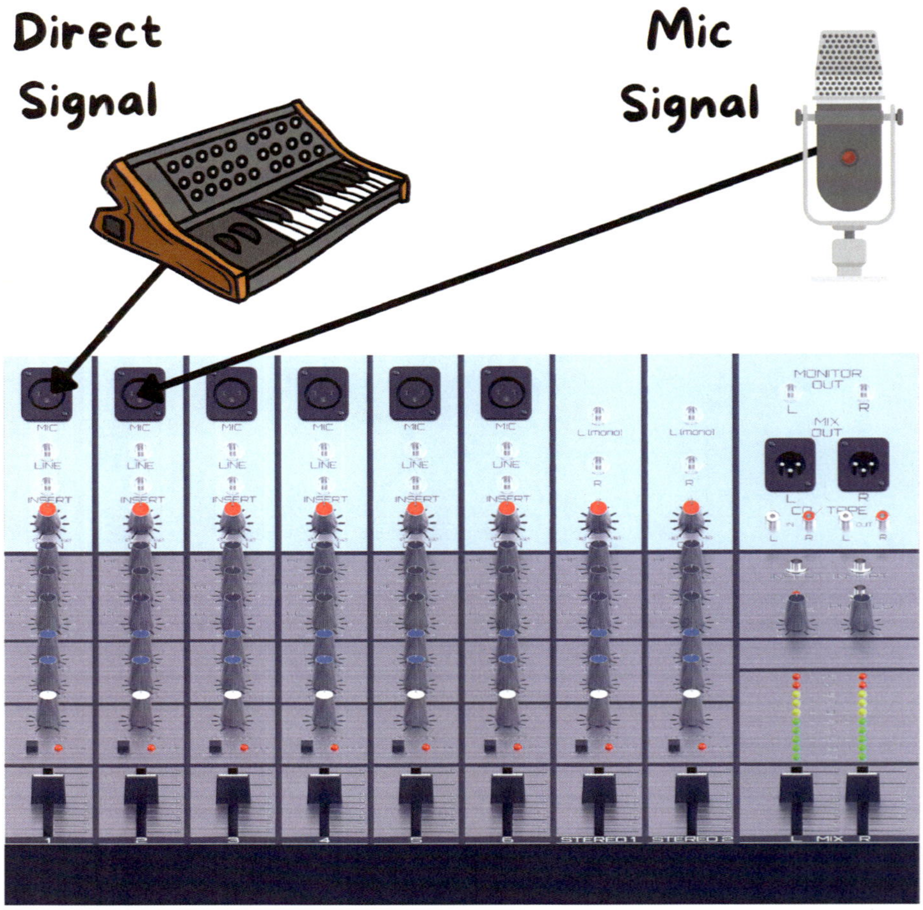

*The physical location audio enters a piece of gear.....

Input Channel "The Channel Strip"

*The input's destination that hosts channel processing, routing, and volume levels.

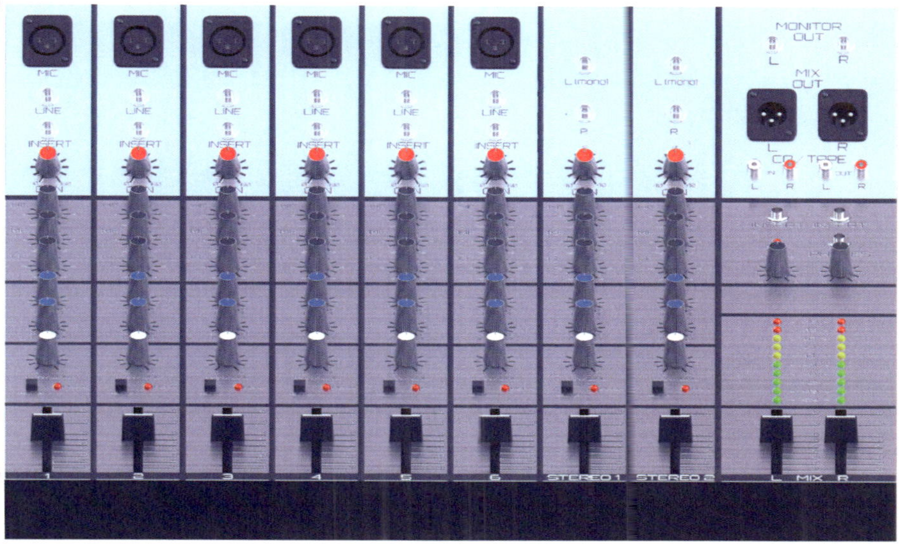

Microphone Pre-Amplifier "Mic Pre"

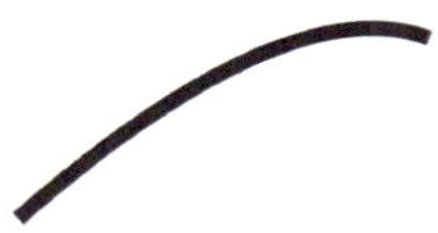

Dynamics "Compression/Gate"

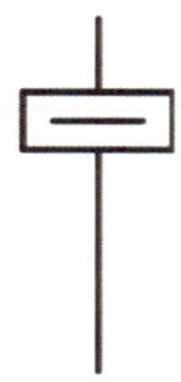

Input Chanel Fader

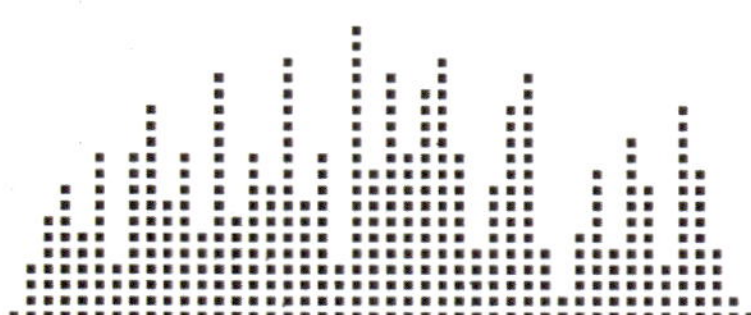

Equalizer "EQ"

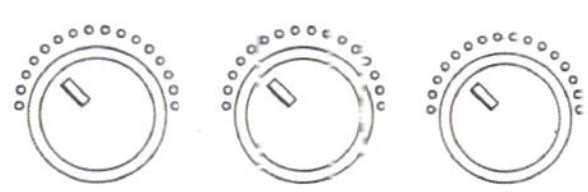

Auxiliary and FX Sends "Sends"

Physical Output

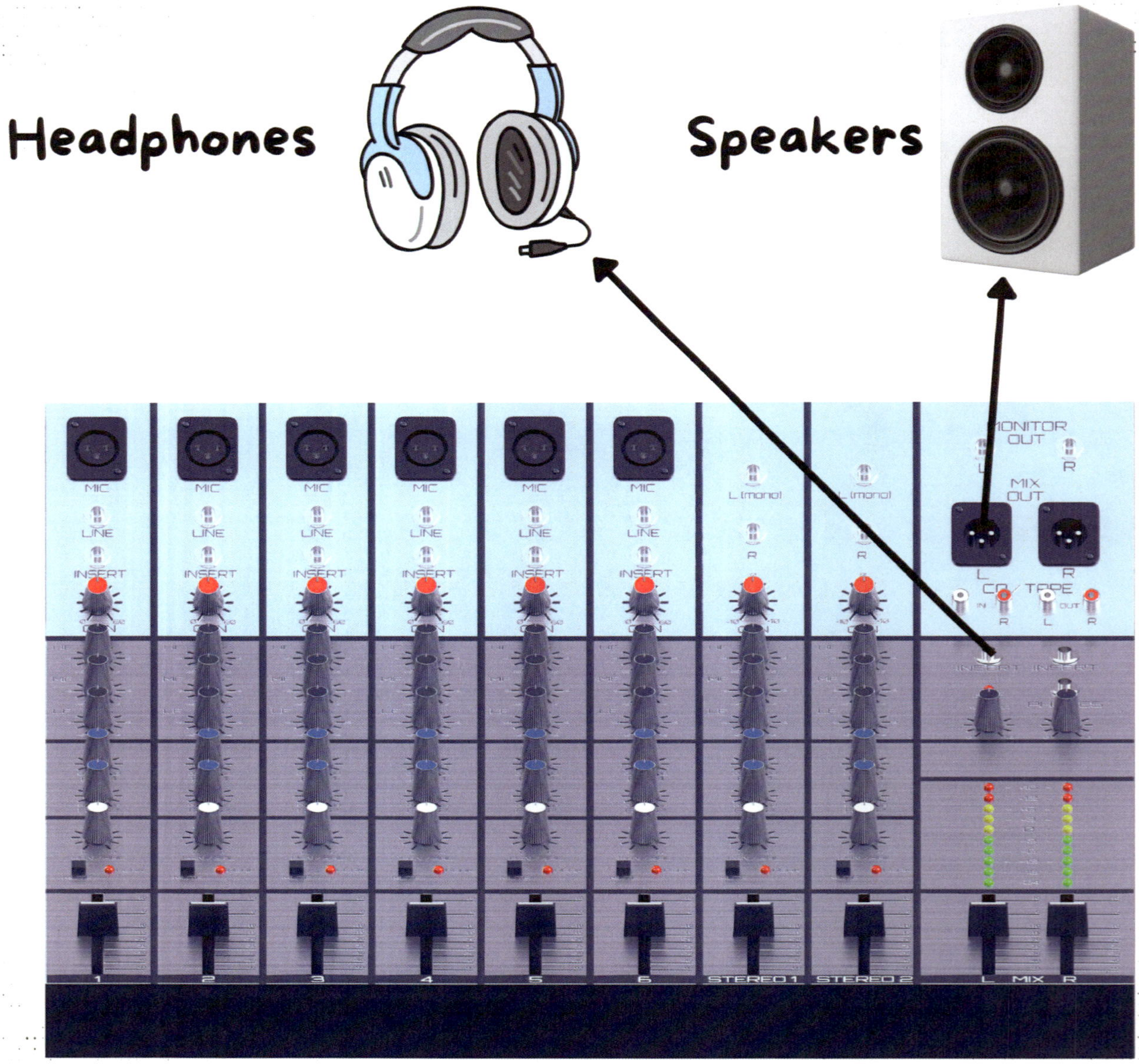

*The physical location audio exits a piece of gear.....

The Mixbus Hierarchy

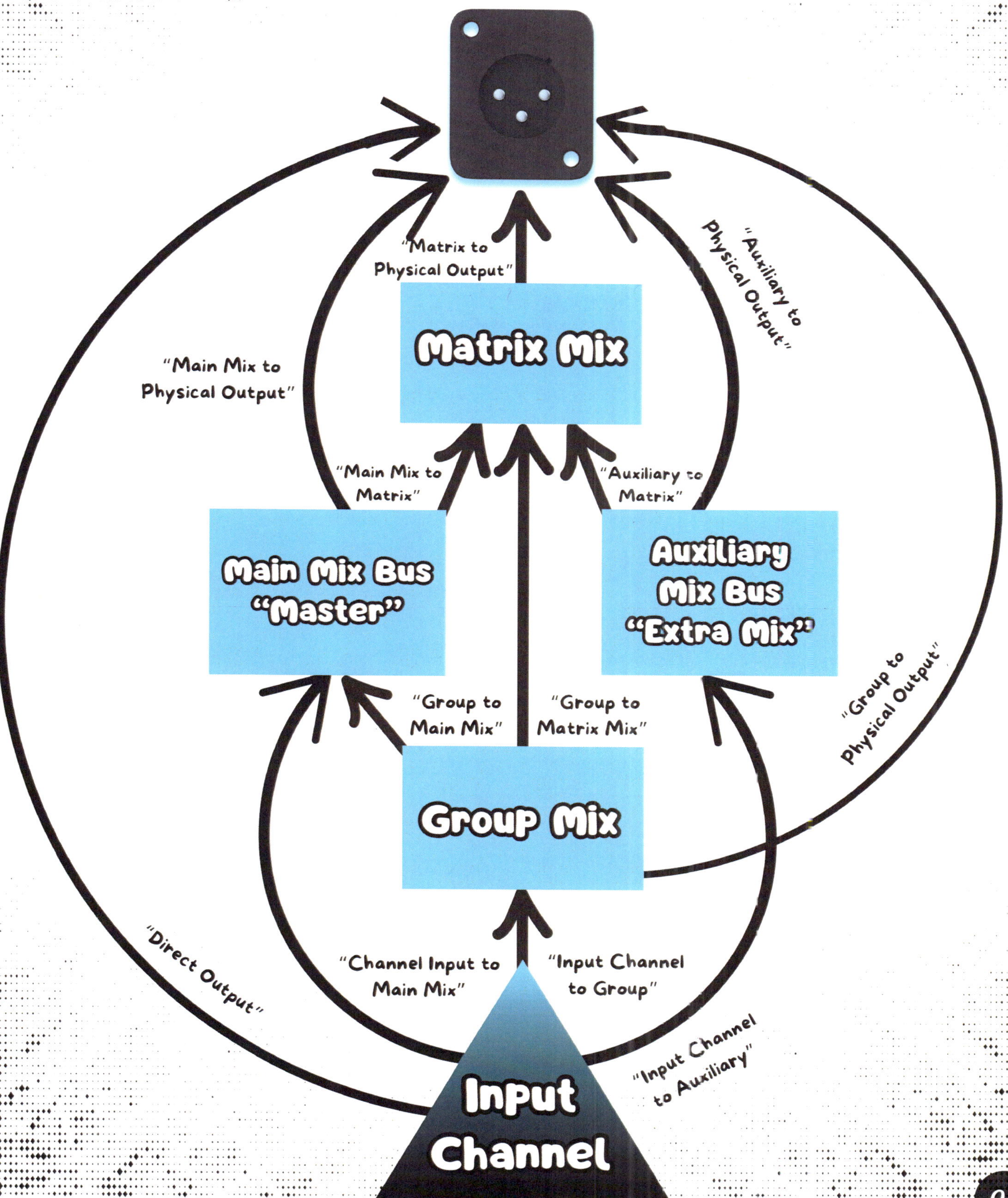

The Direct Output

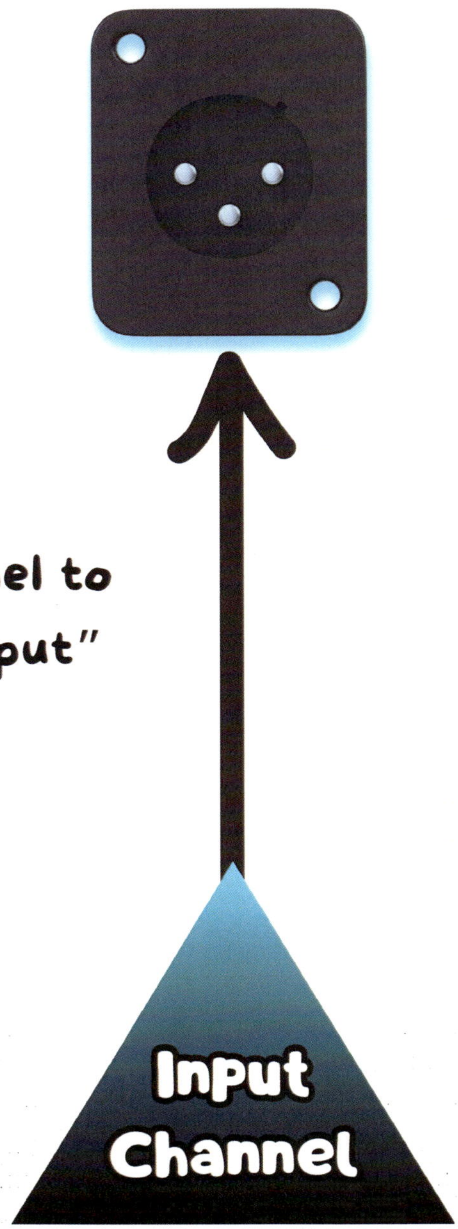

"Input Channel to Physical Output"

The Main Mix

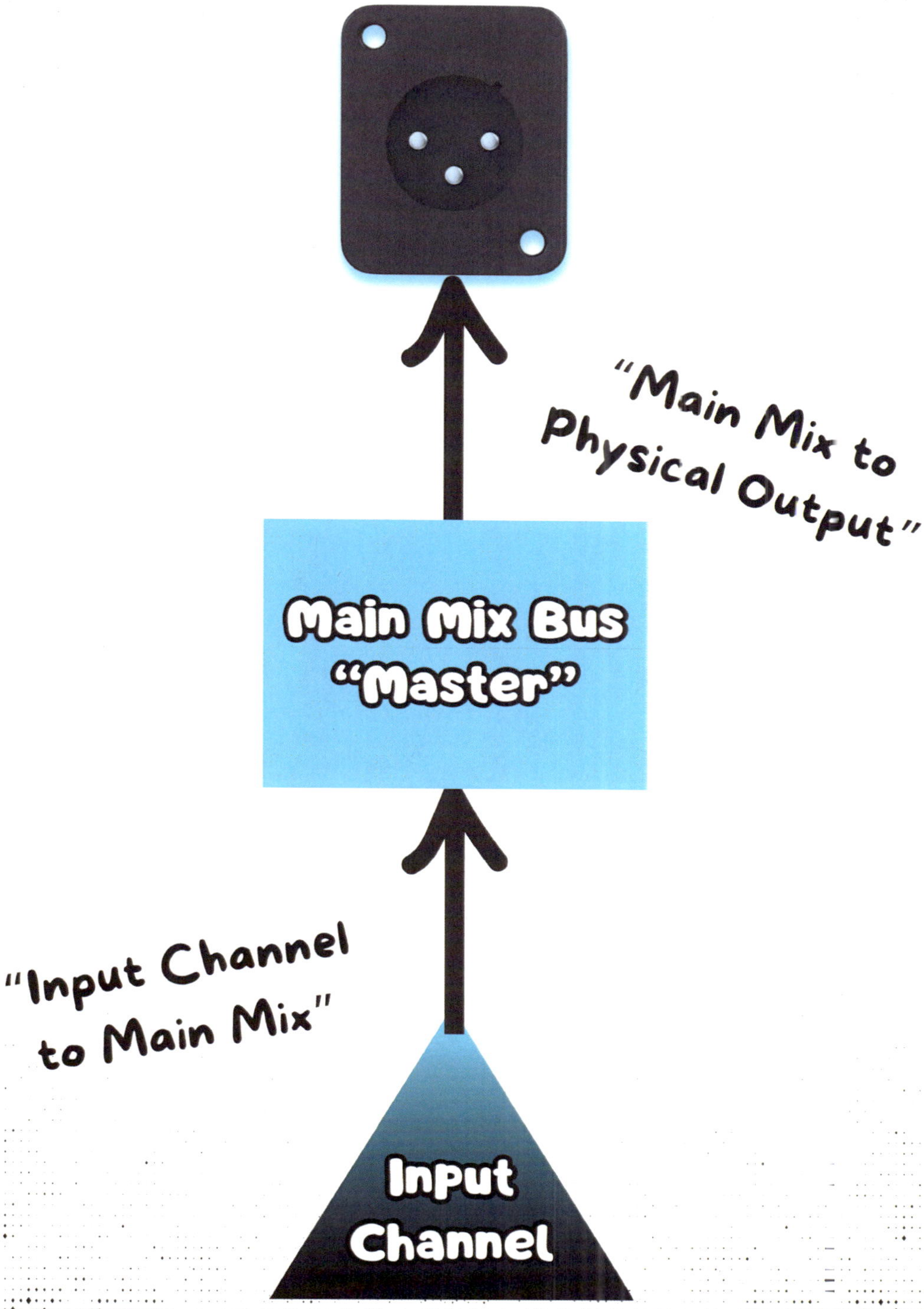
"Main Mix to Physical Output"
Main Mix Bus "Master"
"Input Channel to Main Mix"
Input Channel

The Auxiliary Mix

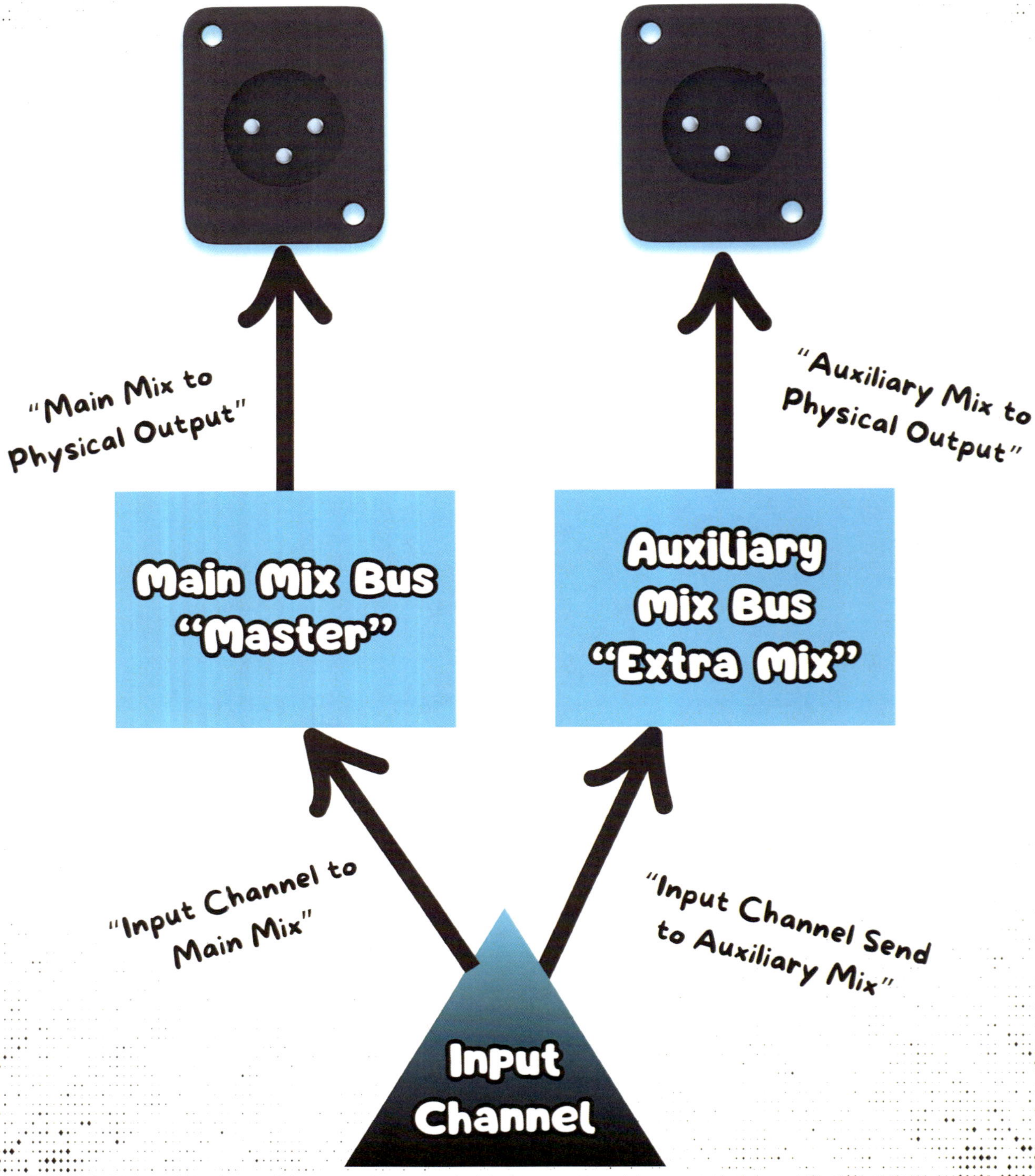

Tap Points

When a signal is sent to an Auxiliary Mix, the point from which the signal is sent within the input channel is referred to as a "Tap Point."

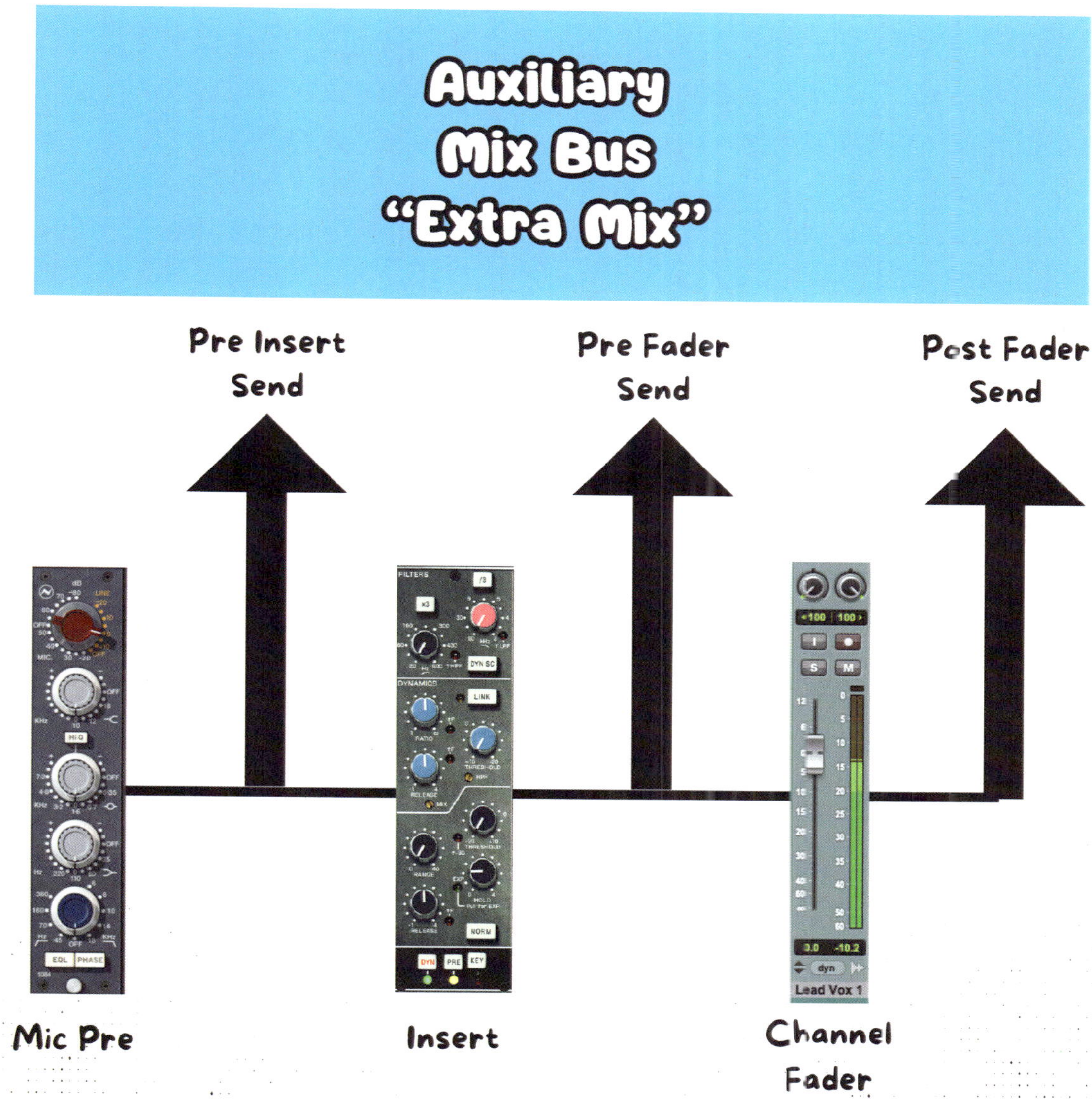

Input Channel Signal Flow

The Group Mix

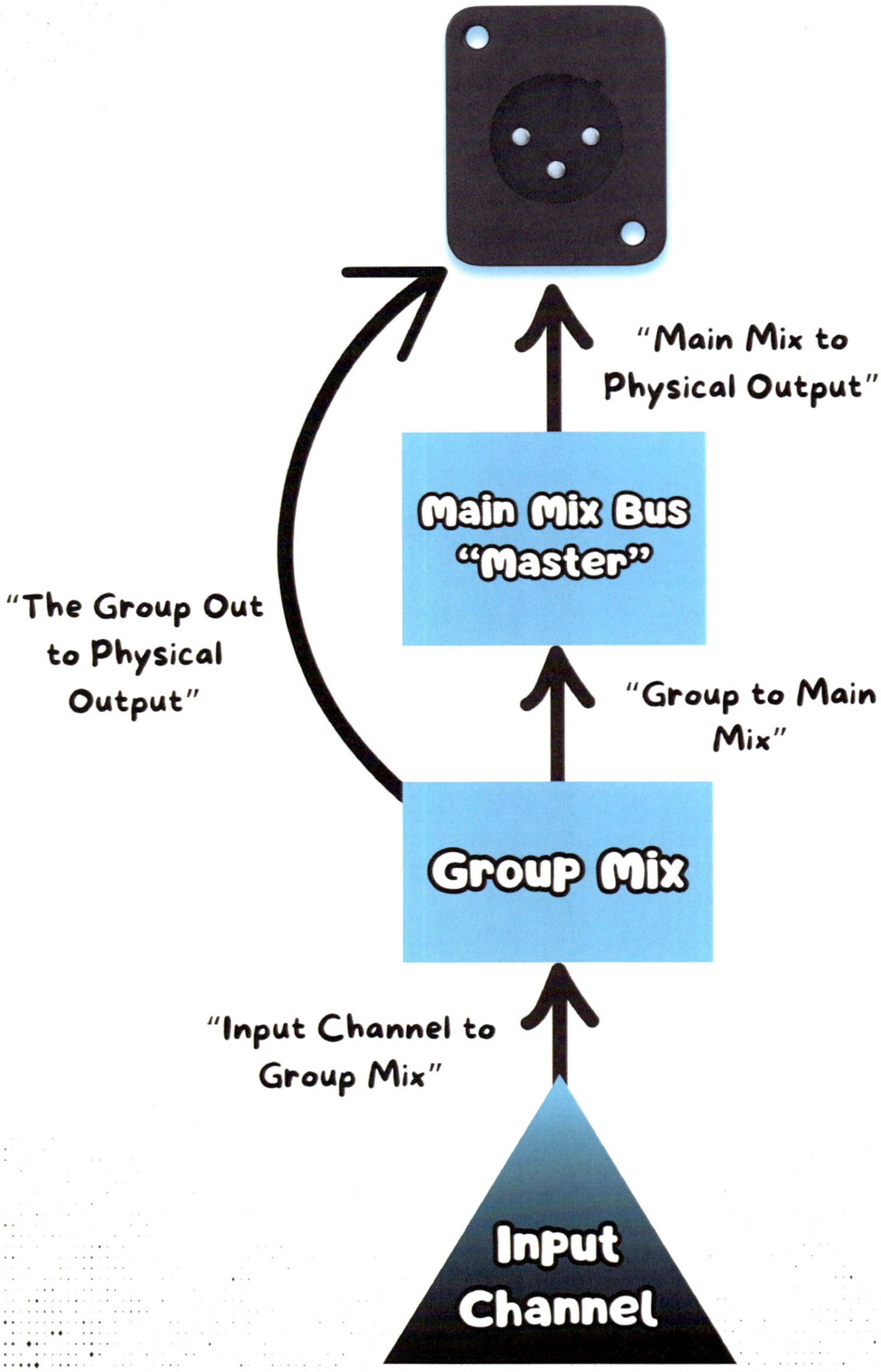

VCA Group

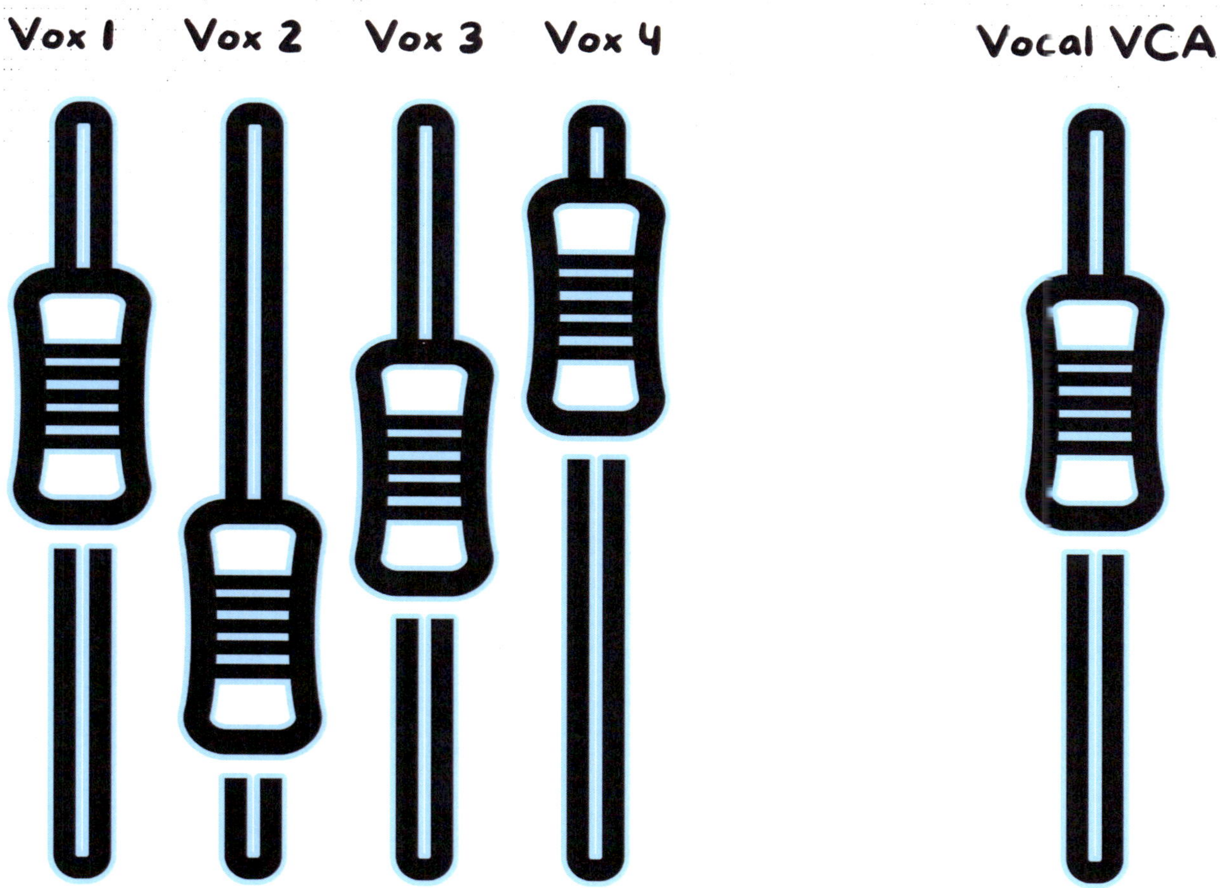

A Voltage Controlled Amplifier, also known as a "VCA," allows volume control of multiple channel faders using a single "VCA" fader. While VCAs can act similarly to a group, they do not pass signal. This means there is no audio processing within a VCA.

The Matrix Mix

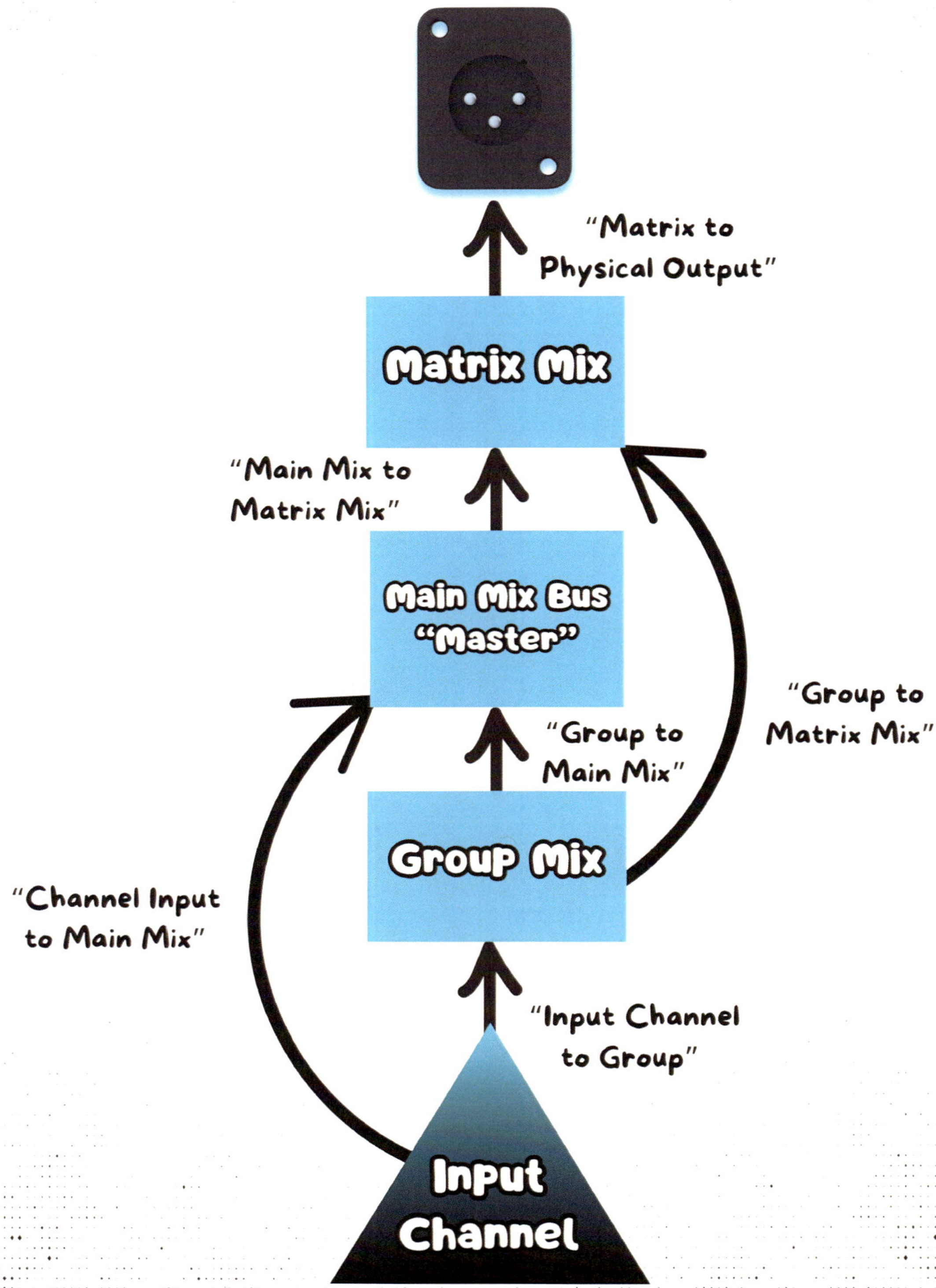

Object Based Audio

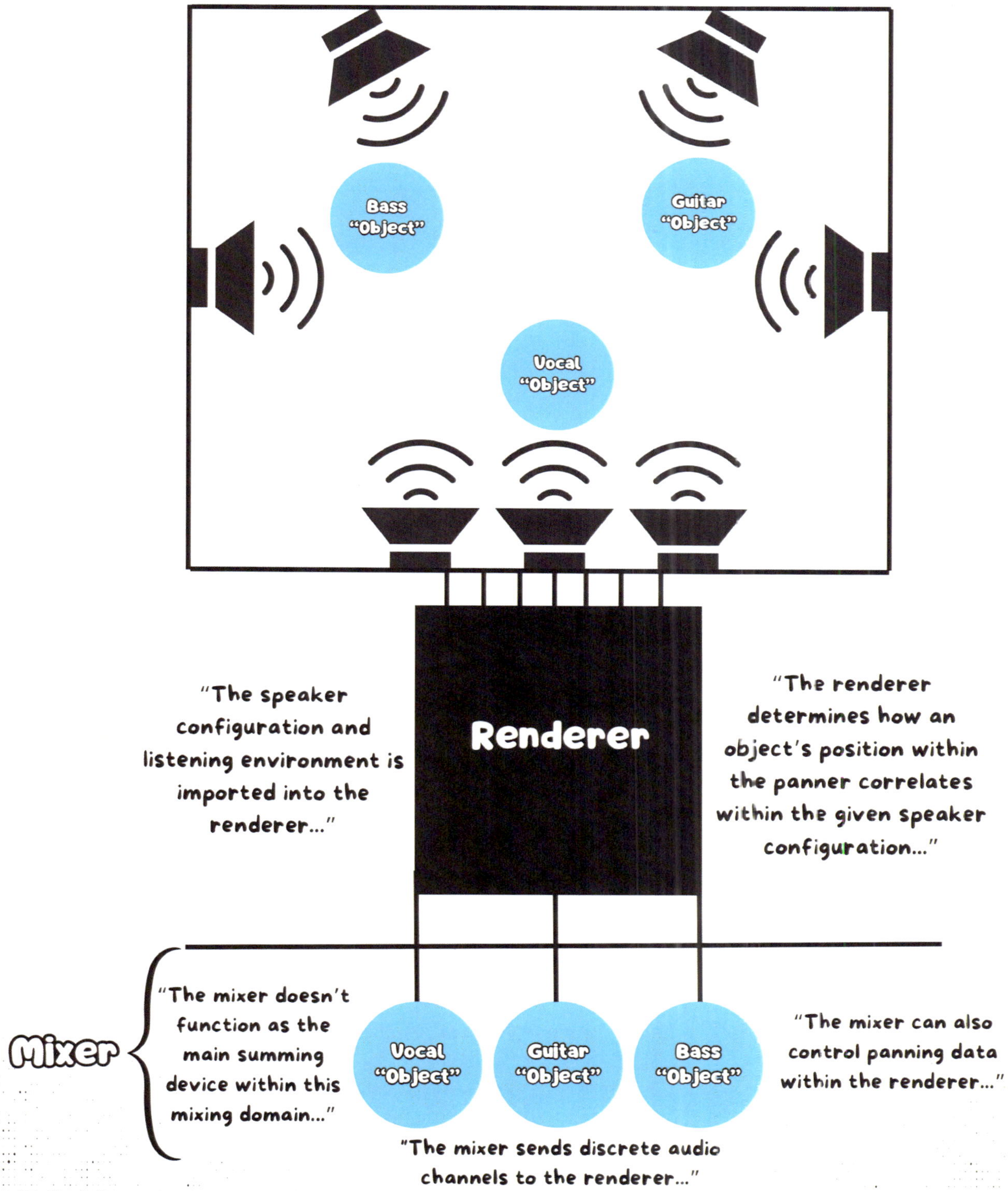

Patchbay

A patchbay is a device that acts as an interconnect between in and out points of audio equipment. Every channel features 4 jacks, 2 in the front (top & bottom), and 2 in the back (top & bottom).

Front

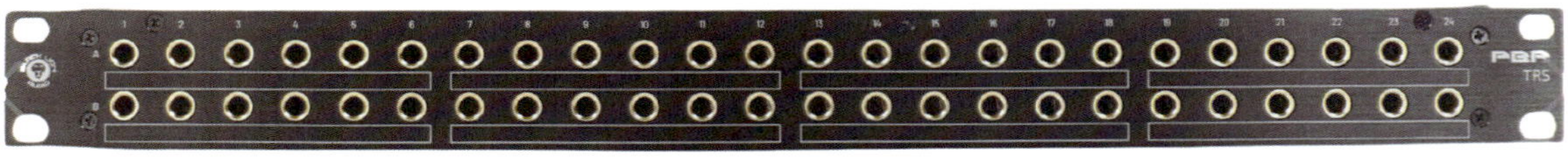

Back

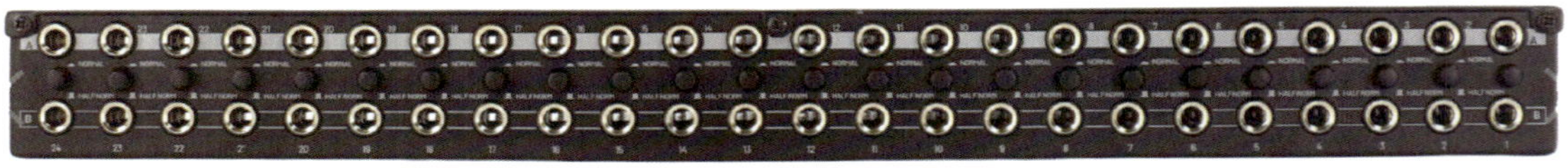

Patchbays are configured to connect cables from the rear position, offering engineers easy access to the connection points from the front position.

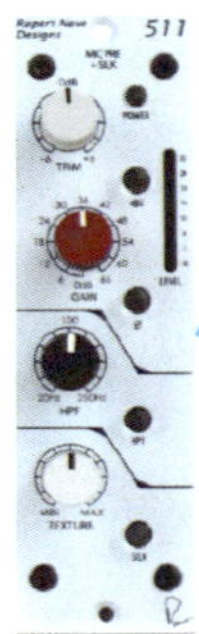

Mic Pre Line Out

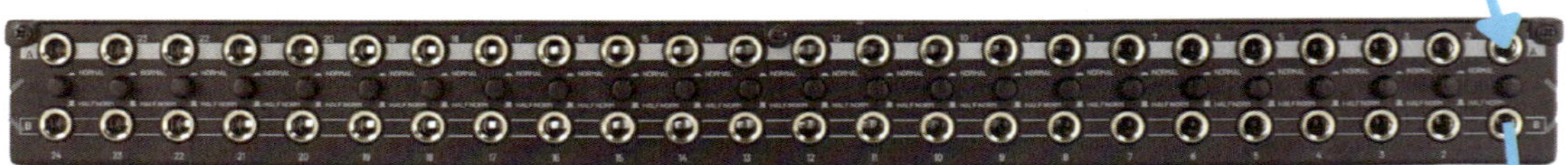

Note* In most cases, the signal of a given patch channel will flow from top to bottom... Notice how the "Mic Pre Line Out" is terminated on the top jack and the "Interface Line Input" is terminated on the bottom jack...

Interface Line In

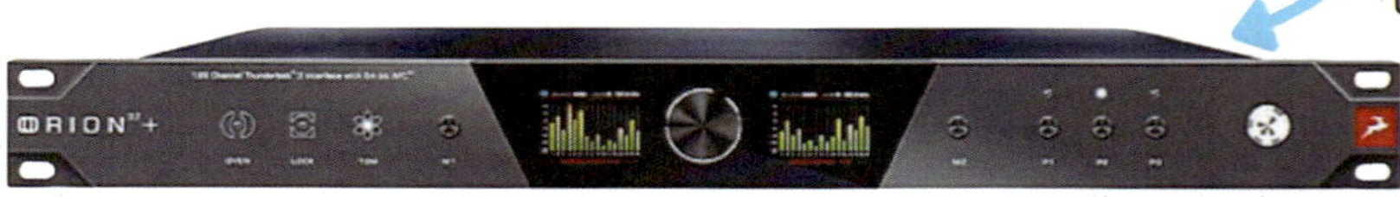

Patchbay Normalling

Non-Normalled "Tie Lines"

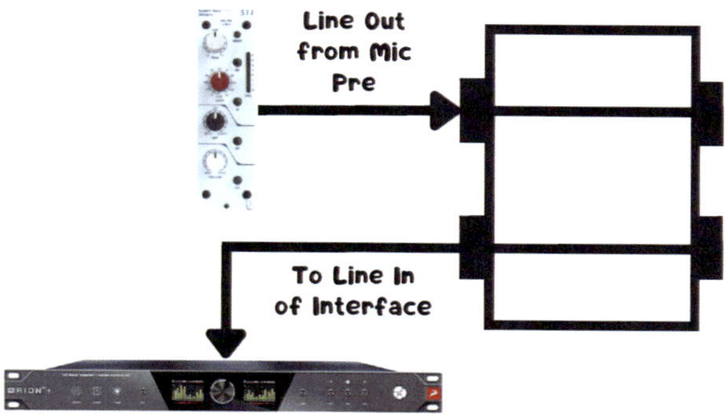

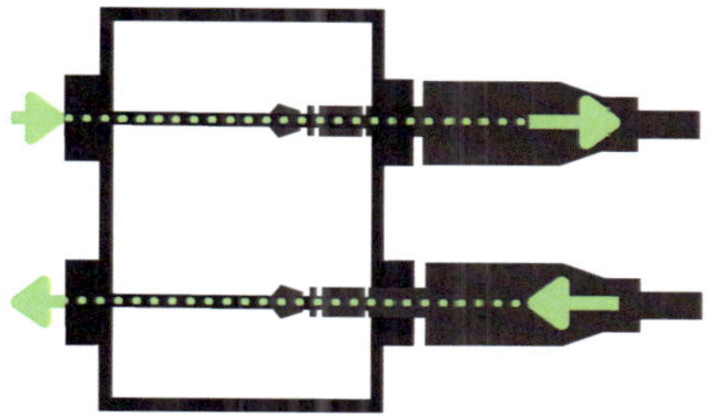

There is no connection between the top and bottom jacks from the rear position. To access the connections, patch cords are needed.

Half-Normalled

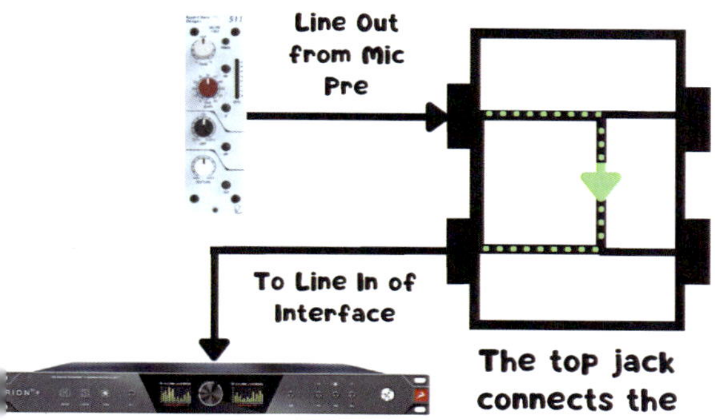

The top jack connects the bottom jack from the rear position... This is called "Normalling the patchbay"..

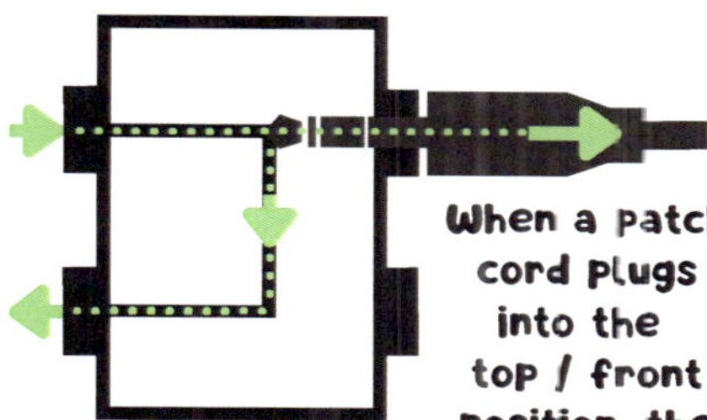

When a patch cord plugs into the top / front position, the signal splits...

The top jack continues to connect with the bottom jack from the rear position...

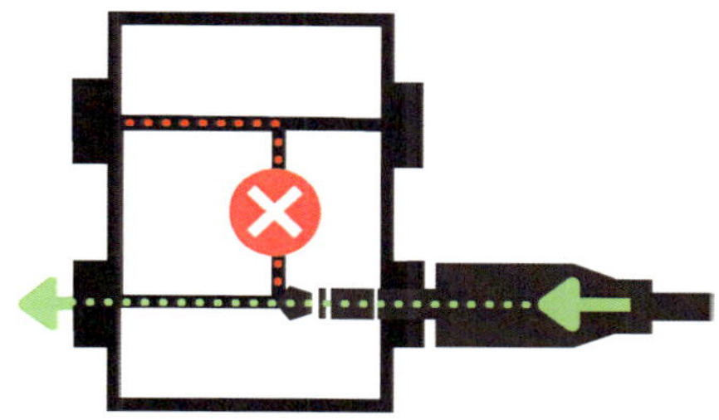

When a patch cord plugs into the bottom / front position, the "Normalled" connection is broken... Remember, the top line can still be accessed by patching an incoming signal...

Fully-Normalled

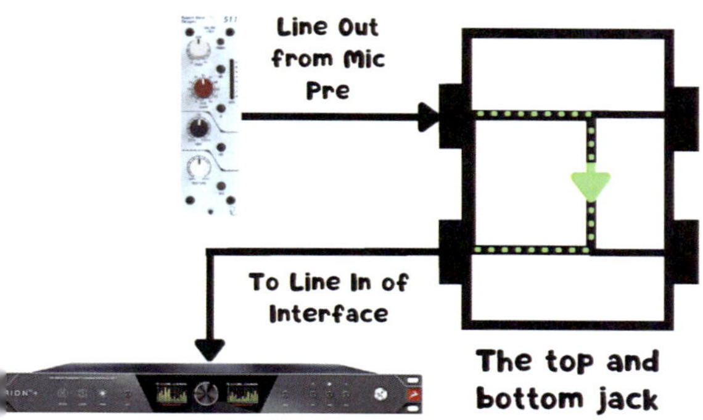

The top and bottom jack are "Normalled" from the rear position...

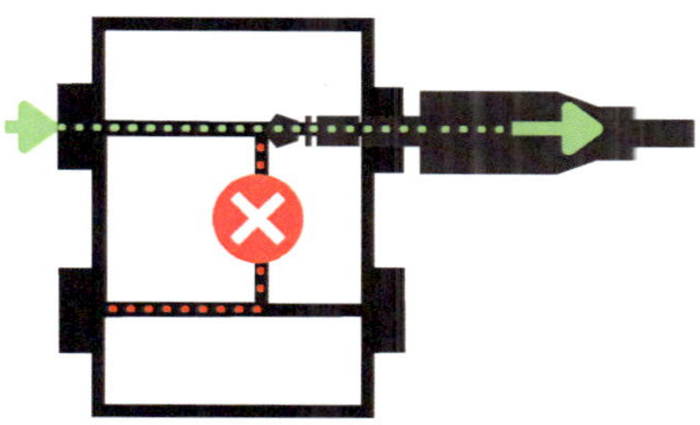

When a patch cord plugs into the top / front position, the "Normalled" connection is broken without splitting... Remember, the bottom line can still be accessed by patching an incoming signal...

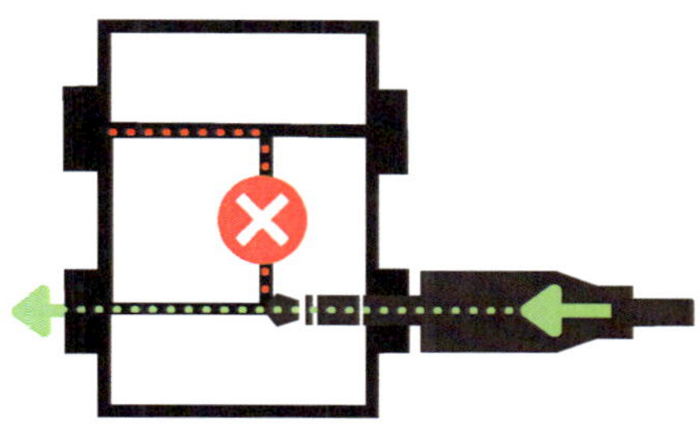

When a patch cord plugs into the bottom / front position, the "Normalled" connection is broken... Remember, the top line can still be accessed by patching an incoming signal...

Patchbay Layout Example

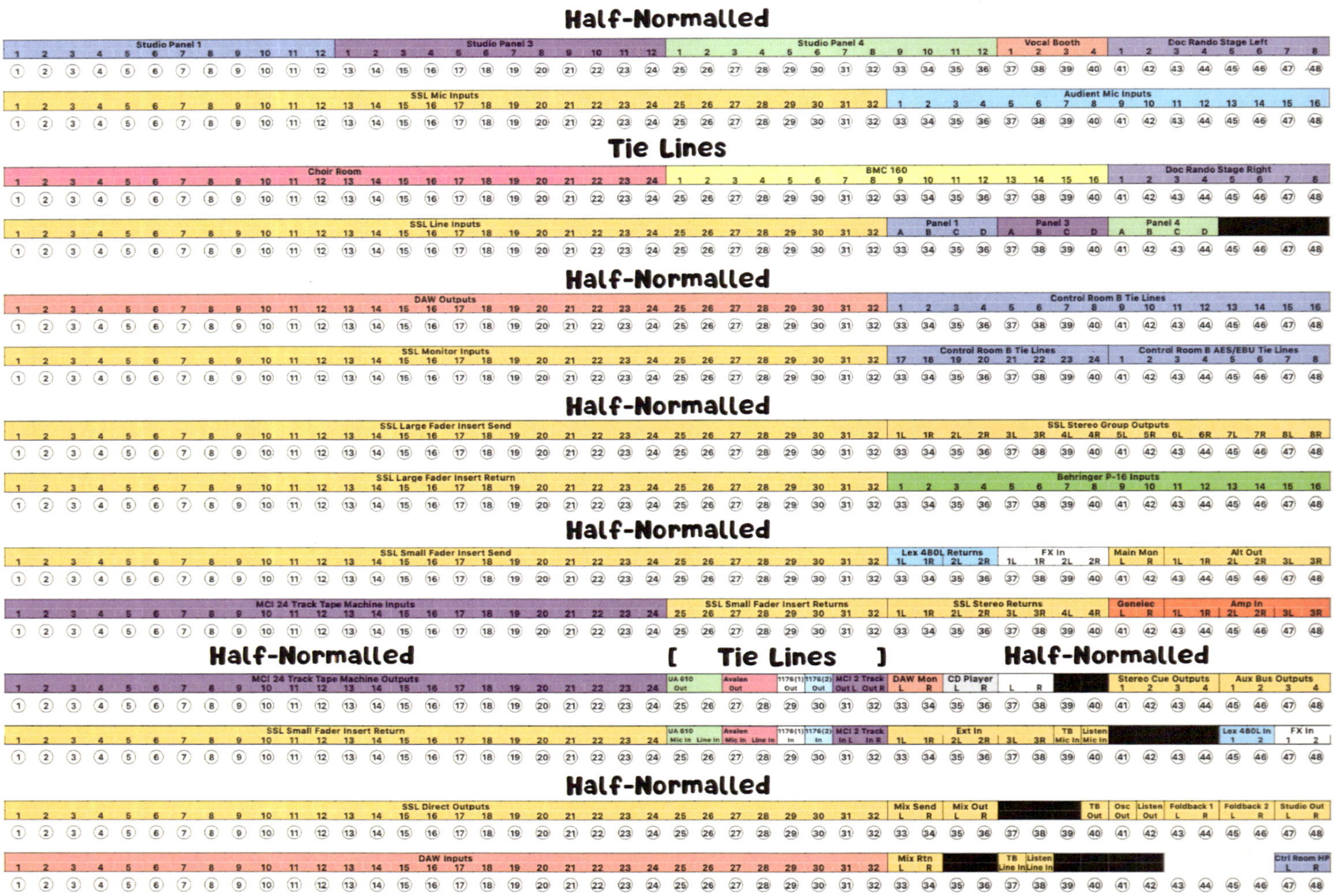

Take a moment to examine the relationship between the top and bottom rows of each patchbay. A patchbay layout can tell you a lot about the connectivity of a studio.

For example... Notice how the SSL Direct Outputs are "Half-Normalled" to the DAW Inputs...

Patching

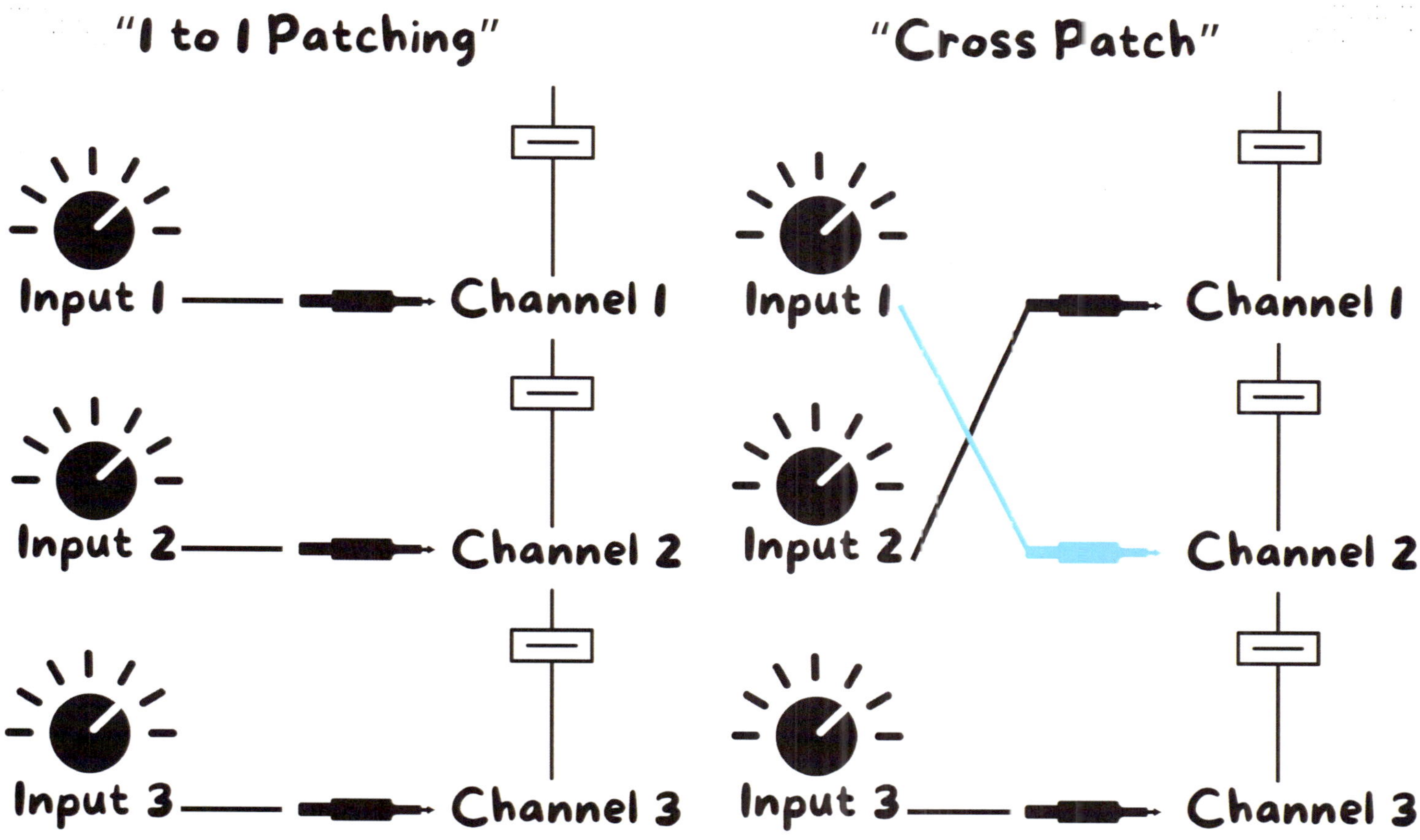
"I to I Patching"
Input 1
Channel 1
Input 2
Channel 2
Input 3
Channel 3
"Cross Patch"
Input 1
Channel 1
Input 2
Channel 2
Input 3
Channel 3

Large Console Signal Flow

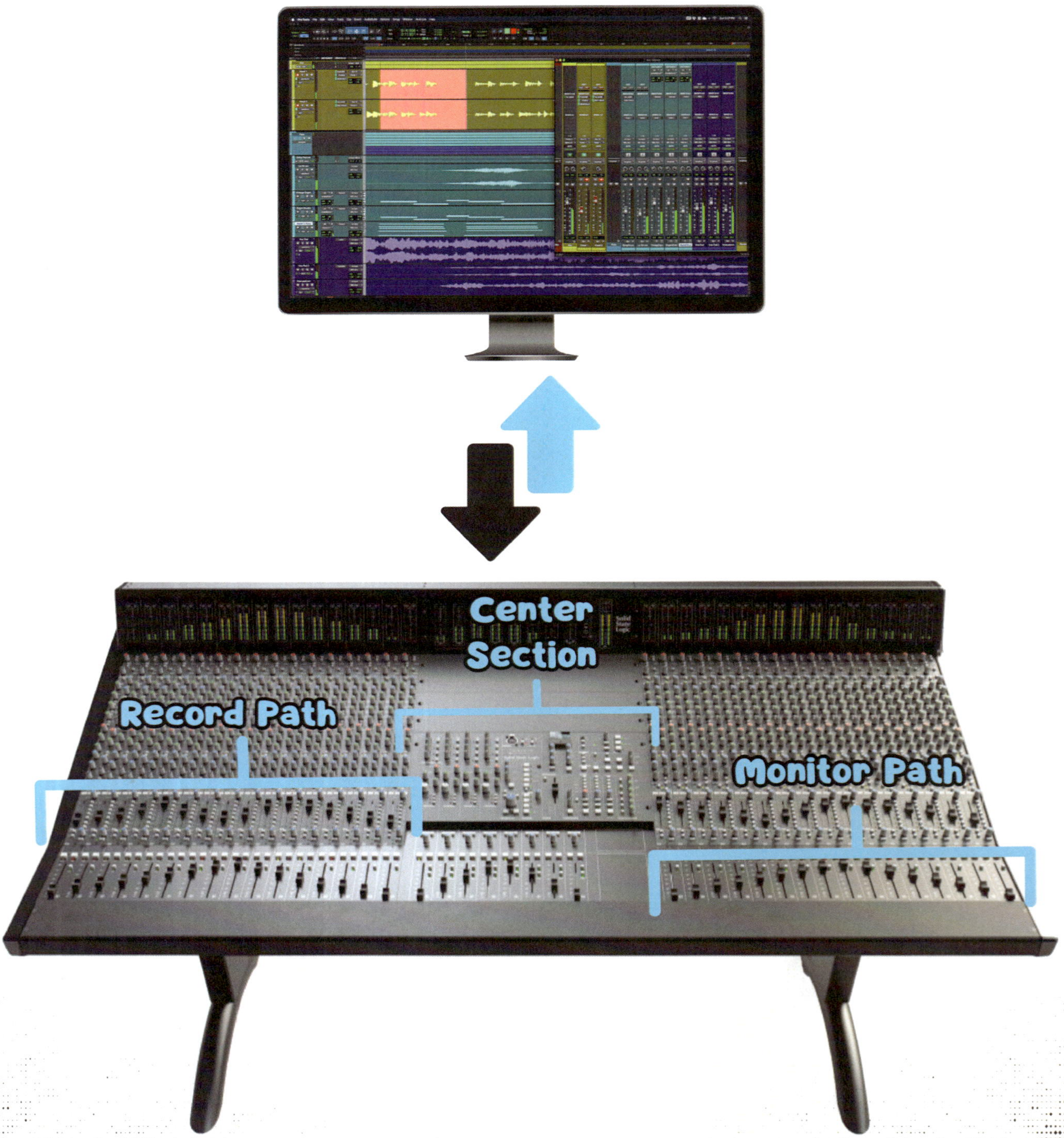

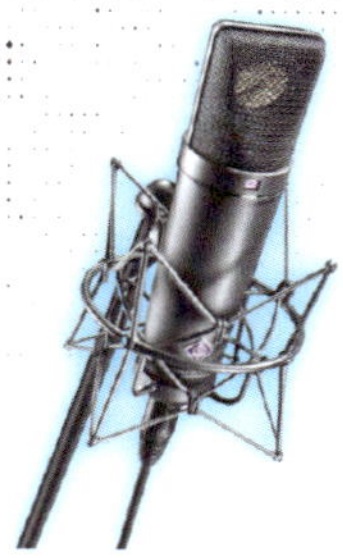

Record Path

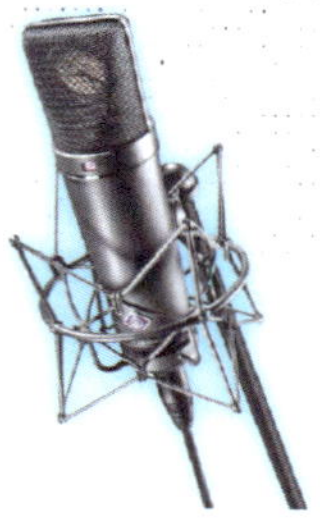

The starting point for a large console begins with the "Record Path"... A microphone can be plugged into the console's mic input, and then sent "Direct Out" to the recording device with level indicated by the "Small Fader"... These "Direct Outputs" are often normalled "1 to 1" with the inputs of the "Recording Device"

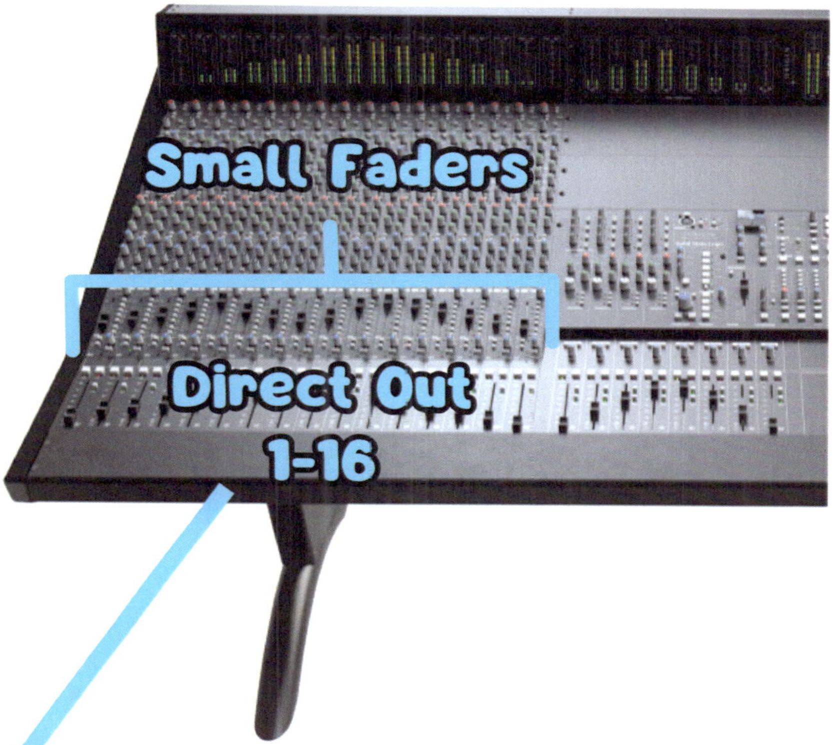

DAW Inputs 1-16

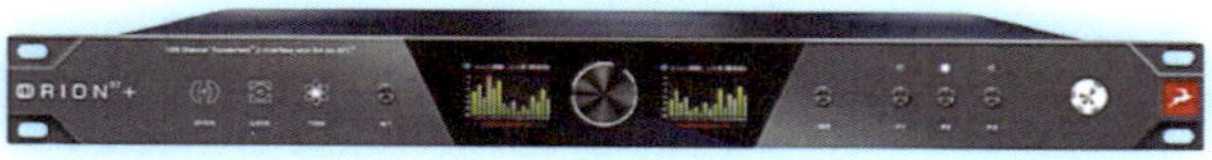

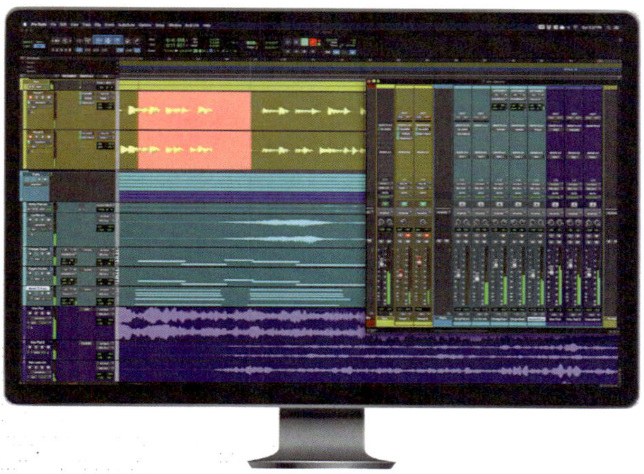

In a modern workflow integrated with a DAW as the recording device, the Direct Outs of the console will normal to the line inputs of the A/D converter (Audio Interfacee).

DAW Returns

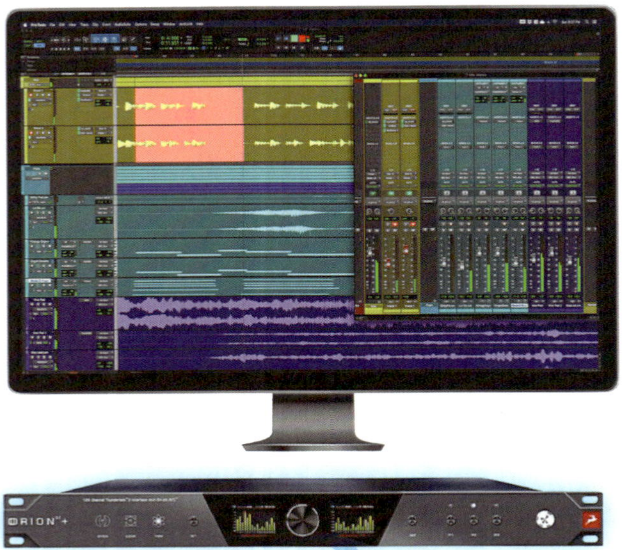

As an audio signal leaves the DAW via the line outputs of the interface, the conventional method is to return the discrete channels back through the console "channel monitor paths". These channel levels shown by the large faders would determine the final balance before summing to the main mix.

Note: The main mix is what feeds the master recording device... Traditionally, this was the master 2 track machine, but now it's simply a dedicated audio track in your DAW... "The Print Track"

DAW Outputs 1-16

In a modern workflow integrated with a DAW as the recording device, the line outs of a D/A converter (Audio interface) will normal to the channel monitor input path or "DAW Returns" of the console...

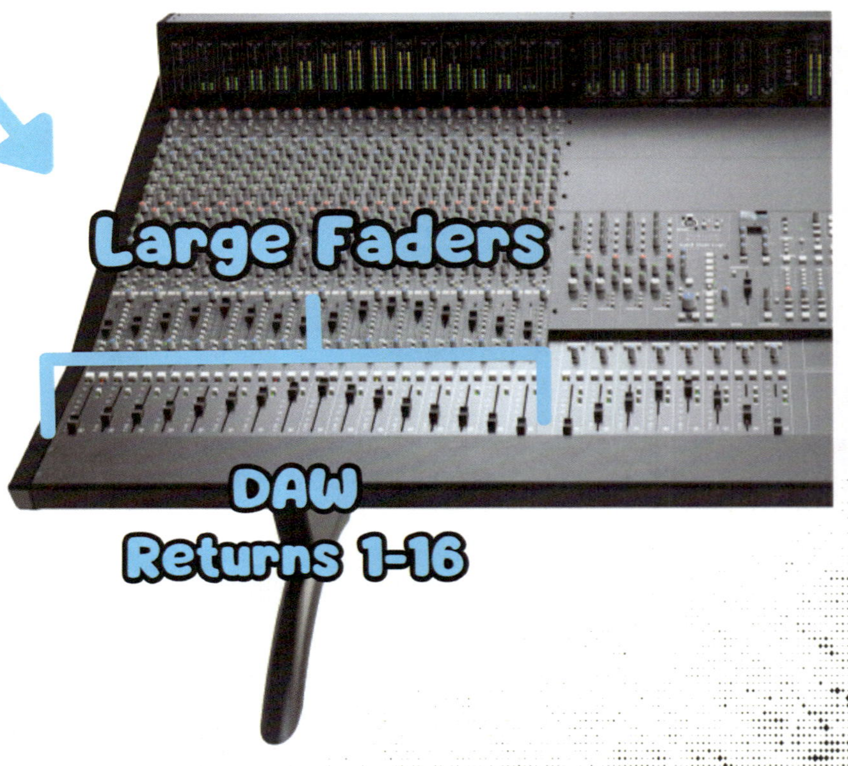

Center Section

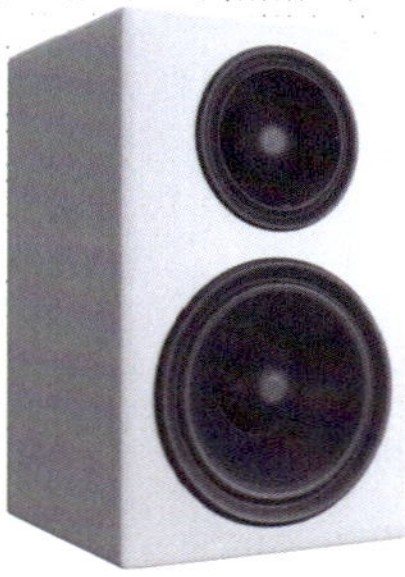

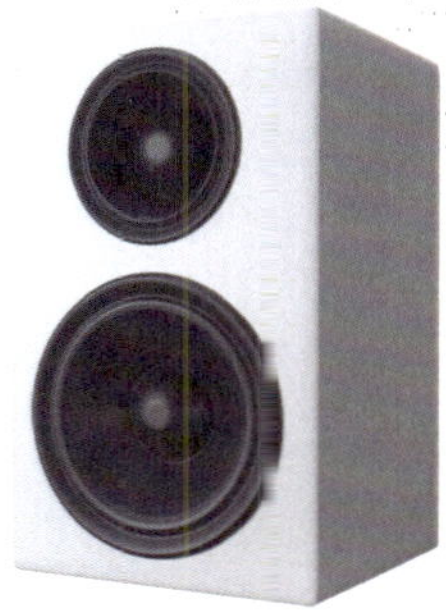

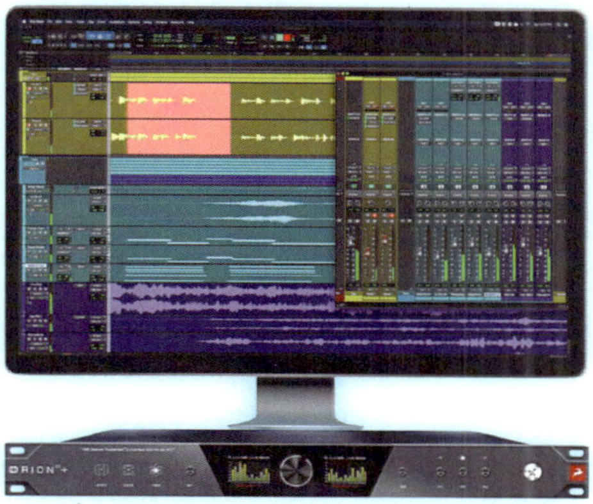

The center section of a console contains numerous operational features. It includes functions such as metering, monitor settings, master levels, auxiliary functions, and speaker distribution. Consider the center section as the primary control hub for the console.

One of the most important concepts in a hybrid DAW / console setup is to establish a "Clean DAW Monitor Path"... Most interfaces have a dedicated monitor output that feeds your speakers. However when speakers are connected to a console, using an "External Input" can assure a path with the least amount of coloration before feeding your speakers. This will prevent additional channel strip processing applied after your DAW. In other words, the path between your DAW and speakers will be as transparent as possible.

DAW Monitor Path

Main Mix Out

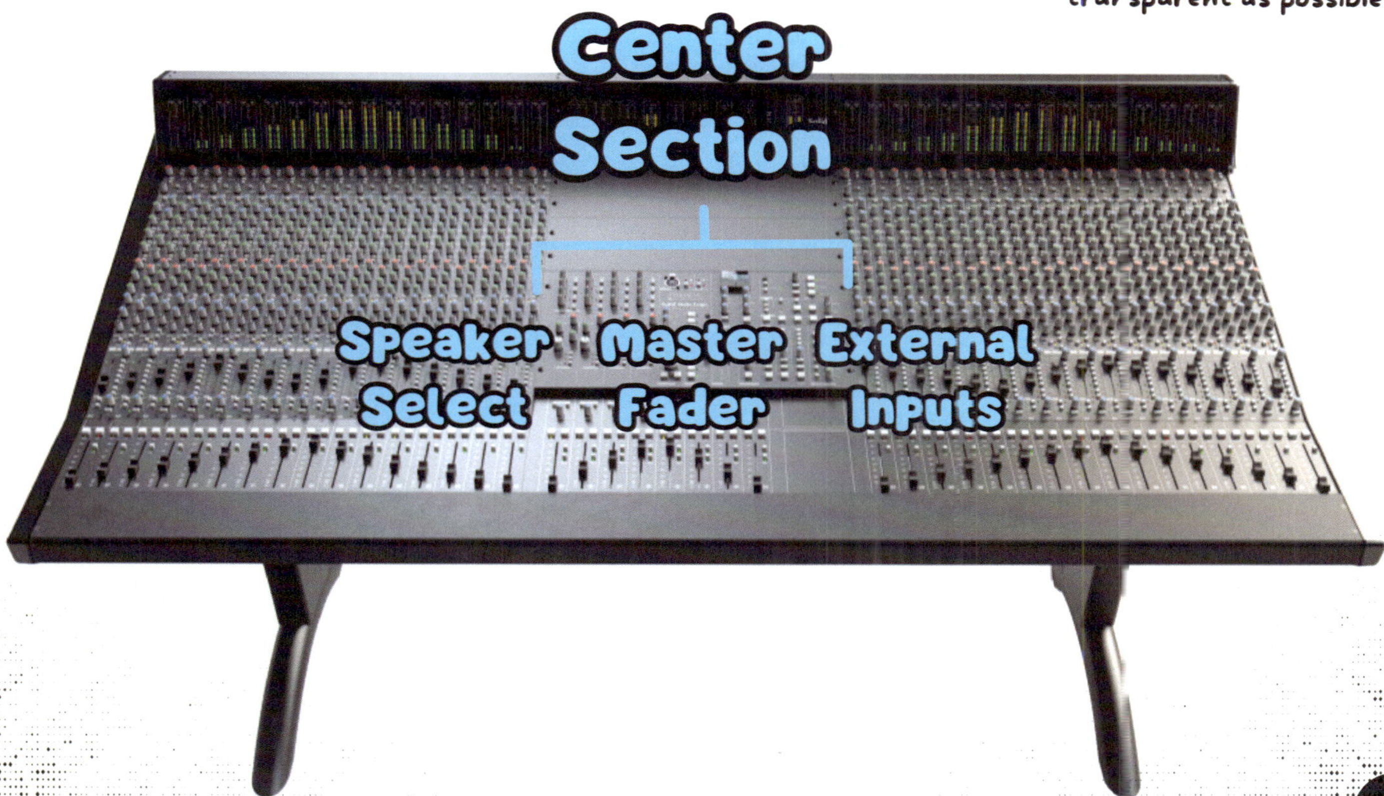

Routing Inputs in the DAW

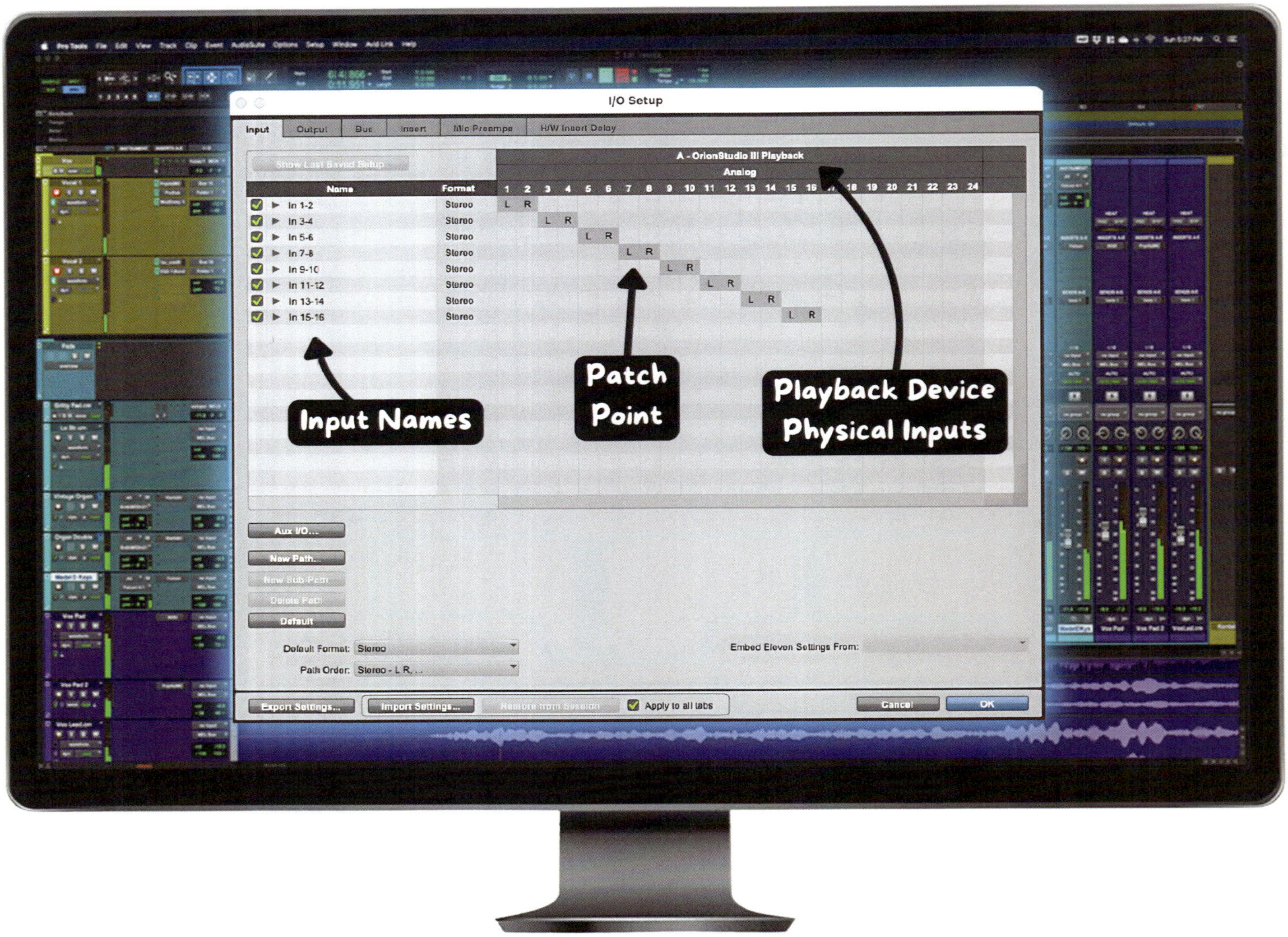

Get Ready! The Input tab is where the relationship between the physical input and input name for the session is established. On the left side of the window, you'll find the names of the inputs, such as "In 1-2" (you can technically label these inputs anything you want). Now, direct your attention to the upper right area of the window, where you'll see information about the connected playback device, physical input number, and audio type, for instance, "Analog." You'll notice a "1 to 1" patch connecting the physical input to its corresponding label. This is considered best practice, as it promotes an intuitive workflow.

Routing Outputs in the DAW

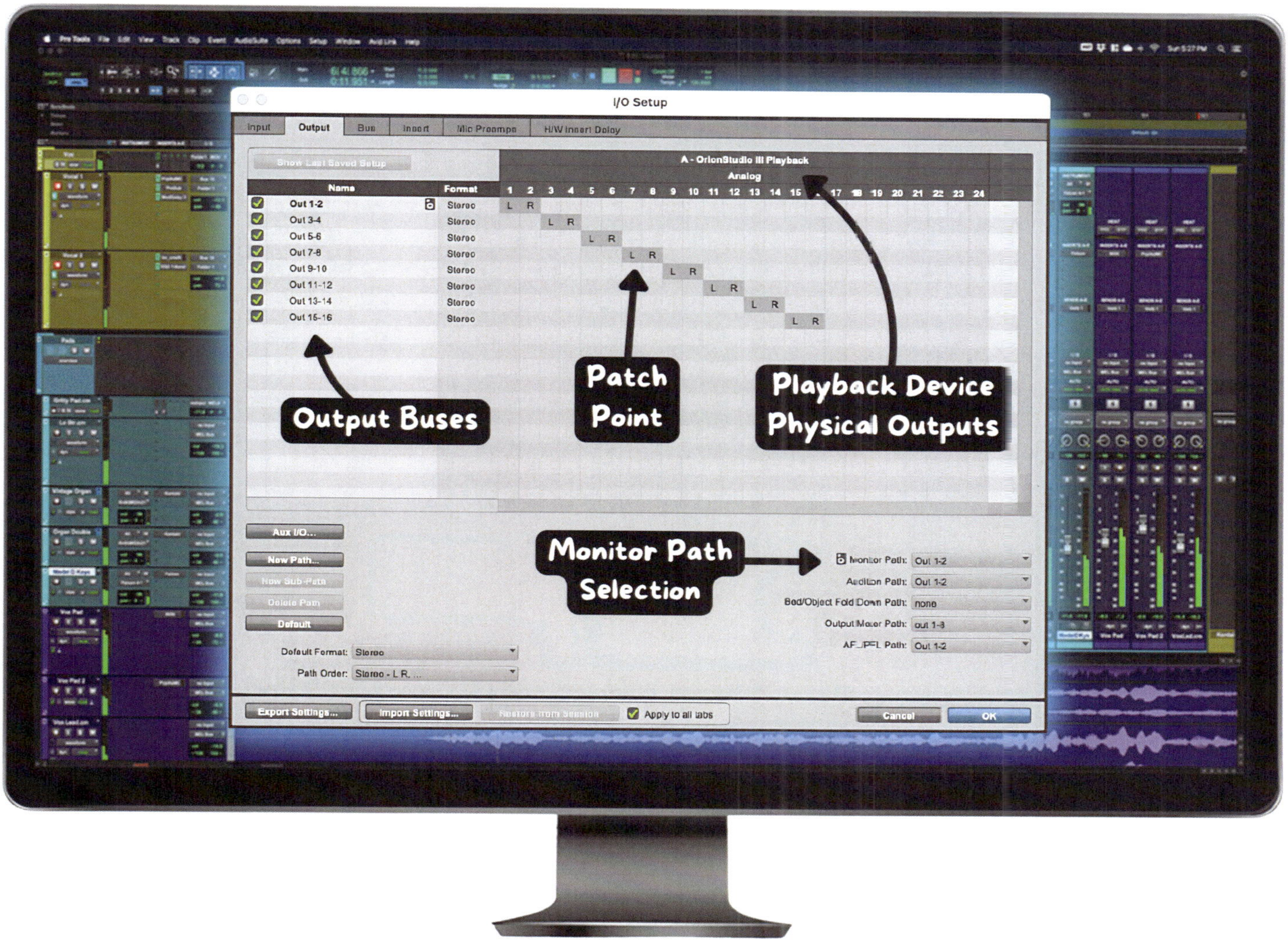

Get Ready Again!! The output tab is where the connection from output bus to physical output is established. On the left side of the window, you'll find the names of the output buses, such as "Out 1-2." Pay attention to the checkmarks, which indicate that the output bus is linked to a physical output on the playback device. Now, direct your attention to the upper right area of the window, where you'll see information about the connected playback device, output number, and audio type, for instance, "Analog." You'll notice a "1 to 1" patch connecting the output bus to the device's physical outputs. This is considered best practice, as it promotes an intuitive workflow.

Routing Buses in the DAW

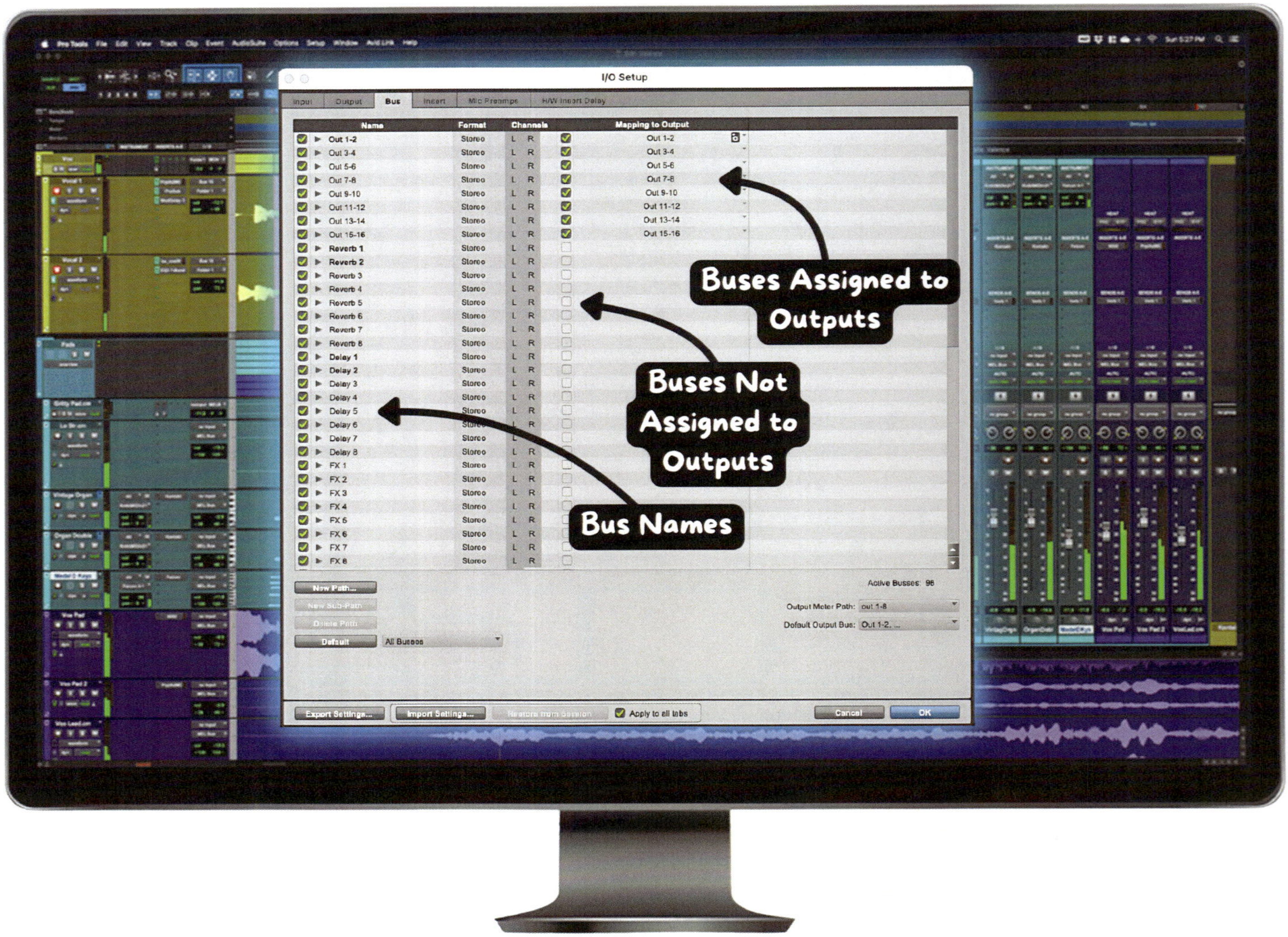

Here We Go!! The "Bus" tab serves as the central hub for all bus paths within the DAW, where they are both routed and named. On the left side of the window, you'll find the bus names, such as "Out 1-2," "Reverb 1," or "FX1." Keep in mind that not every bus is routed to a physical output, which is marked by a checkmark. It's essential to remember that while outputs are fed by buses, not all buses feed outputs. You can name the buses as you wish, but it's wise to label them according to their functions.

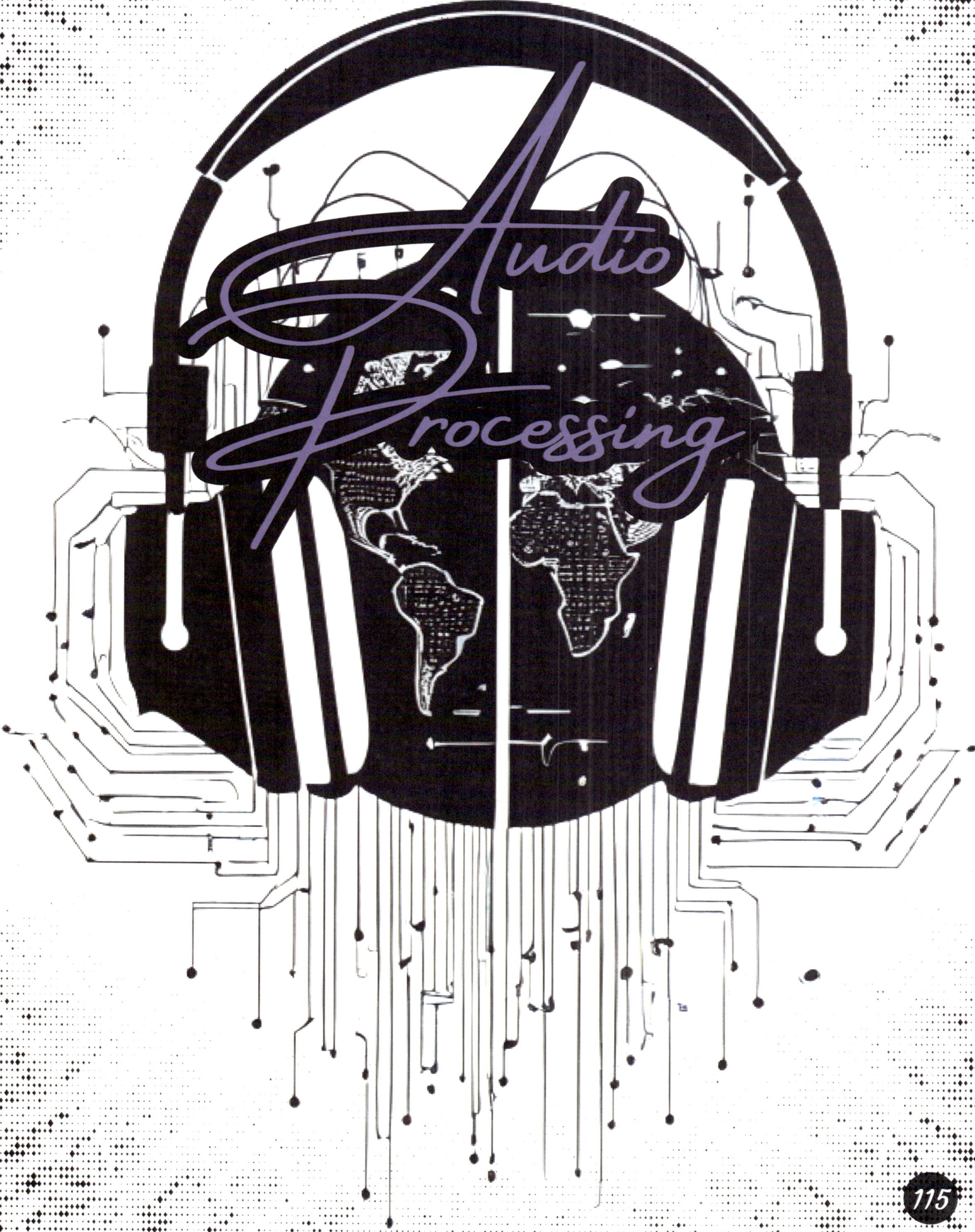
Audio
Processing

Inline Processing

Refers to modifying an audio signal by passing the entire signal through a signal processor... This is also commonly known as "Inserting" an effect...

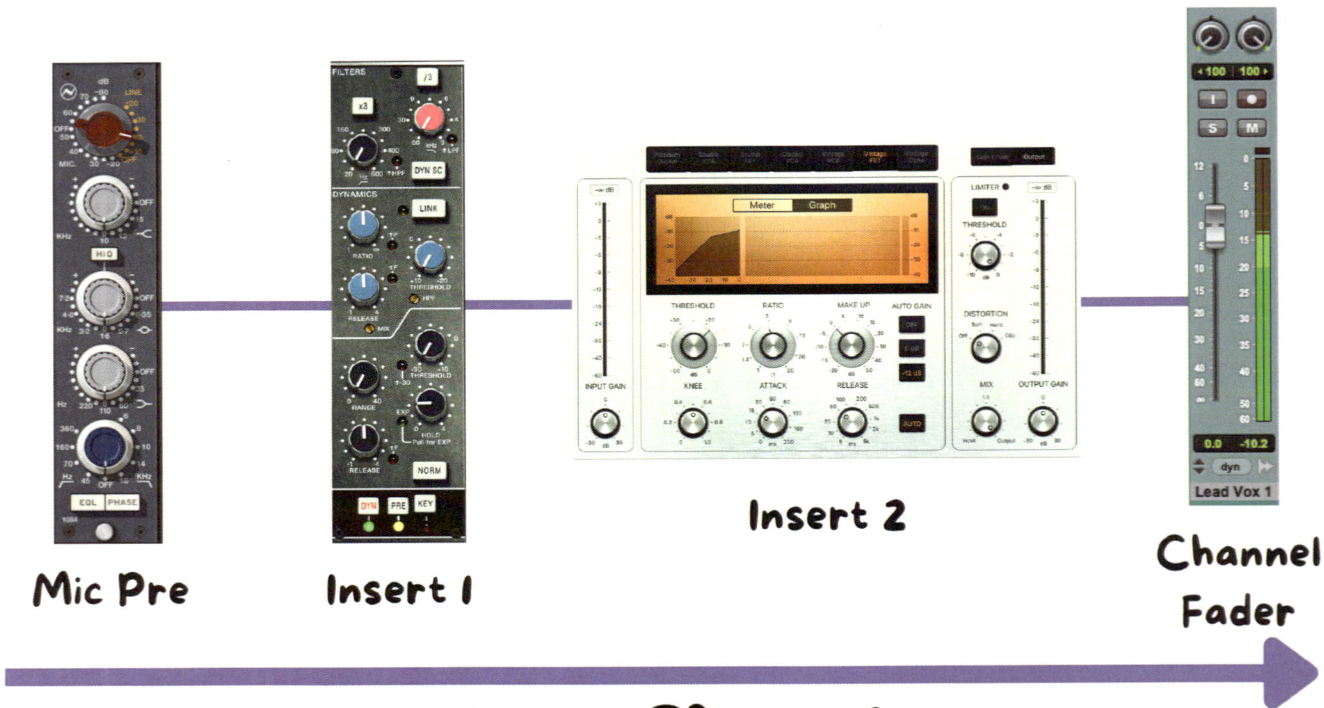

Input Channel Signal Flow

Parallel Processing

Refers to duplicating a signal and modifying that duplicate signal... then mixing the altered track alongside the original track in parallel.

Original "Dry" Signal | Parallel "Wet" Signal

"Balancing special effects such as Reverb is made more intuitive through parallel processing."

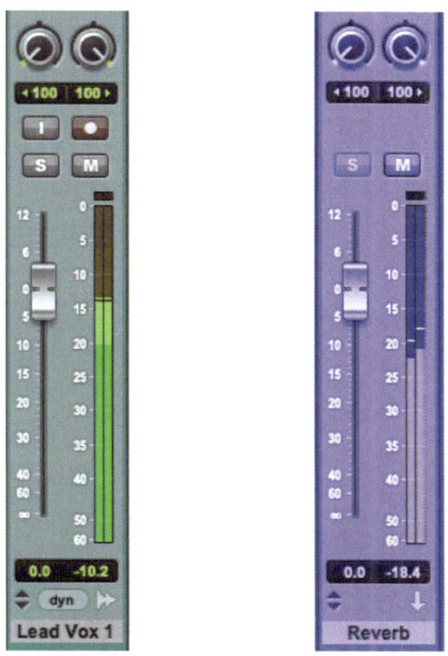

Channel Fader | FX Return

FX Send & Return

This refers to the process of sending a particular amount of an original signal to another channel, typically an auxiliary input that serves as the FX Return... The FX Return track is then mixed in parallel with the original track...

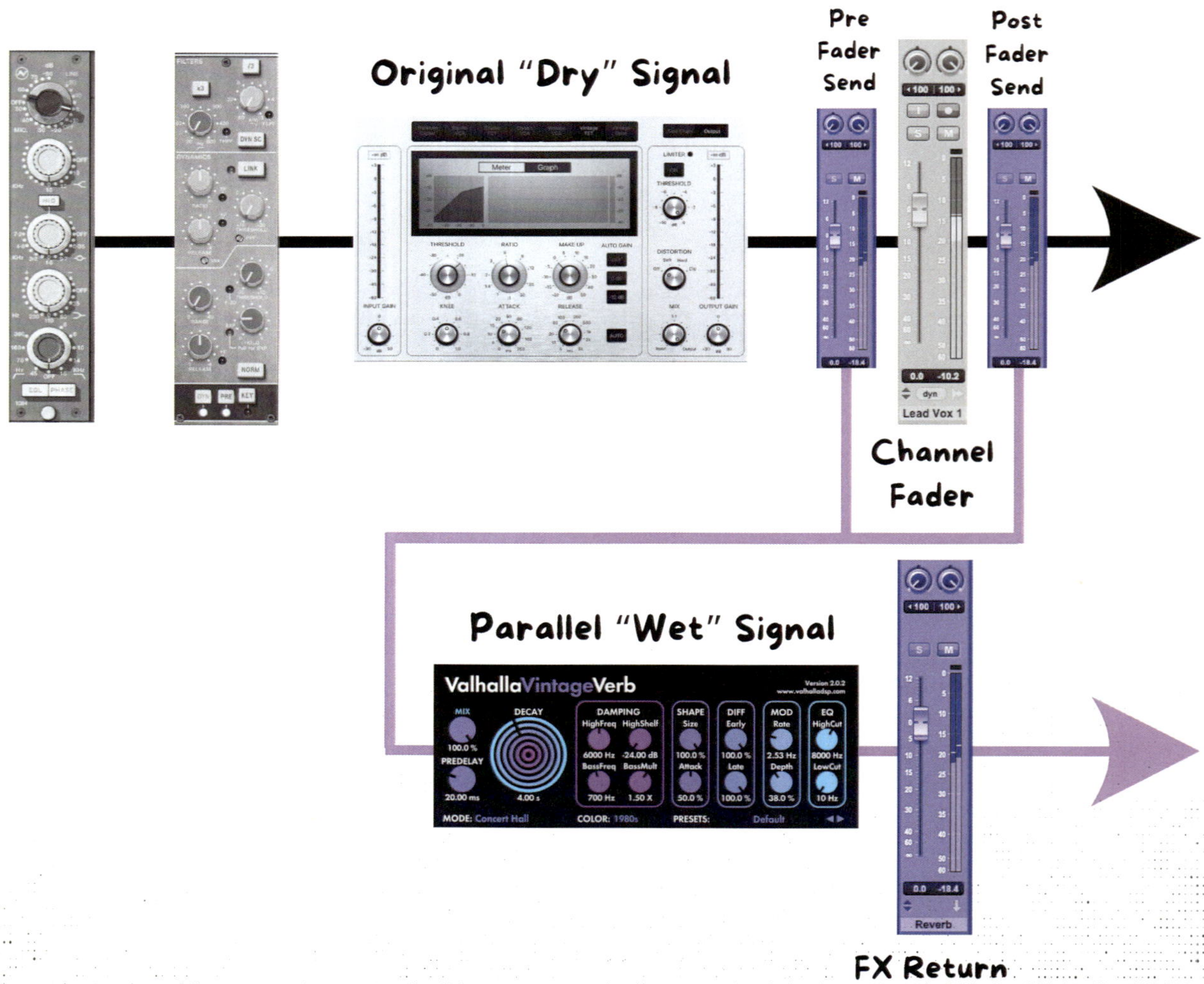

The Human Ear (Fletcher Munson Curve)

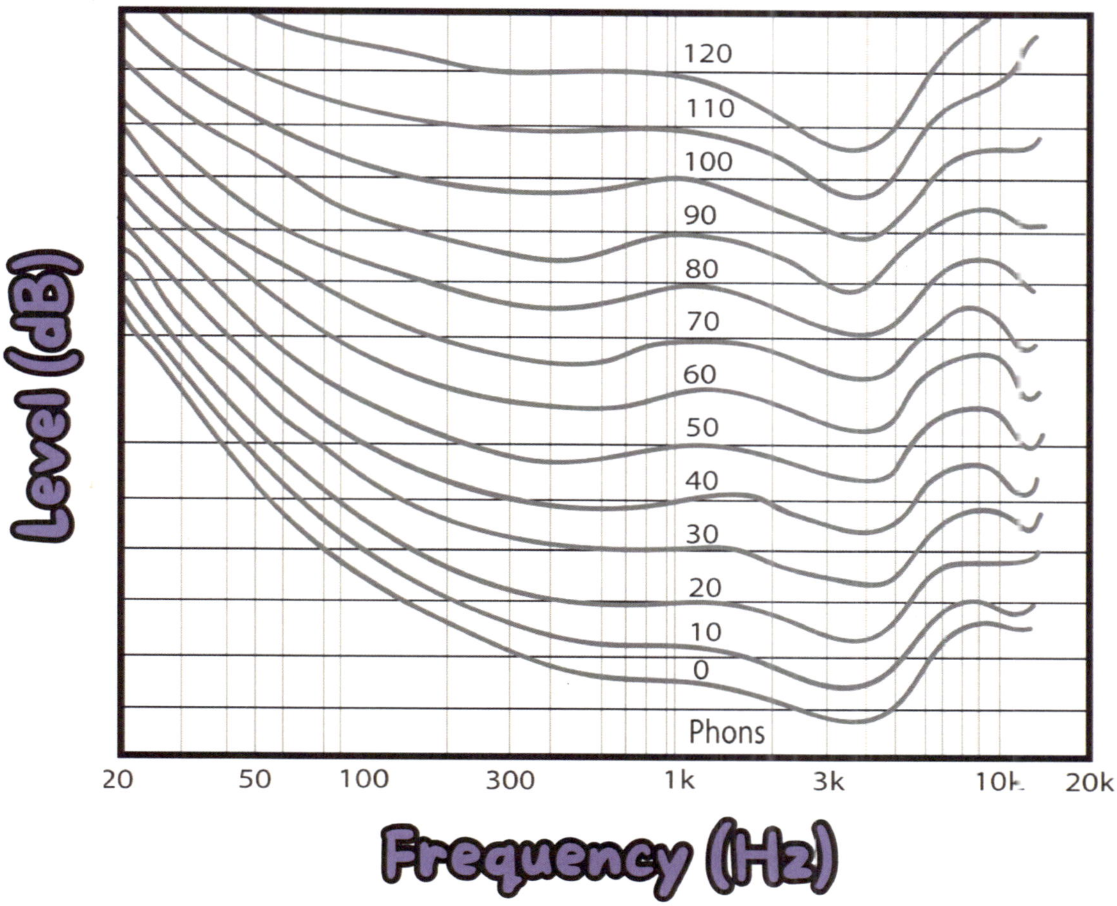

Humans perceive the frequency spectrum differently depending on the volume level. Additionally, varying volumes are required at different frequencies for us to perceive the spectrum as balanced. Notice the volume dip between 2k-6k and the amplified volume at the lower and higher ends of the spectrum. This occurs because humans are most sensitive to upper mid frequencies, and we need more volume to perceive the outermost frequencies in the spectrum. Additionally, the most accurate or neutral representation of the spectrum is typically perceived at approximately 85dB.

(This is the ideal volume for mixing)

Higher Frequency Reference Points

Low Teeth "Sh" — 4-6 kHz

Too much
4-6 kHz can be harsh on the ears. Transient details are frequently found in this frequency range.

Not enough
4-6 kHz can lead to a lack of energy, resulting in a dull sound.

Teeth "Ss" — 6-12 kHz

Too much
6-12 kHz may result in excessive sibilance and a lack of smoothness.

Not enough
6-12 kHz can lead to a lack of clarity and articulation.

Tip of the Teeth "Sss!" — 12-16 kHz

Too much
12-16 kHz can take on a "hissing" quality which over time can lead to ear fatigue.

Not enough
12-16 kHz can cause a lack of brightness, resulting in the ear perceiving a "muddy" sound.

Air "Hhh" — 16-20 kHz

Too much
16-20 kHz can create a tiring atmosphere that can eventually cause ear fatigue.

Not enough
16-20 kHz can lead to a diminished sense of "openness" and cause the ear to focus more on lower, harsher frequencies.

Midrange Frequency Reference Points

Low "Howl" — 400-800 Hz

Too much
400-800 Hz can result in "Cloudy" or "Howling" midrange.

Not enough
400-800 Hz can lead the ear to hear resonances in the surrounding midrange frequencies.

Low Comfortable "Whistle" — 800Hz - 1 kHz

Too much
800 Hz - 1 kHz can create "Papery" or "Hollow" resonance in the midrange.

Not enough
800 Hz - 1 kHz will create a lack of midrange harmonics that are fundamental for many instruments and voices.

High Comfortable "Whistle" — 1-2 kHz

Too much
1-2 kHz can lead to an unpleasant "Ambient" or "Whistling" resonance, which may be harsh on the ears.

Not enough
1-2 kHz can lead to a lack of intelligibility in sounds such dialogue.

High "Whistle" — 2-4 kHz

Too much
2-4 kHz can cause an unpleasant "Biting" harshness that triggers the sensitive range of human hearing.

Not enough
2-4 kHz may result in a lack of articulation or "punch" in the sound.

Lower Frequency Reference Points

"Bass too low to sing"

20-60 Hz

Too much

20-60 Hz can overwhelm the clarity of the overall frequency spectrum. It can also potentially lead to negative physical responses.

Not enough

20-60 Hz can shift the ear's attention from the low end to the lower mid range, creating an underwhelming experience.

"Bass you can almost sing"

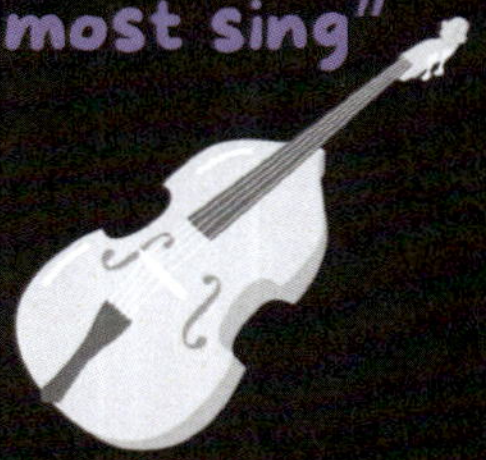

60-150 Hz

Too much

60-150 Hz can create a sense of "boominess" that can overpower a mix.

Not enough

60-150 can result in a "thin" sound that lacks low end "punch". This can also result in a "mid-rangy"

"Muffled mouth sound"

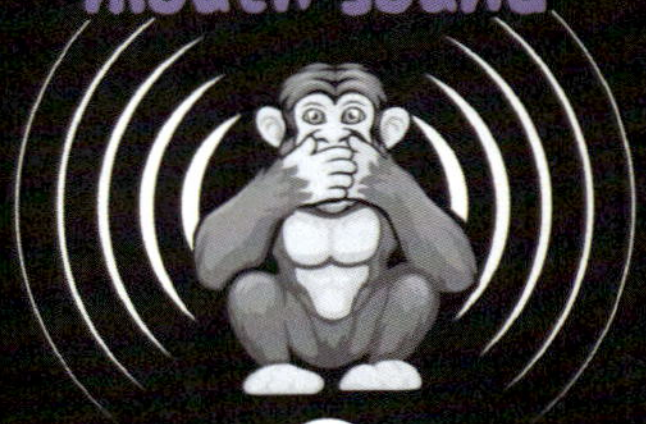

150-300 Hz

Too much

150-300 Hz can create a sense of "muffling" or "muddiness" that fights clarity in the sound.

Not enough

150-300 Hz may lack "warmth," leading to a "thin" sound.

"Howling sound"

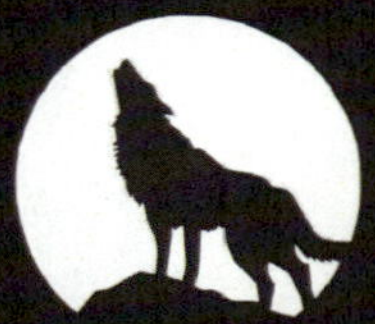

300-400 Hz

Too much

300-400 Hz may lead to a "howling" effect, diminishing definition in other frequency ranges.

Not enough

300-400 Hz can create a "scooped mid-range effect, exaggerating the "boominess" and "harshness" in a mi

EQ

Equalizing or "EQ" for short is the type of audio processing that involves boosting or reducing specific areas of the frequency spectrum... Below demonstrates the 3 fundamental EQ filter types...

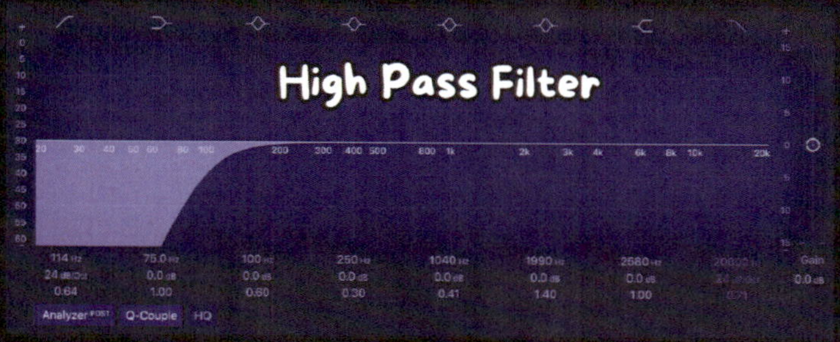

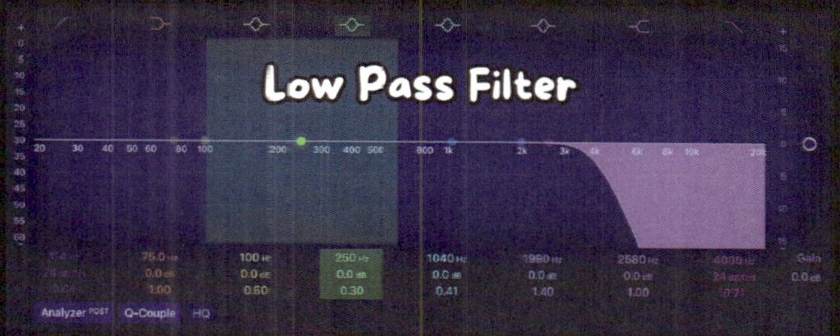

High and low pass filters roll off the extended ends of the frequency spectrum. The two fundamental parameters of these filters are the frequency and slope measured in dB/Octave (the higher the dB/Octave the steaper the slope).

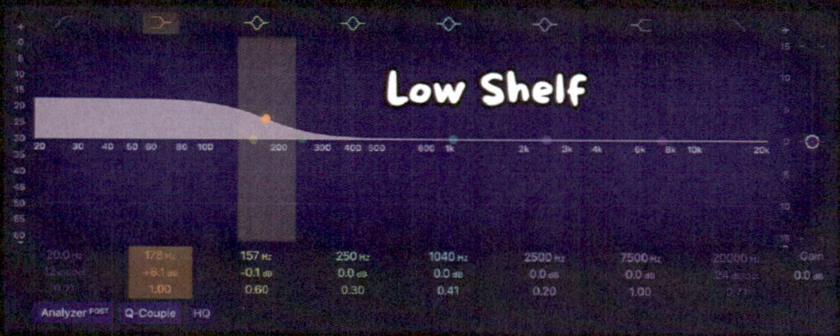

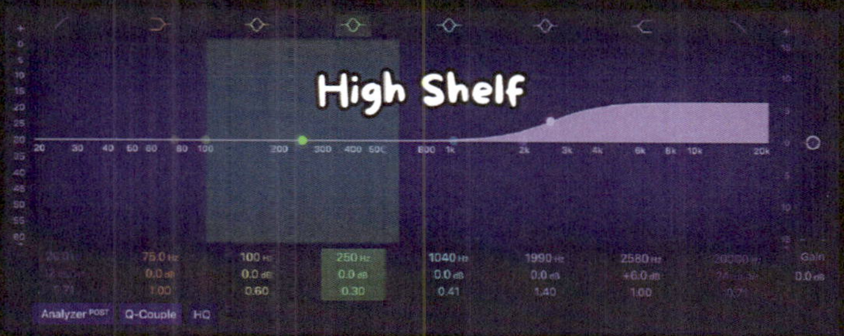

High and Low Shelves are particular EQ bands that allow frequencies beyond a particular frequency to be amplified or attenuated equally (like a shelf)...
The main parameters of this band are frequency, gain, and Q (the slope of the shelf).

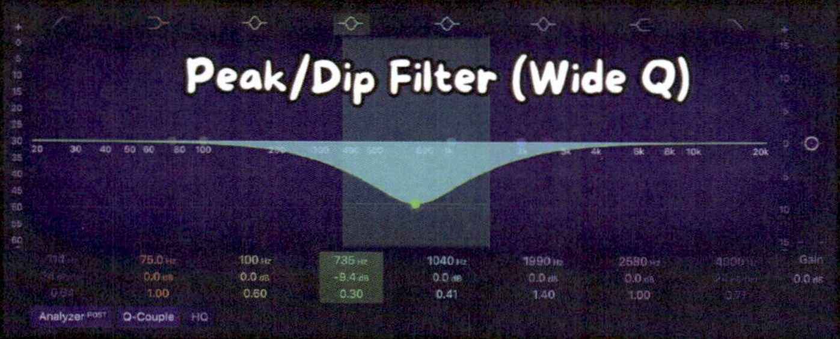

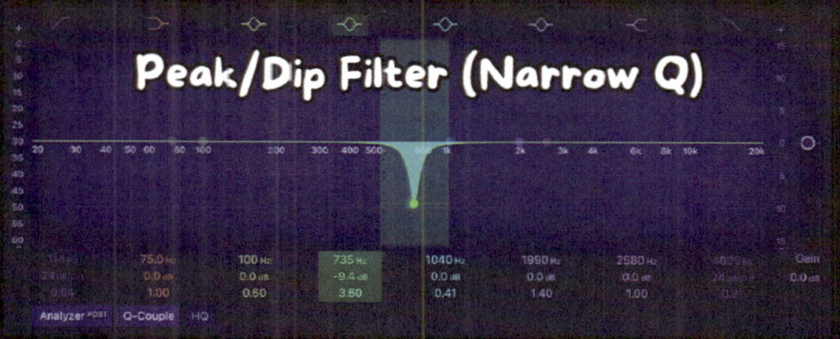

A Peak/Dip Filter allows amplification or attenuation at a specific frequency point in the spectrum. The main parameters of this band are frequency, gain, and Q (how wide or narrow the band is).

Corrective EQ

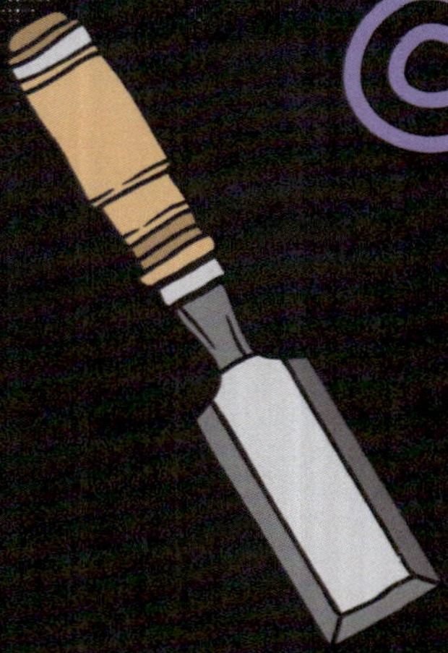

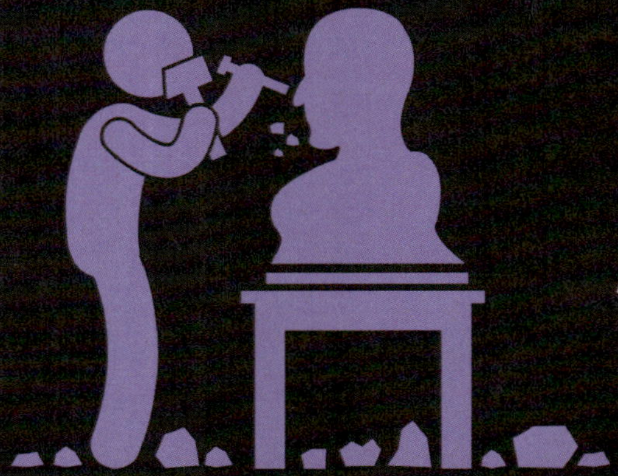

When addressing issues in the frequency spectrum, Corrective EQ is one the most effective approaches in this pursuit. By simply removing undesirable resonances in the sound, the audio will gain clarity. Similar to a sculptor shaping a rock, corrective EQ can unveil a refined piece of audio.

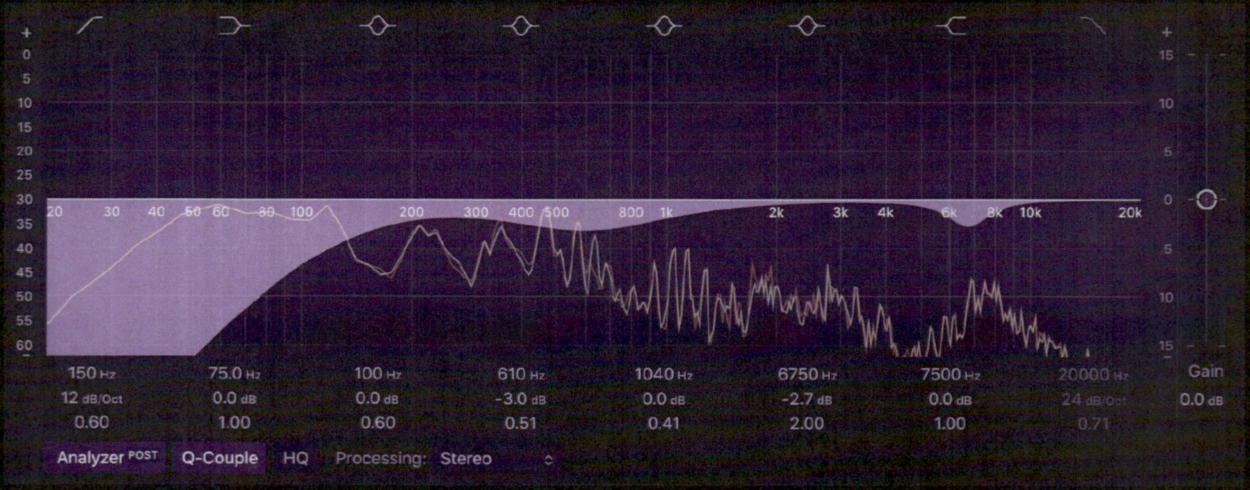

Notice the use of subtractive EQ to refine the audio signal... There is a high-pass filter to reduce the extended low end, a broad dip to eliminate some mid-range buildup, and a narrow dip to control a harsh high frequency.

Coloration EQ

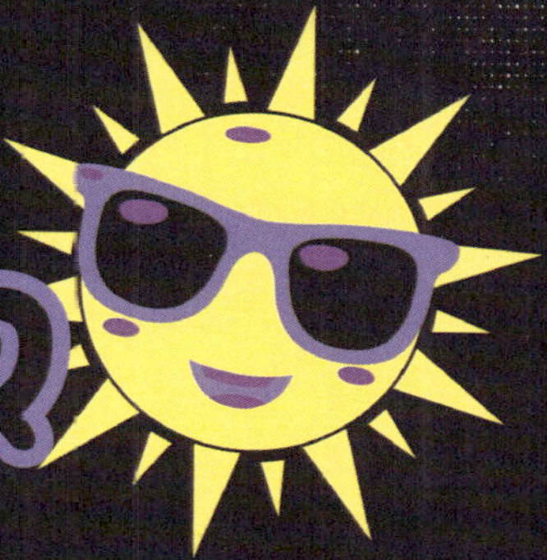

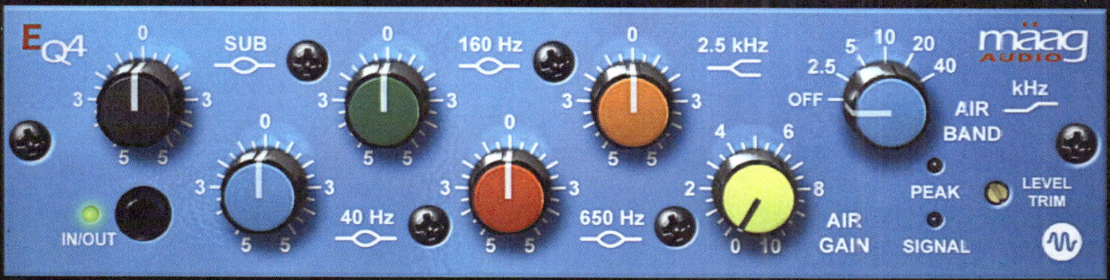

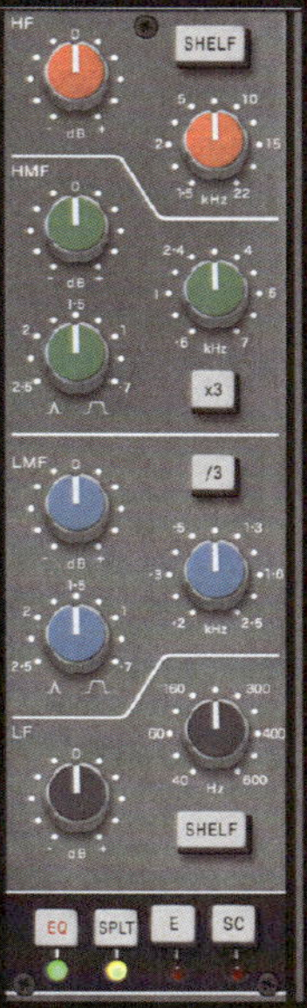

When aiming to enhance or "Add" EQ to your sound, opt for an EQ that offers "Harmonic Content" to elevate the excitement in your mix. It's like adding "Vitamins" to your sound. Many of these plugins are emulations of analog gear. Analog equipment is valued for the lush harmonic content meticulously integrated into their design by these manufacturers.

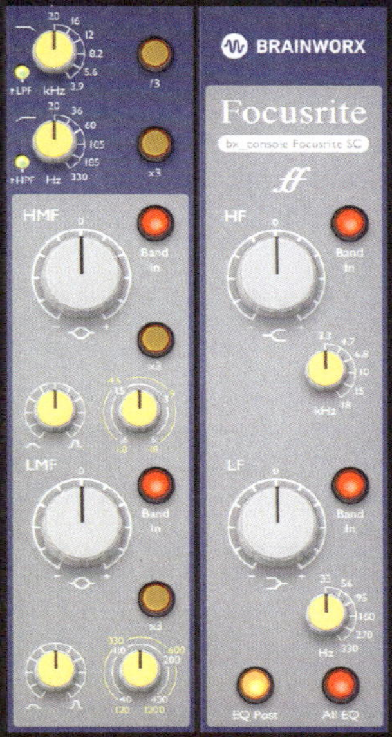

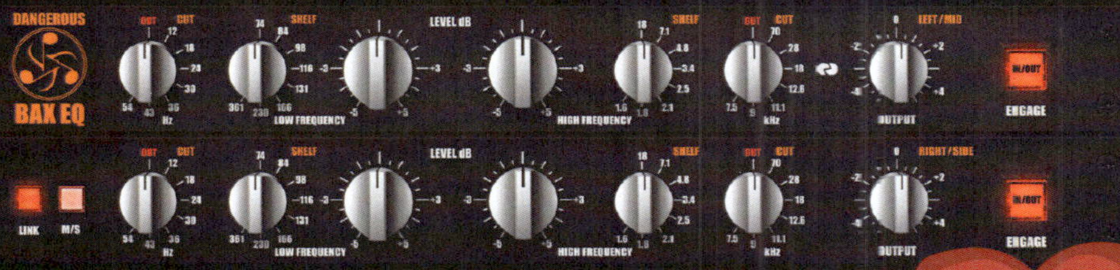

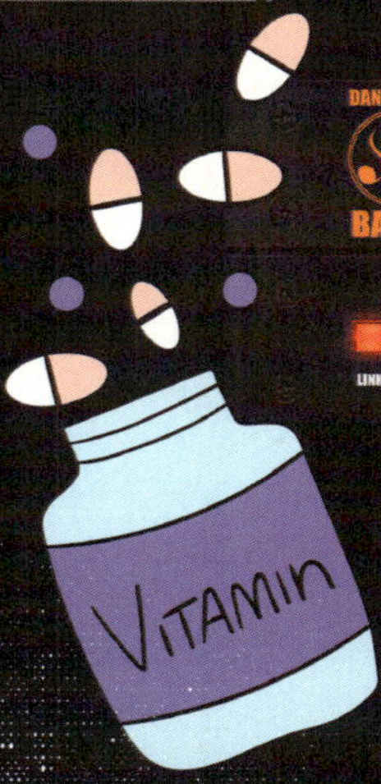

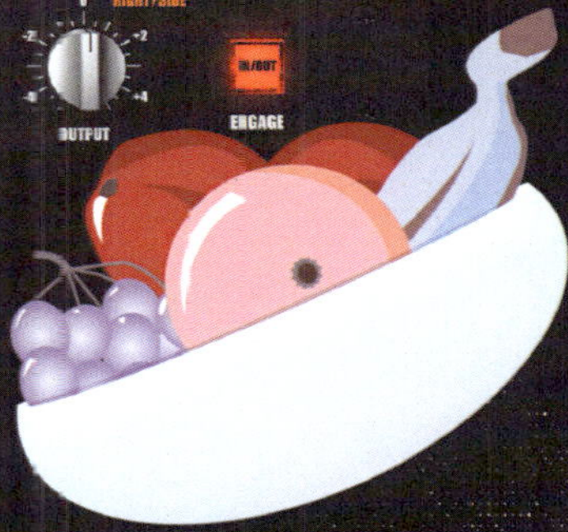

Compressor

A compressor's purpose is to reduce the dynamic range of an audio signal. This means it will simultaneously reduce loud level sound, while increasing low level sound. There are a handful of fundamental parameters that can modify a compressor's behavior.

Threshold: The dB level at which compression will engage

Ratio: The amount of reduction once the threshold is met

Gain Reduction Meter: A visual reference that shows the amount of reduction to a signal in decibels (dB)

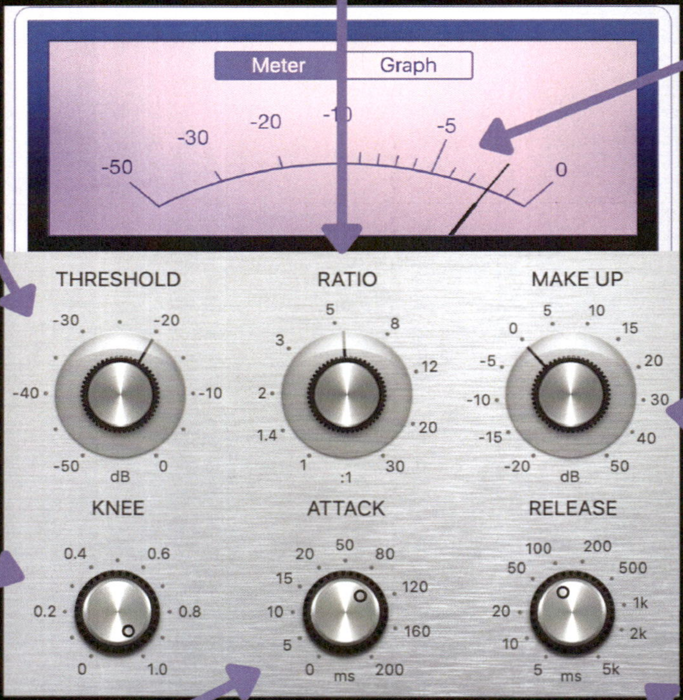

Knee: Determines the immediacy of the ratio when the threshold is met

Make Up Gain: Allows the compressor to amplify the signal to make up for the loss of signal.

Attack: The speed at which the compressor reacts to the incoming signal

Release: The speed at which the compressor returns to its un-compressed state.

Flavors of Compression

Optical (Opto)

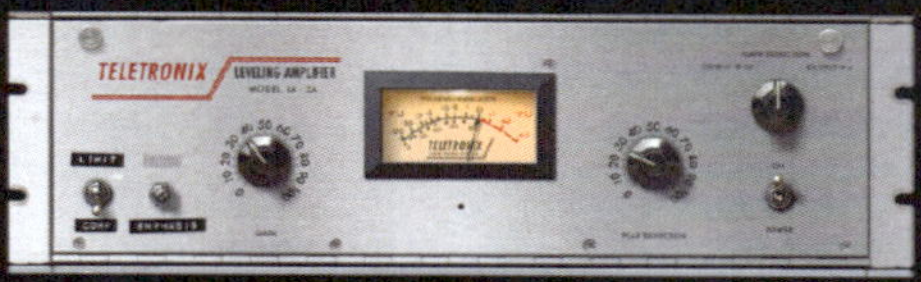

Smooth.. Great on Vocals and Bass

FET

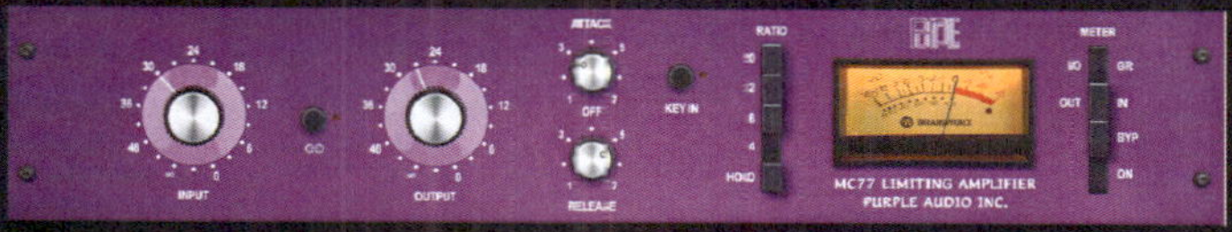

Bold... Great on Drums and anything needing Energy!

Fully Variable

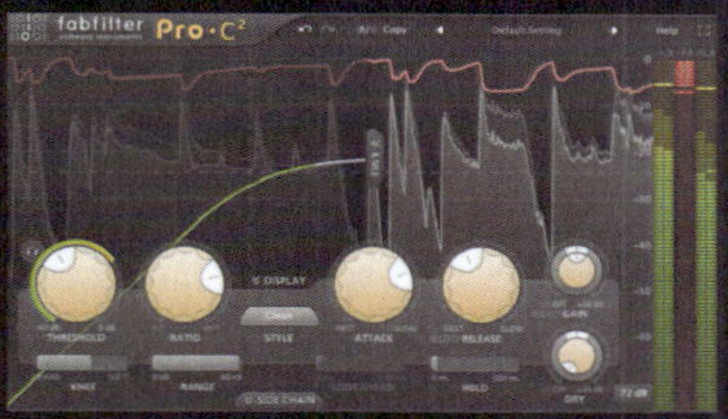

Versitile... Allows full flexibility on compressor parameters

VCA

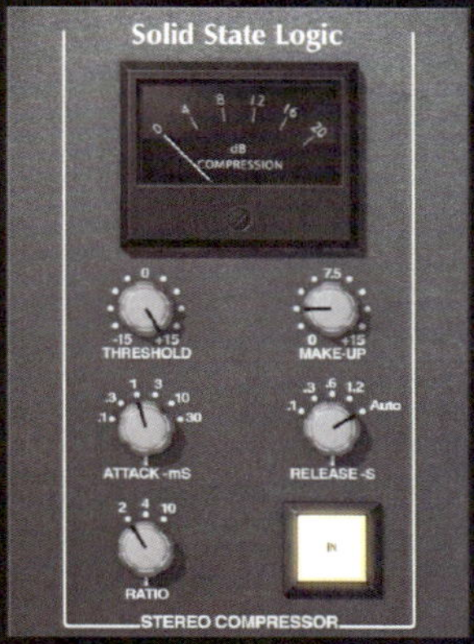

Transparent... Great for Mix Buses

Tube (Vari-Mu)

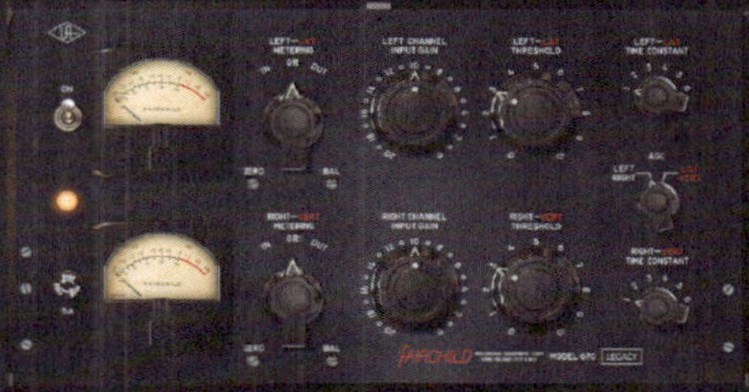

Rich... Great at preserving transients while adding warmth

Limiter

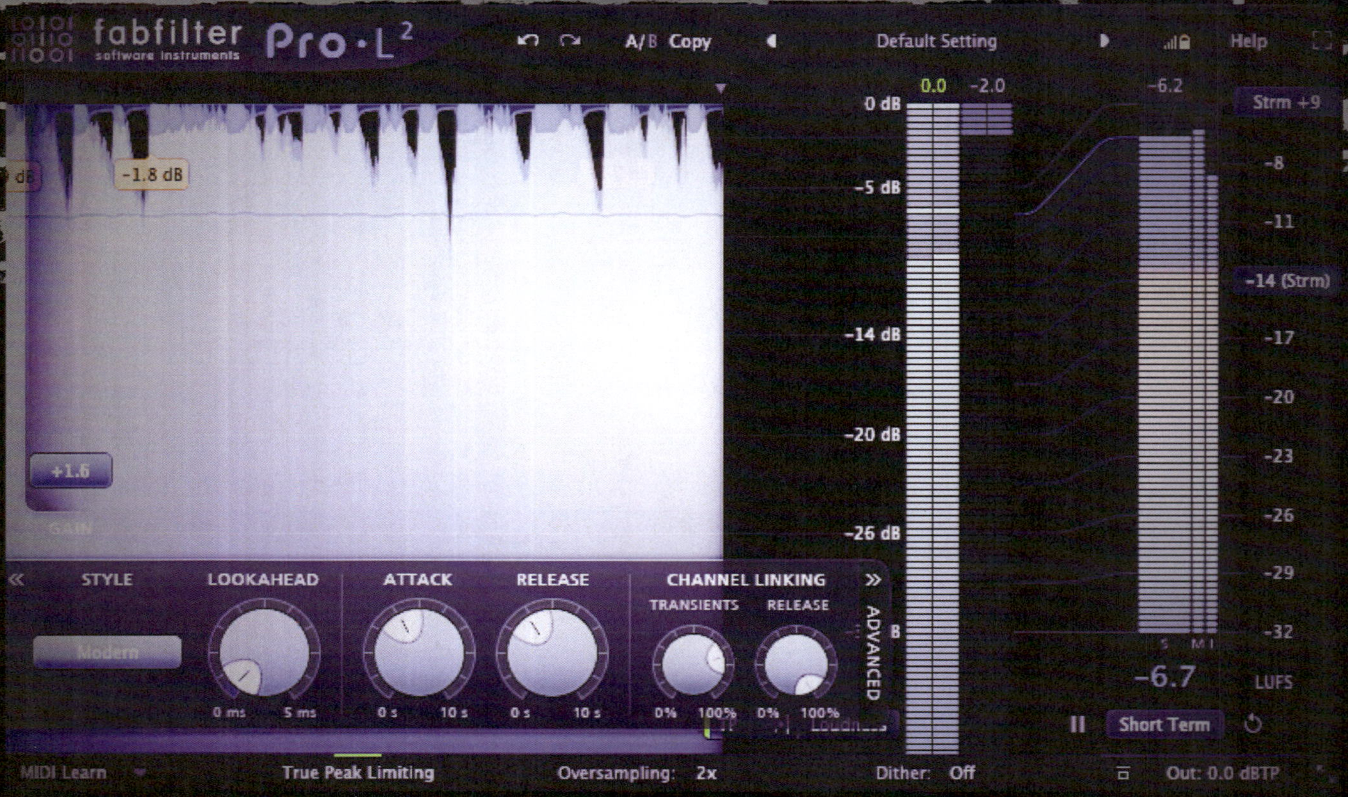

It's a compressor with a very high ratio...

Much like a compressor, a limiter's job is to reduce the level o sound as it reaches a particular dB level. Limiters are often used to protect the point of unwanted distortion within an audio channel. In other words, we can push the level of the audio up, while the limiter prevents the signal from "clipping". Commonly, limiters are used at the end of a mix bus, but can be used anywhere the function is needed.

Transients
"Punchy vs. Smooth"

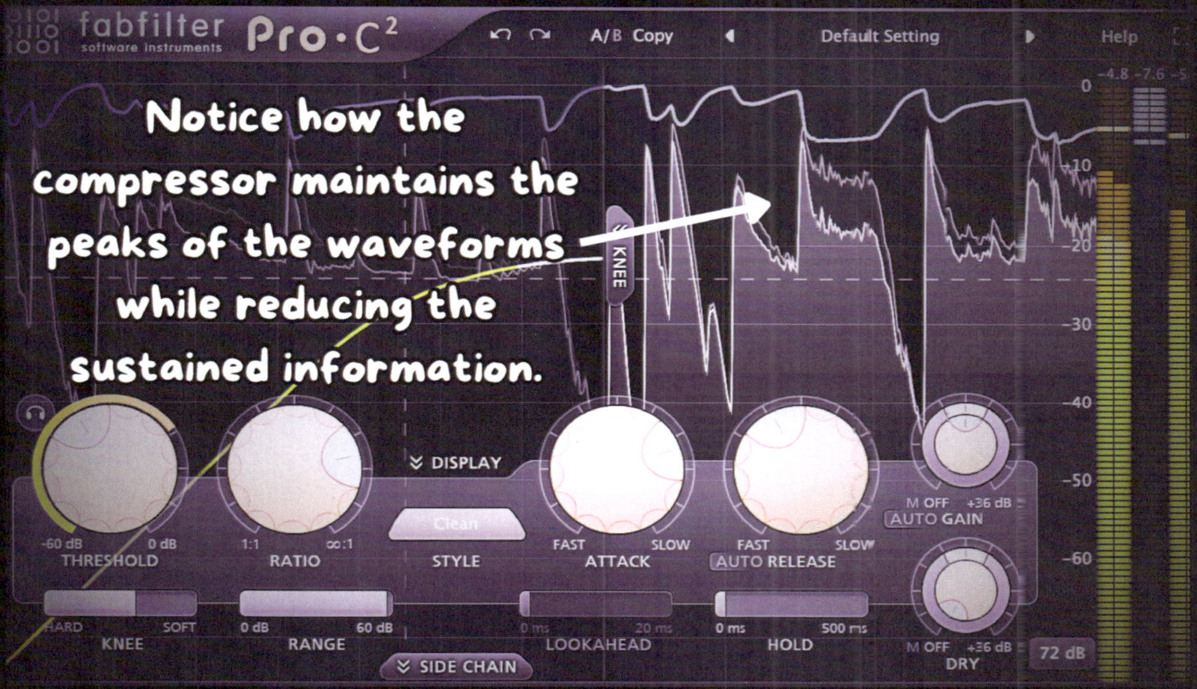

With a slower approach to compression (Slower attack time), you can preserve the transient and compress resonance. This is a great approach for "tightening" transient information like drums.

With a faster approach to compression (faster attack time), you can limit the transient, resulting in a smoother sound. This is a great approach to "smoothen" an overly transient performance.

Transient Designer

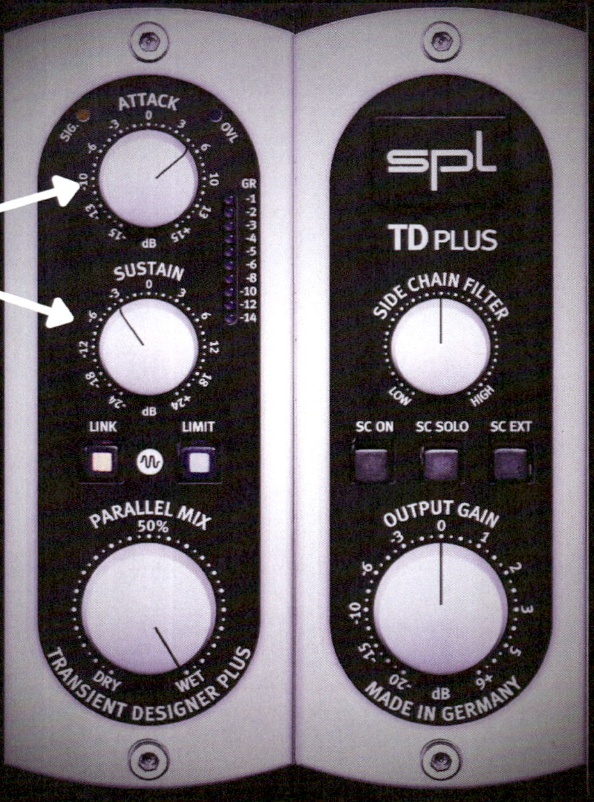

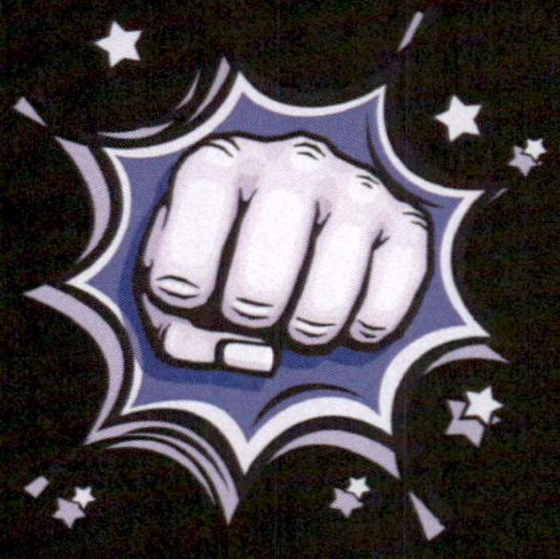

A "transient designer," also referred to as a "wave designer" or "wave shaper," is designed to refine the envelope of a specific sound. This tool works wonders with percussive instruments like drums or rhythmic synths.

By increasing the attack and decreasing the sustain, you can achieve a "punchier" and "less resonant" sound. Conversely, reducing the attack and increasing the sustain can help soften a "stabbing transient" or enrich tonal details.

Contradicting EQ Dilemma

What if I love the clarity in the top end of the bass guitar, but the "popping" of the pickup feels like a dagger to my ears? I don't want to sacrifice clarity by EQ'ing out the harsh frequencies created only by the "popping". What do I do?

What if a singer has a dark voice, but strong sibilance? I'd love to brighten her voice without turning up the sibilance. What do I do?

What if I love the warmth when this cello plays high notes, but when low notes are played, the lower mid frequencies become too extreme? How do I tame down the low end without sacrificing warmth on the high notes?

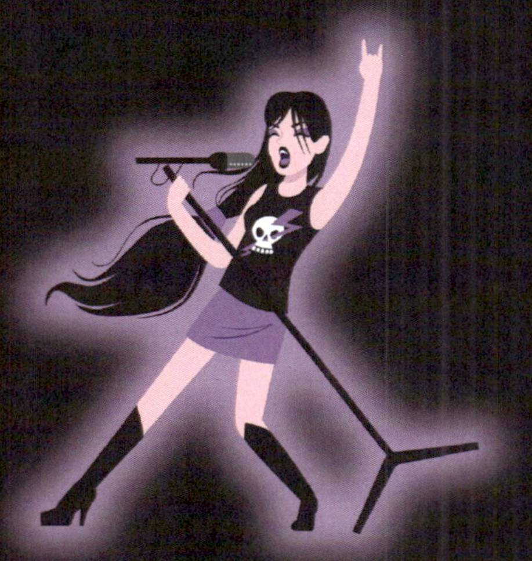

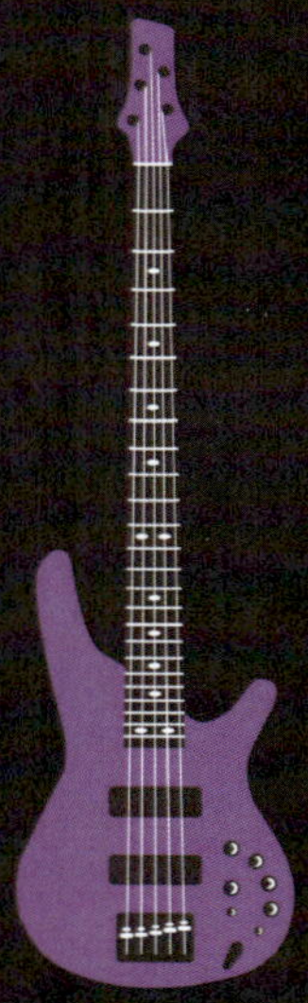

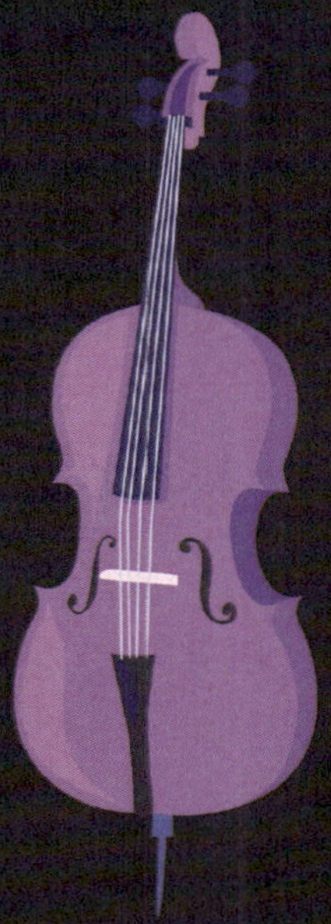

We can fix it!!

During the mixing process, it's common to encounter "Contradicting EQ" challenges like these... Resort to a "De-Esser", "Multiband Compressor", or "Dynamic EQ" to fix these tricky problems...

De-Esser

A "De-Esser" is an FX processor that is specifically designed to manage sibilance in a vocal. It functions much like a compressor focused in the 4-12kHz range. Many of the parameters are like a typical compressor, however you'll see parameters that specify the focused range of frequencies.

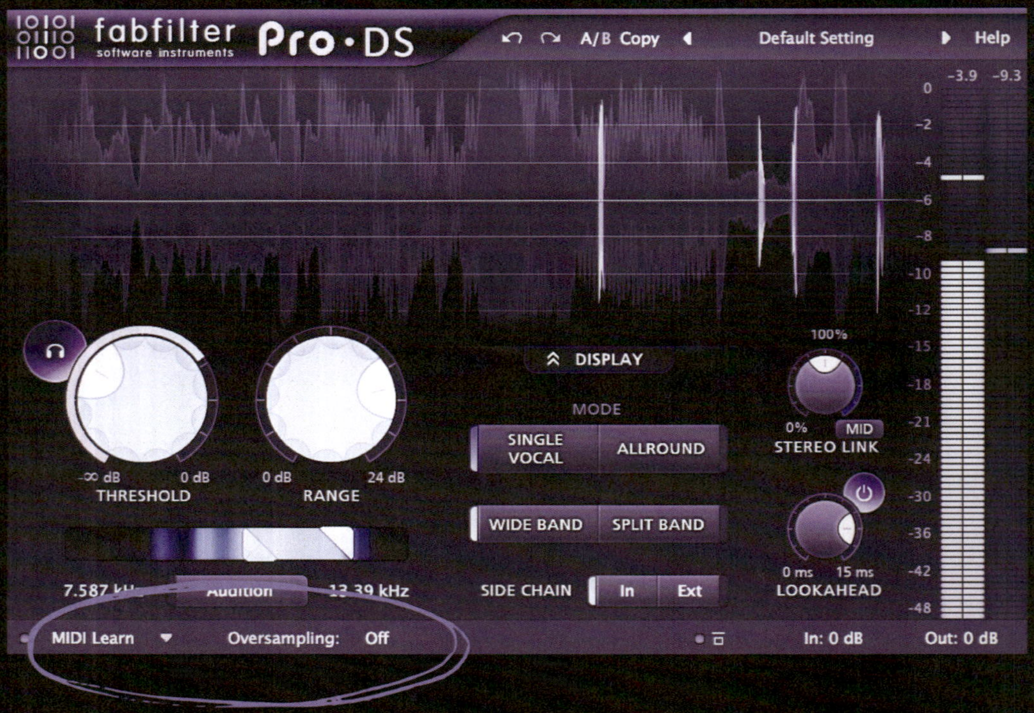

Tip* Be sure to pinpoint the harsh frequencies in the sibilance and set the threshold where the reduction only applies when the "S" appears.

Multiband Compressor

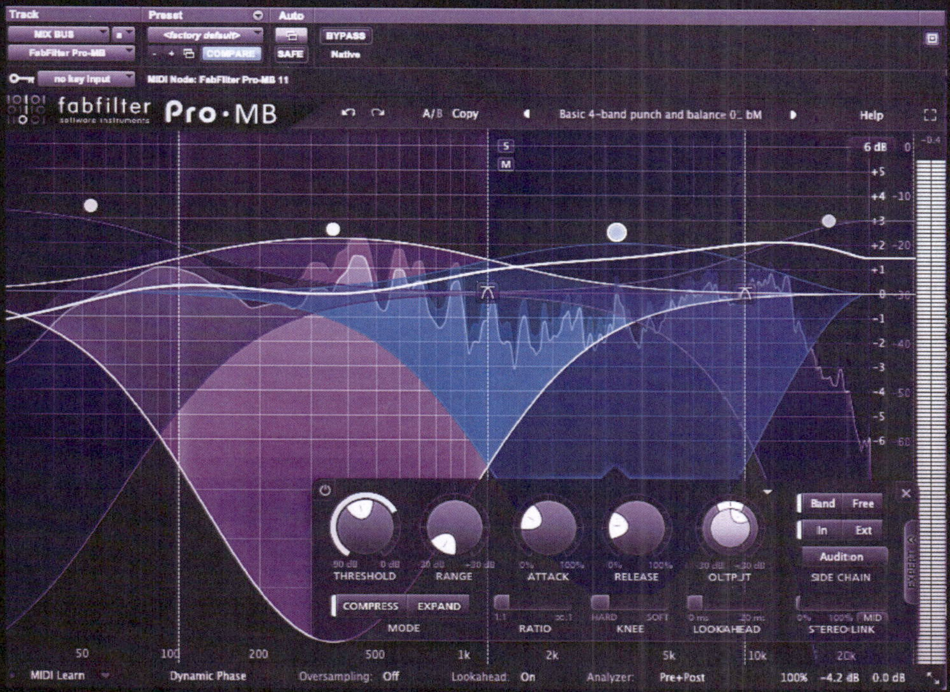

Consider utilizing a multiband compressor similarly to how you would use an EQ. This tool balances the frequency spectrum through compression instead of adjusting volume. Instead of relying on a single compressor for the entire spectrum, it allows several compressors to collaborate, each targeting specific frequency ranges. While multiband compressors demand a bit more finesse, when properly adjusted, they can deliver a dynamic sound that reacts with the performance.

Multiband Technique

1. Separate the bands by function in context with the sound...

Ex: Piano

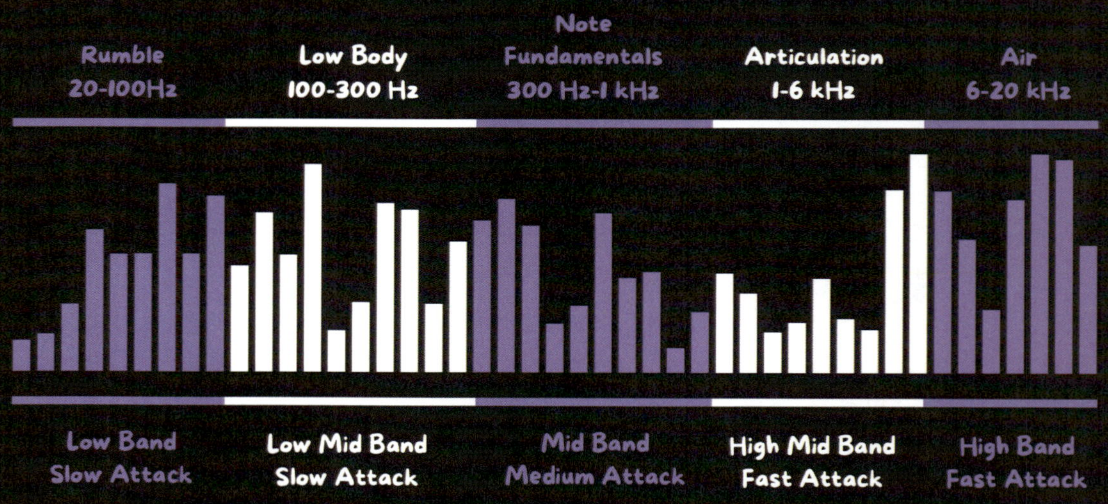

2. Think of each band like it's own sound with it's own discrete compressor settings...

3. Pay attention to transients and whether or not you want to "tighten" or "smooth out" that part of the sound...

Remember "fast attack = limiting transients" and "slow attack = lighten resonance"...

4. Use your ears as if you are EQ'ing, but use your knowledge of compression to EQ the sound...

Tip* As a starting point for controlling the dynamics of a complex sound, try using slow attack times for the low frequencies, medium attack times for mid frequencies, and faster attack times for the high frequencies...
Try tightening the low end and midrange, while creating a smooth top end...

Dynamic EQ

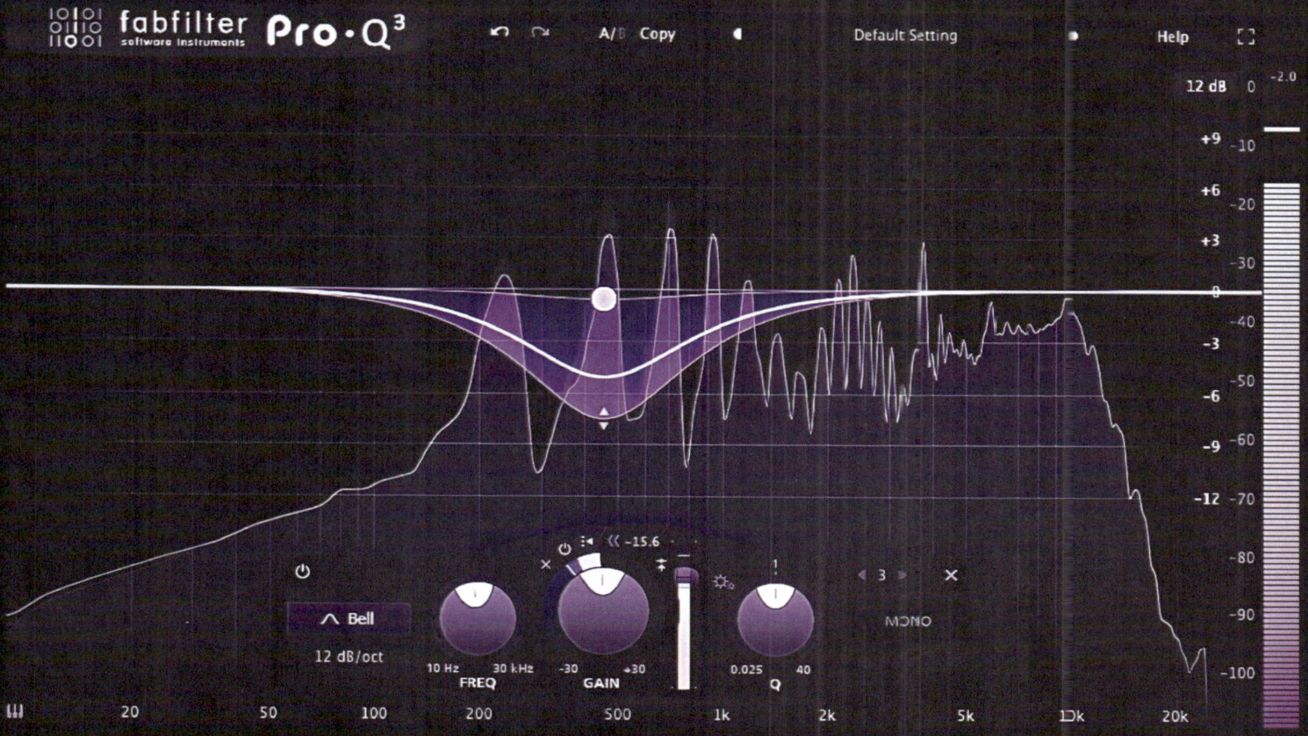

A Dynamic EQ is an adaptive form of equalization that lowers the volume when the signal surpasses a specific threshold. In contrast to a multi-band compressor, which applies compression at a ratio, a Dynamic EQ simply applies volume adjustments similar to volume automation. Much like a goalie in Hockey, a Dynamic EQ serves as a protective barrier against sporadic spikes or temporary imbalances in volume across the frequency range.

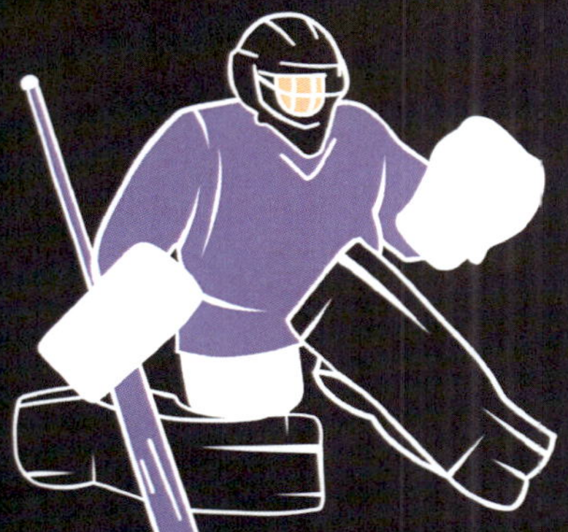

Gate / Expander

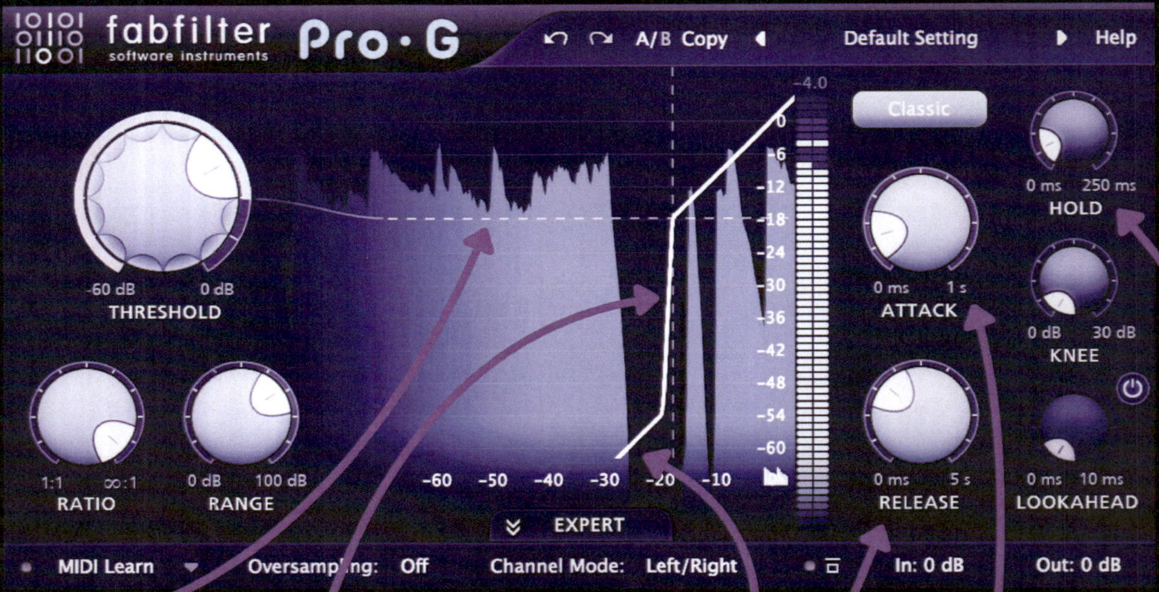

Threshold - The level at which the gate opens...

Range - the level of audio when the gate is in the closed position...

Attack - The speed at which the gate opens...

Ratio - the amount of signal passing through the gate when the threshold is hit...

Release - The speed at which the gate closes when the signal falls below the threshold...

Hold - The amount of time the gate will remain open when signal falls below the threshold...

Both gates and expanders function as dynamic processors that enhance dynamic range. A gate accomplishes this by using a reduction ratio, allowing audio to be audible above a certain threshold while lowering the volume below that ratio. In simpler terms, the sound remains silent or diminished until the dB threshold is reached, at which point the sound is released, effectively "opening the gate." On the other hand, an expander similarly increases dynamic range, but instead of reducing volume at the threshold, it incorporates an increase in volume.

Saturation

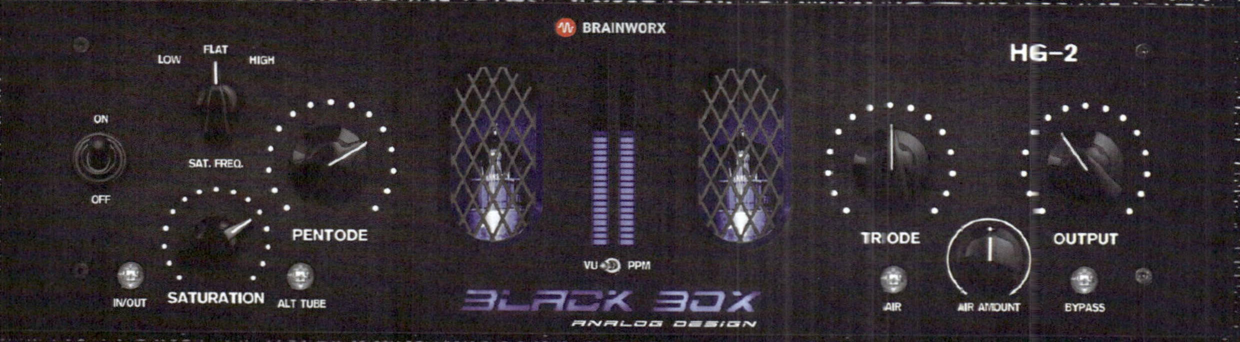

When analog equipment is pushed beyond its nominal limits, saturation starts to take effect. This saturation can vary from subtle harmonic coloration to intense distortion. Many leading audio manufacturers emphasize this harmonic content to achieve their signature sound. While this electronic noise is a natural characteristic of analog devices, it must be artificially created in the digital realm. Although developing an ear for saturation can be challenging, experimenting with various flavors can open up endless possibilities for creating unique vibes and finishing touches on mixes.

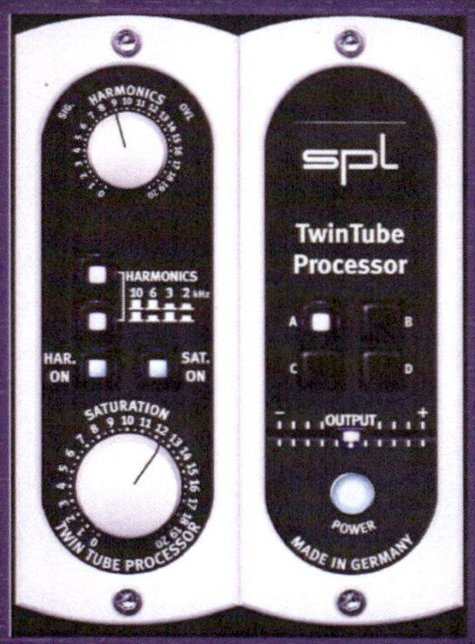

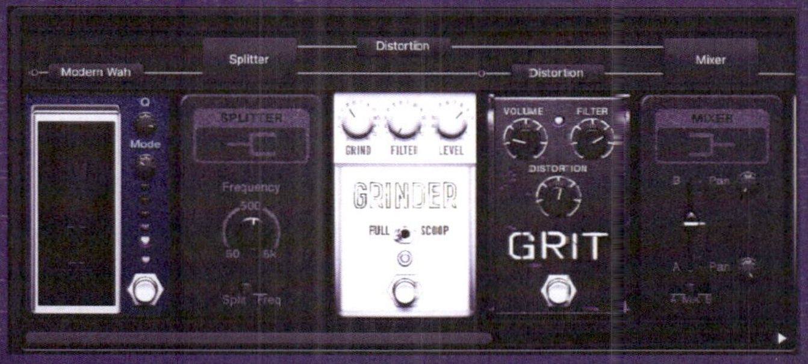

Odd and Even Harmonics

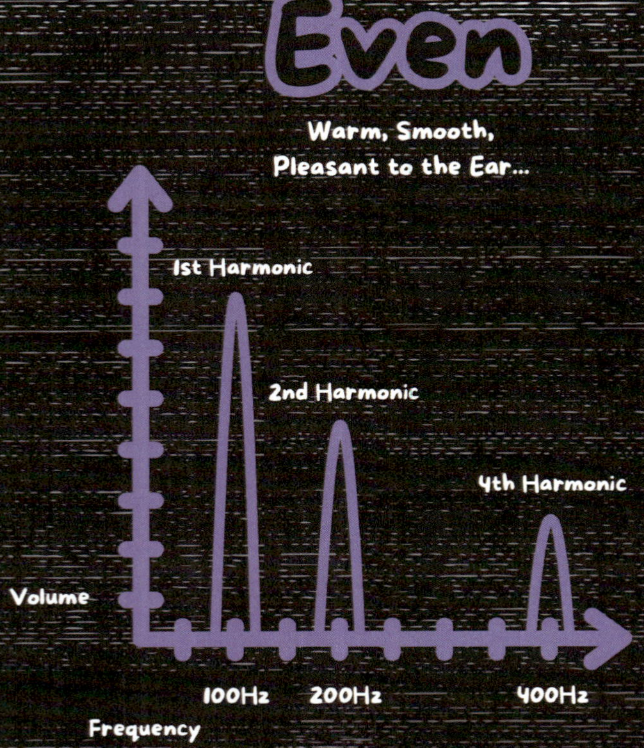

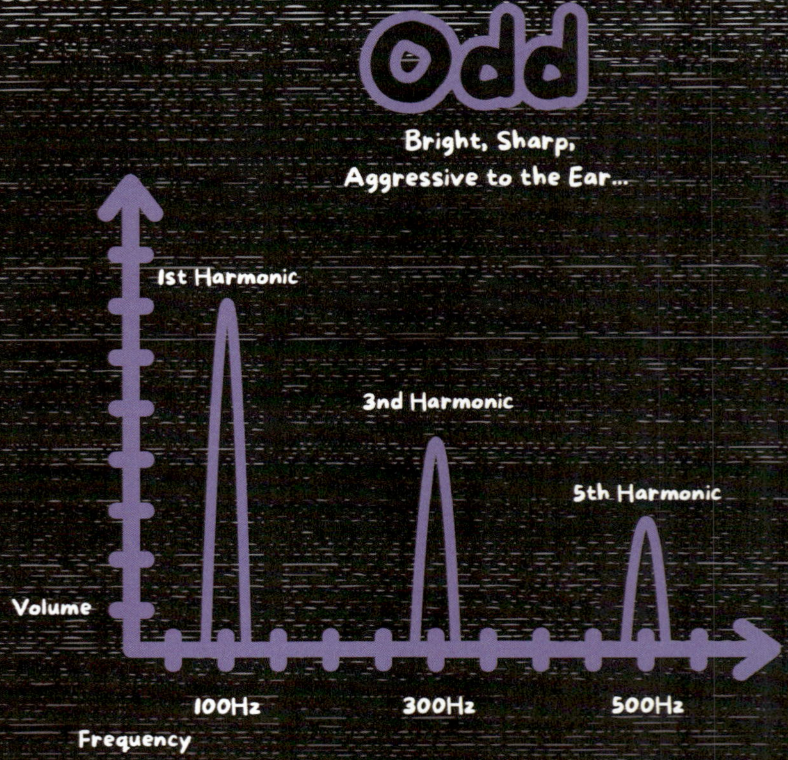

When a piece of equipment produces harmonics, envision a tiny harmony for each fundamental frequency. Above, you can see 2 visuals of both even and odd harmonics stemming from the initial 100Hz "1st Harmonic." Equipment that generates even harmonics typically offers a smoother, warmer sound that is pleasing to the ear. This quality can help sweeten sounds that may be overly harsh, such as cymbals, brass, or electric guitars. Conversely, gear that produces odd harmonics tends to infuse a unique energy, brightness, and vibrancy into the sound. This can be an excellent way to inject some excitement into a dull mix that may be lacking in liveliness.

Modulation

Phaser

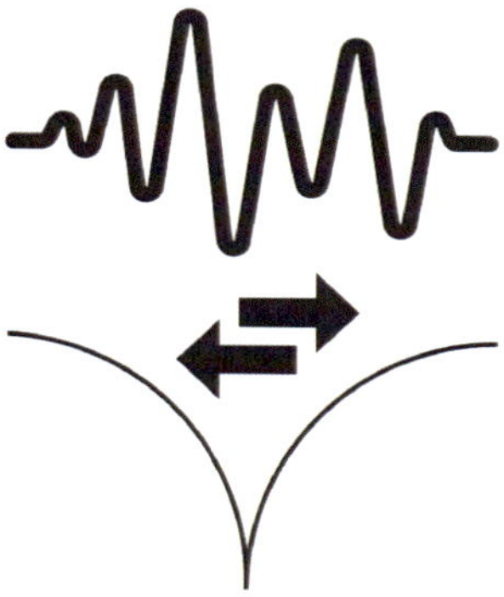

A phaser effect is created by applying a polarity flip to a particular or series of frequencies. Unlike a flanger which achieves it's sound with delay, a phaser gets its "sweeping sound" when the inverted frequency or frequencies oscillate up and down the frequency spectrum.

Chorus

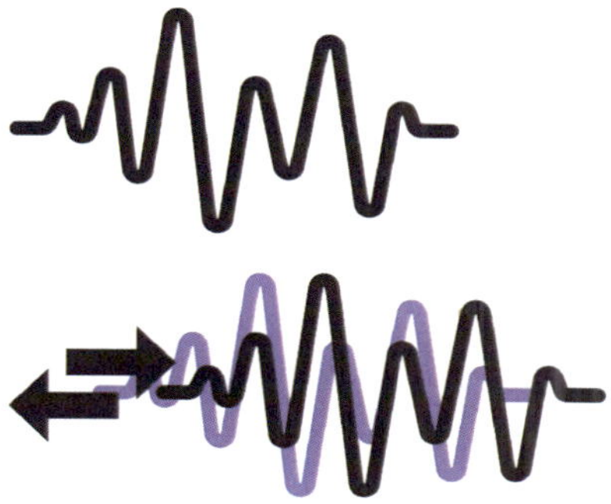

A chorus effect is produced by a duplicated waveform that varies in both time and pitch. The speed of the pitch variation is referred to as the rate, while the amount of pitch modulation is known as the depth. These two characteristics are represented as parameters in a chorus effect.

Flanger

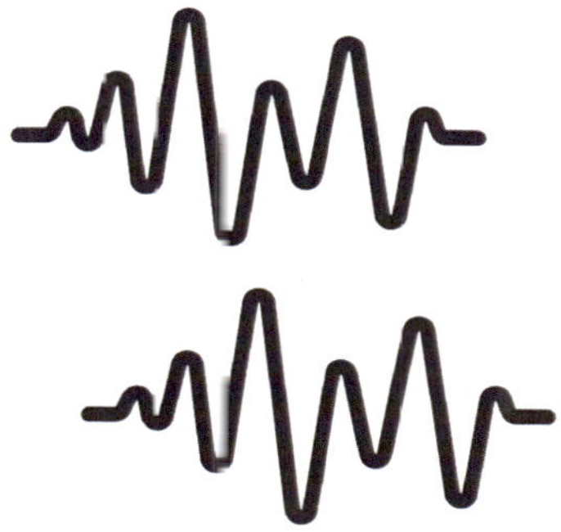

A flanger effect is created by a duplicate waveform that is slightly delayed in time. The relationship between the duplicates create both phase cancelation and summation, resulting in the distinctive swishing sound of flanging.

Pitch Shifting

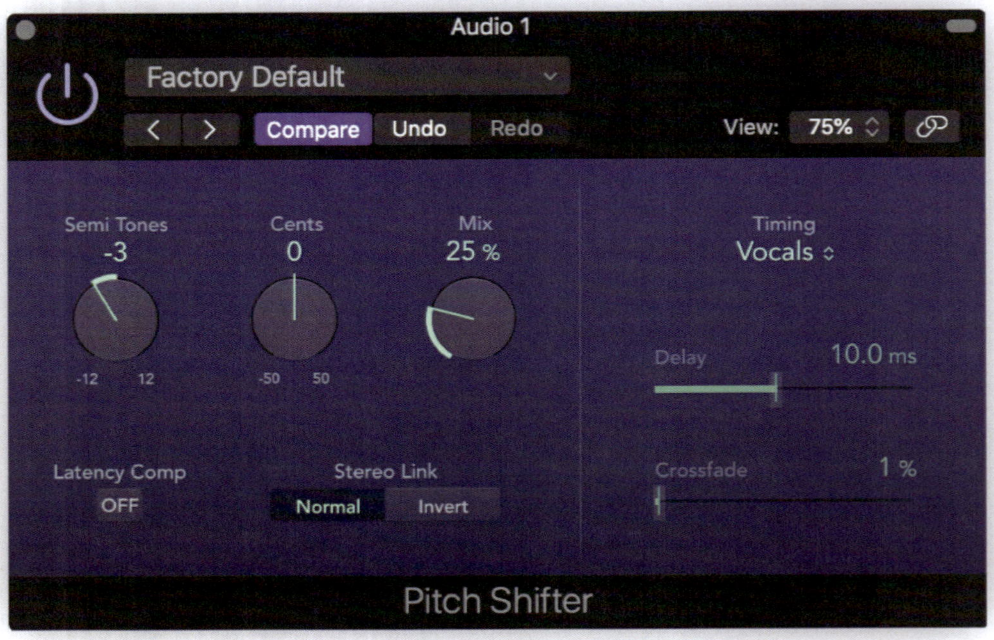

One method for modifying a waveform is to adjust or shift its pitch. This adjustment is typically measured in semitones (the half steps used in music) or cents (for micro-tuning within a single semitone). A helpful guideline is to utilize either subtle "cents" shifting on "track doubles" or octaves (12 semitones) when working with melodic components. Non-melodic instruments allow for more experimentation. Keep in mind that the more you alter a digital waveform, the greater the risk of introducing unwanted artifacts. When applied tastefully within the song's context, pitch shifting can enhance the weight and density of your mix.

Auto-Tune

Determine the key and scale of the song

Determine the type of voice... This helps optimize the effect...

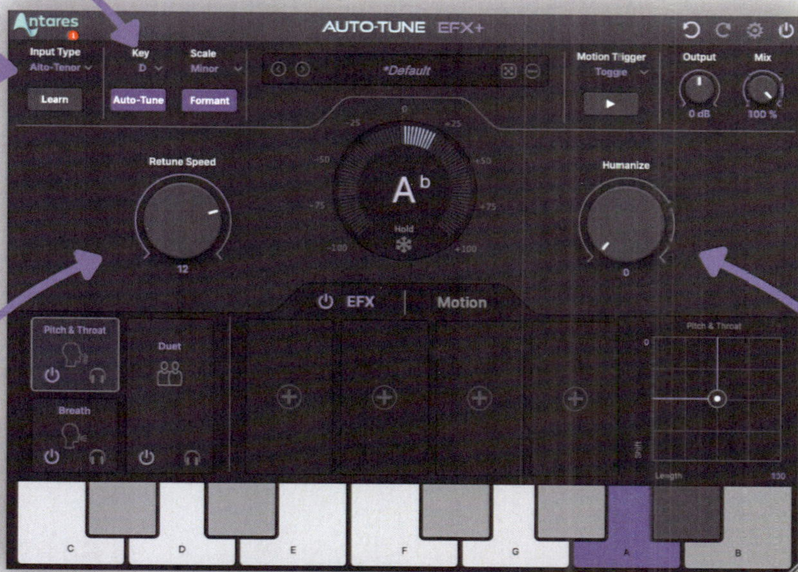

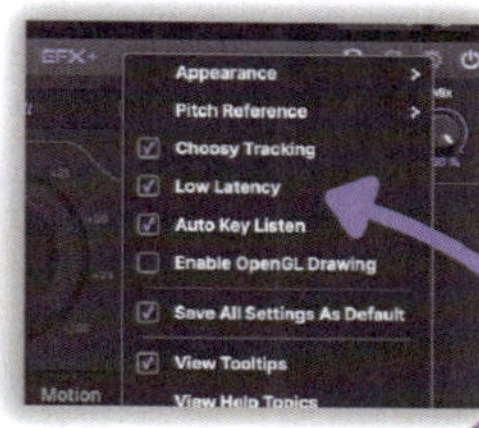

Engage "Low Latency" mode while Recording

Adjust the "Retune Speed"... Determines how fast the note gets corrected... "0" is "T-Pain" and you can increase the number to make the effect more subtle...

Adjust "Humanize"... Determines how much the note is corrected... "0" corrects to perfect pitch...

Ok time to Preach!! Auto-Tune has become a signature sonic thumbprint for modern music... It is a powerful tool, and one of the most controversial effects processors ever created for obvious reasons. If you choose to take this approach, treat Auto-Tune like an instrument, and record with the effect engaged... It will influence your artistic decisions in the same way distortion and delay pedals influence a guitar player... At the end of the day, there is more to music than hitting notes. In conclusion, you still have to create great music and rest assured there are no shortcuts...

Reverb Basics

Mix refers to the balance of original or "dry" signal to the reverb or "wet" signal...

Damping a reverb refers to the tone shaping of the reflections...

Diffusion refers to the amount of randomizing the reverb adds to the source's frequency content...

Using a high cut filter helps eliminate high frequencies in the reverb...

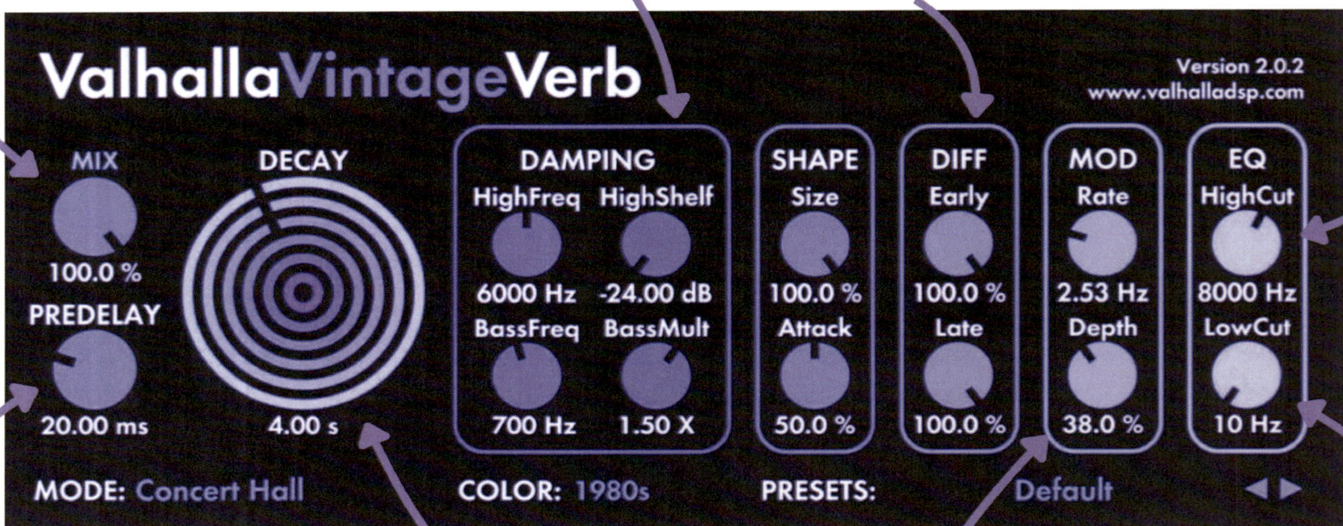

Pre-Delay indicates the amount of delay time before the reverb begins to produce sound...

Decay refers to the length of the reverb tail...

Modulation can add variation to the reverb's frequency content...

Using a low cut filter helps eliminate low frequencies in the reverb...

Room Reverb

Room reverbs are an excellent method to enrich the atmosphere of a space while avoiding an overt "effect." Typically, microphones are positioned at relatively close ranges, particularly when the recording environment lacks ample space. A helpful tip for room reverbs is to perceive them as ambiance or color instead of just "Big Tailed" reverbs.

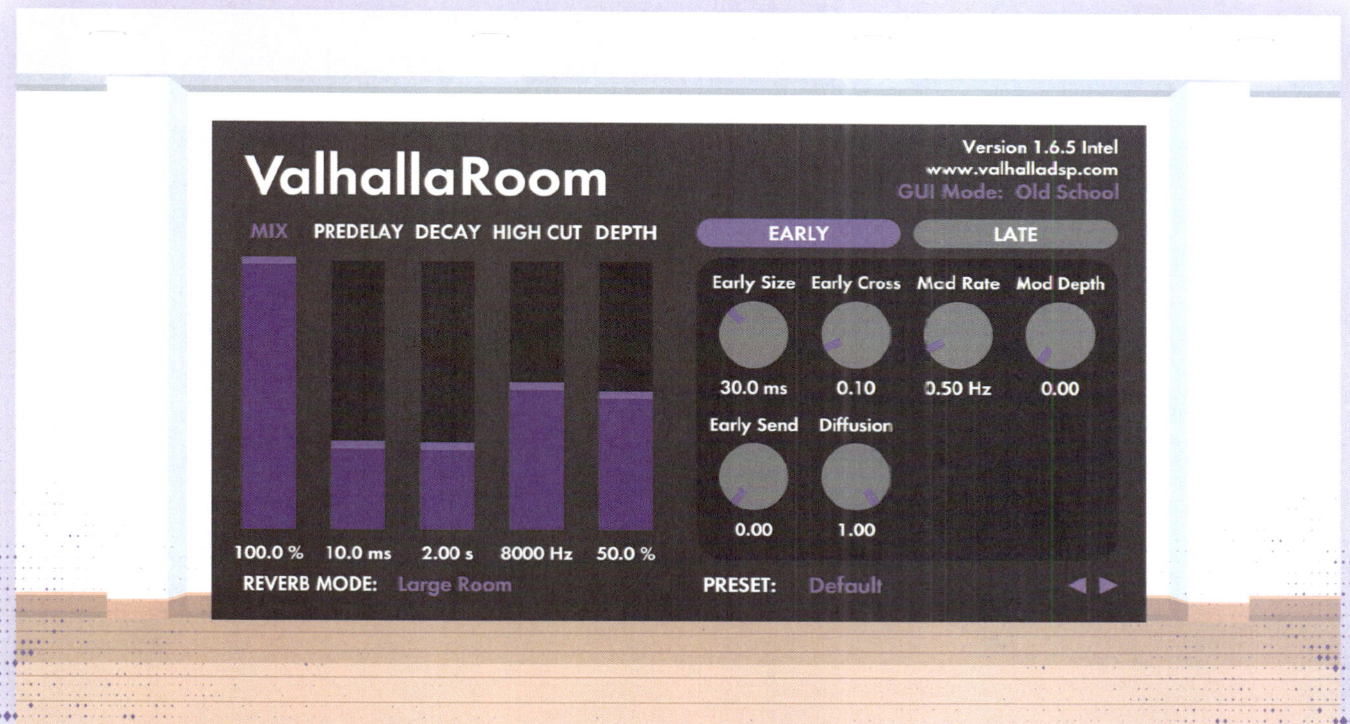

Hall Reverb

Hall reverbs offer a bold and distinctive atmosphere. These reverbs typically have longer tails, and their size and acoustic treatment contribute to a warmer tonal quality. Halls are particularly effective for acoustic instruments and any sound that you wish to enhance with a "grandiose" ambiance.

Tip: Be mindful of the reverb's decay time. Aim to establish a sense of cohesiveness between the musical phrasing of the song and the reverb tails.

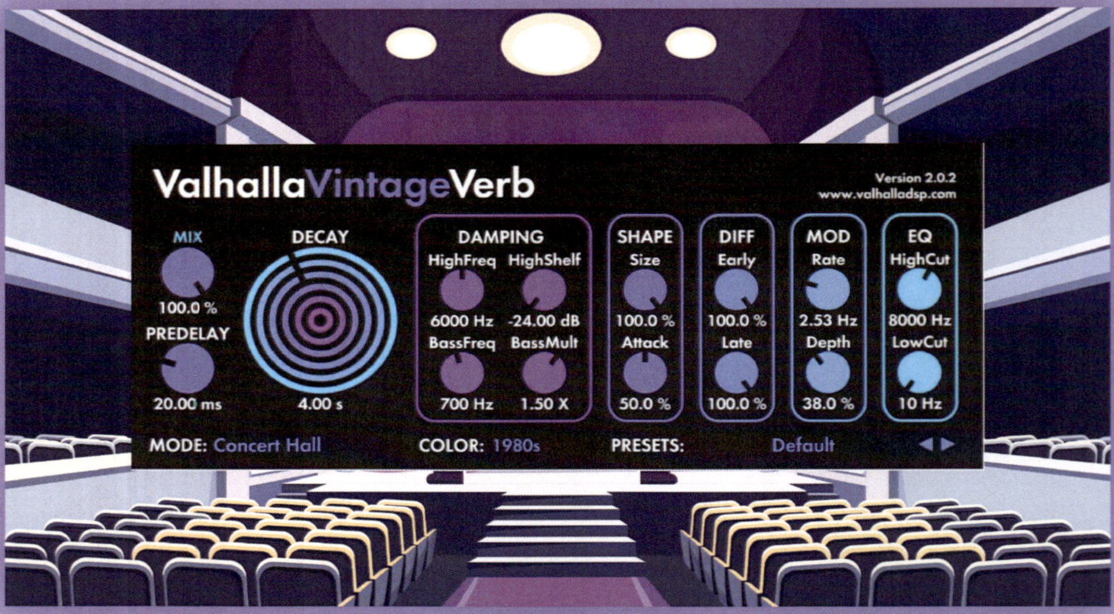

Tip: Experiment with longer pre-delay times... This can create a sense of clarity to the original sound while adding some tasteful rhythmic textures similar to the early reflections in a large space.

Plate Reverb

Plate reverbs stand out because they don't replicate the sound of a room. Instead, they generate a distinct ambiance through the vibrations of a metal plate. These reverbs offer a smooth yet bright quality, making them ideal for anything that requires enhanced clarity. It's for this reason plates are often a favorite on vocals and drums.

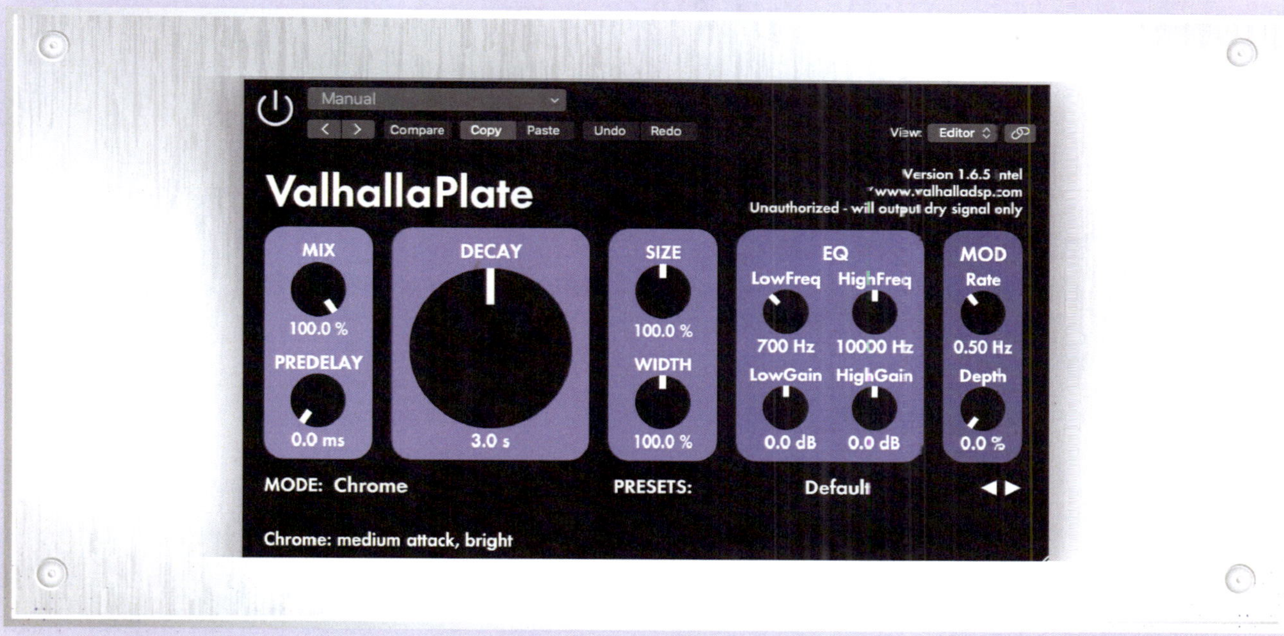

Spring Reverb

Spring reverbs produce a unique "fluttery/metallic" sound that creates a substantial amount of modulation into the reverb tail. While this ambiance may be too pronounced for certain situations, it can provide the perfect touch in others. Spring reverbs are particularly effective on guitars or vocals when used for special effect. Even if applied to just one section rather than the entire track, it can serve as a powerful reverb tool.

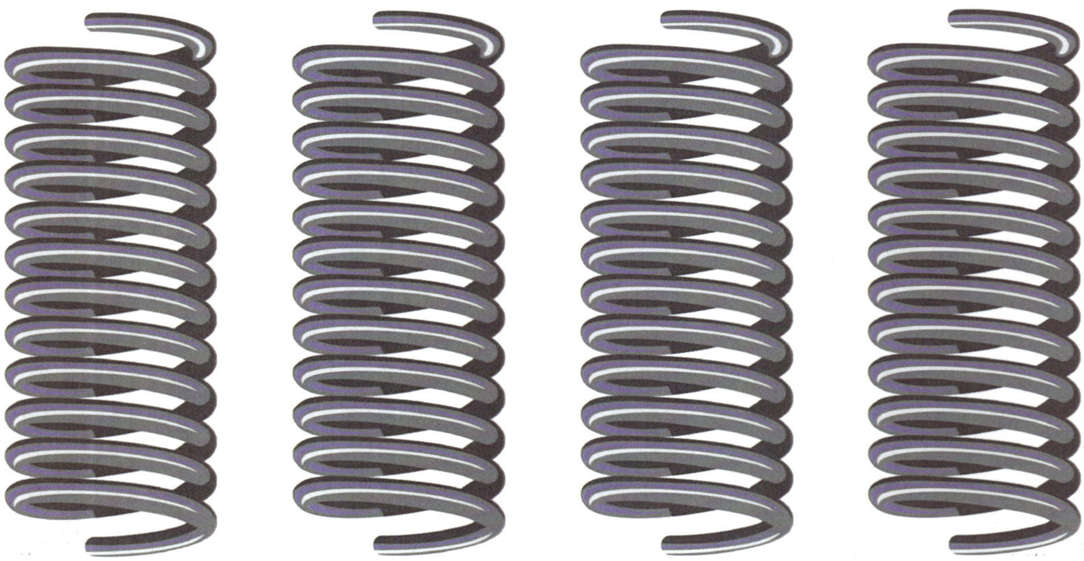

3 Reverbs Trick

A great technique to help achieve rich, luscious ambience is to use a combination of 3 different reverbs. While there's no definitive right or wrong choice of reverb, feel free to experiment and build upon this idea until you find the perfect atmosphere for your song.

Short

- Set up a reverb with a short pre-delay and decay time. Treat this reverb as a subtle enhancement to the original sound. Room reverbs typically work very well for this purpose.

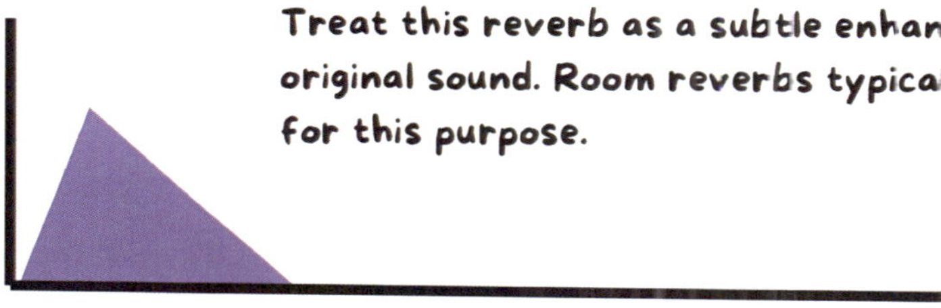

Medium

- Set up a reverb with a moderate pre-delay and decay time. A good musical reference would be to keep the decay time within 1 or 2 beats of the song. This reverb serves as an excellent balance between short and long reverbs.

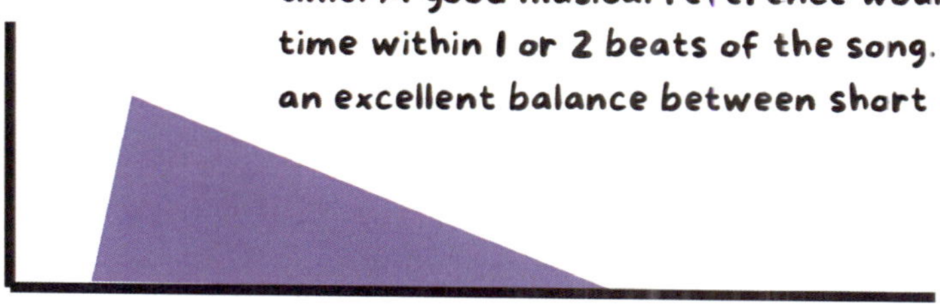

Long

- Create a reverb with a long pre-delay and decay time. This reverb will serve as the most prominent and lengthy effect in your track. For reference, the decay can last up to 1 or 2 bars of the song.

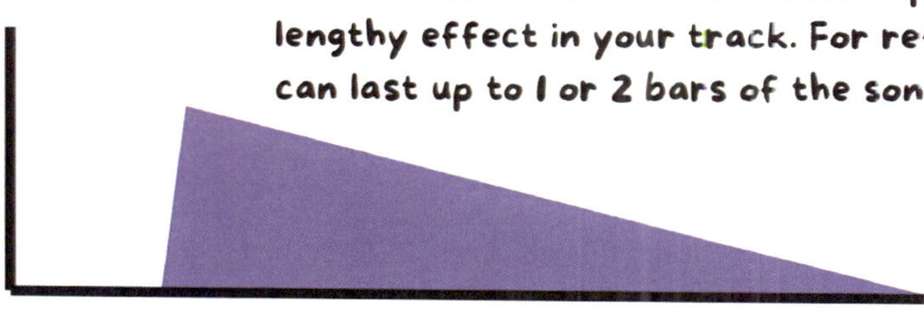

Delay Basics

Mix refers to the balance of original or "dry" signal to the delay or "wet" signal...

The delay time pertains to the rate at which echoes occur within the delay effect. This can be measured in milliseconds or in note values relative to the tempo of the music.

Feedback determines the number of echoes within the delay...

Using a high cut filter helps eliminate high frequencies in the delay...

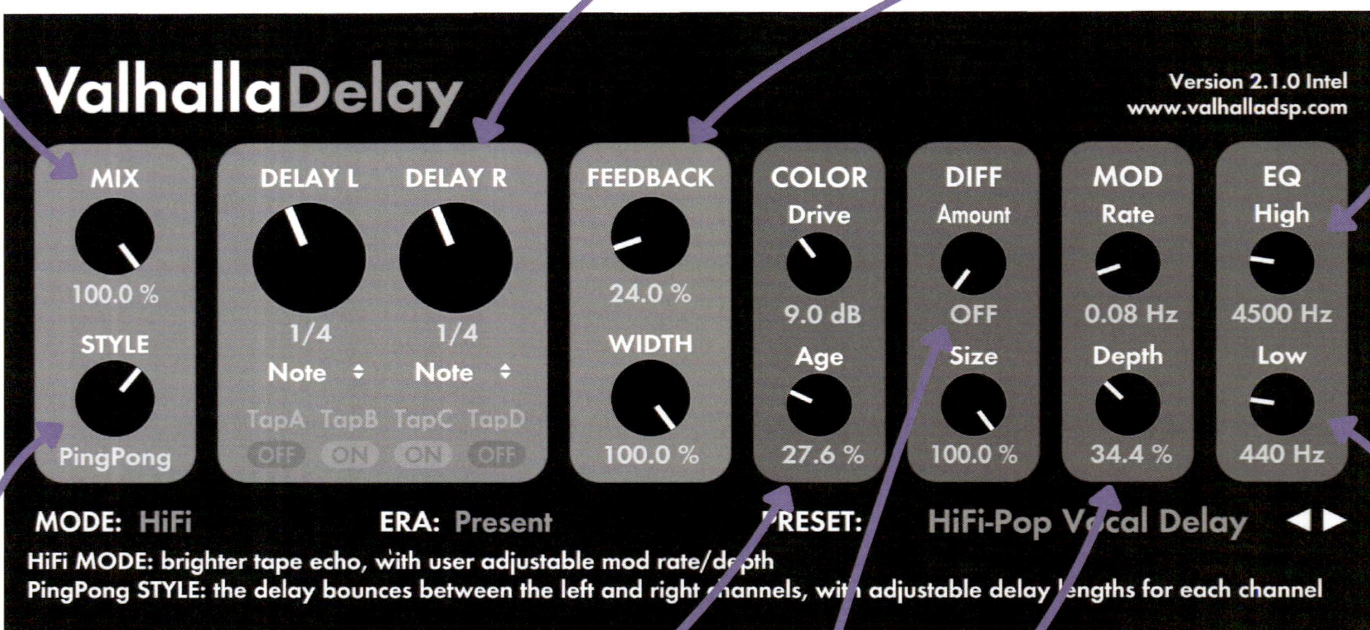

The type of delay can influence its behavior. In this instance, the delay style is referred to as a "Ping Pong" delay, which produces an alternating echo effect that moves from left to right.

Numerous delay processors provide various types of harmonic content. In this context, "color" refers to the amount and character of distortion that is incorporated into the delay effect.

Modulation can add variation to the delay's frequency content...

Diffusion pushes the delay into the background, creating a reverb type delay effect.

Using a low cut filter helps eliminate low frequencies in the delay...

Slap Delay

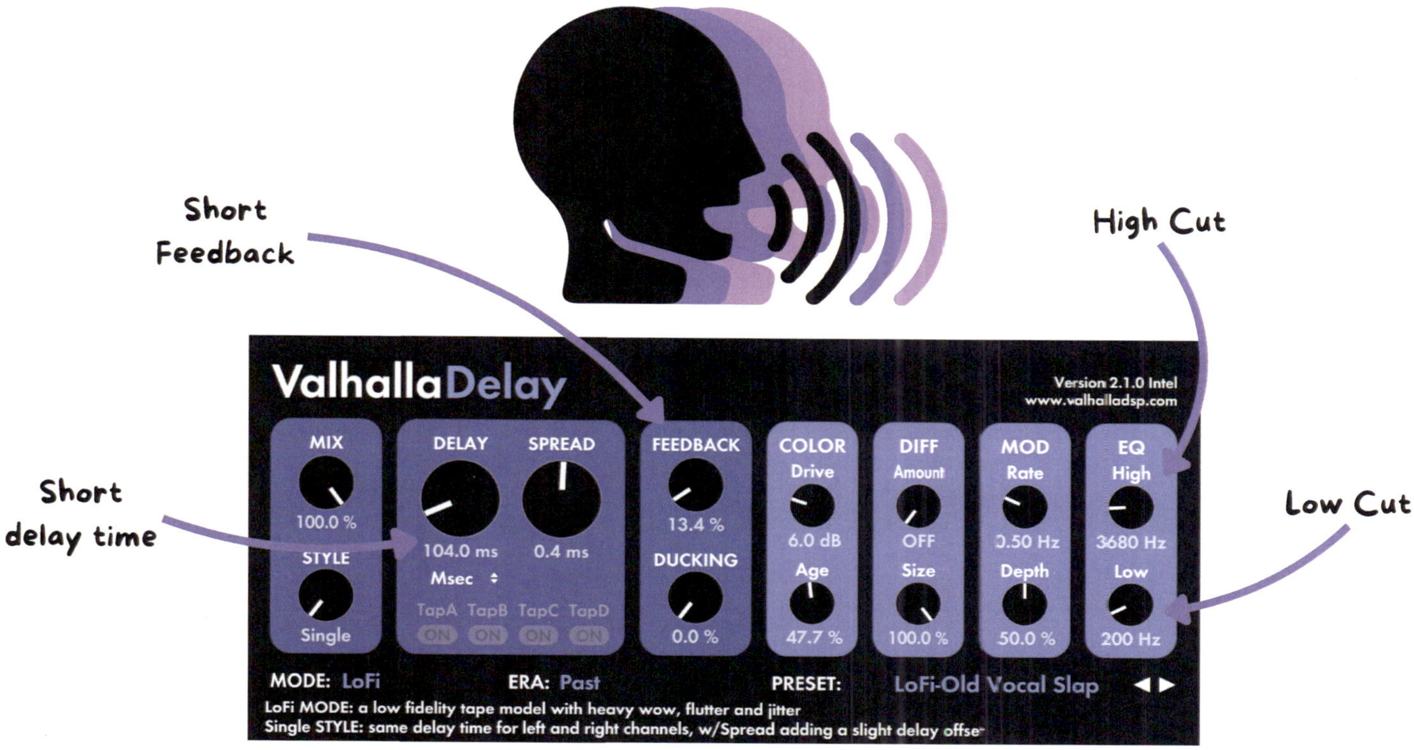

One of the simplest yet most effective methods for creating a delay is the "slap delay," which mimics the slap echo heard in live performances. To achieve this effect, keep the delay time short—ideally no more than 250ms or a 16th note within the musical context. Additionally, adjust the feedback level to produce around 2-3 echoes. Slap delays work wonderfully with a touch of high and low-pass filtering to create a "lo-fi" sound. This also helps to blend the effect. The "slap" is a timeless sound that gained iconic status in the 1950s and 1960s, and it continues to hold its popularity even today.

Timed "Tap" Delay

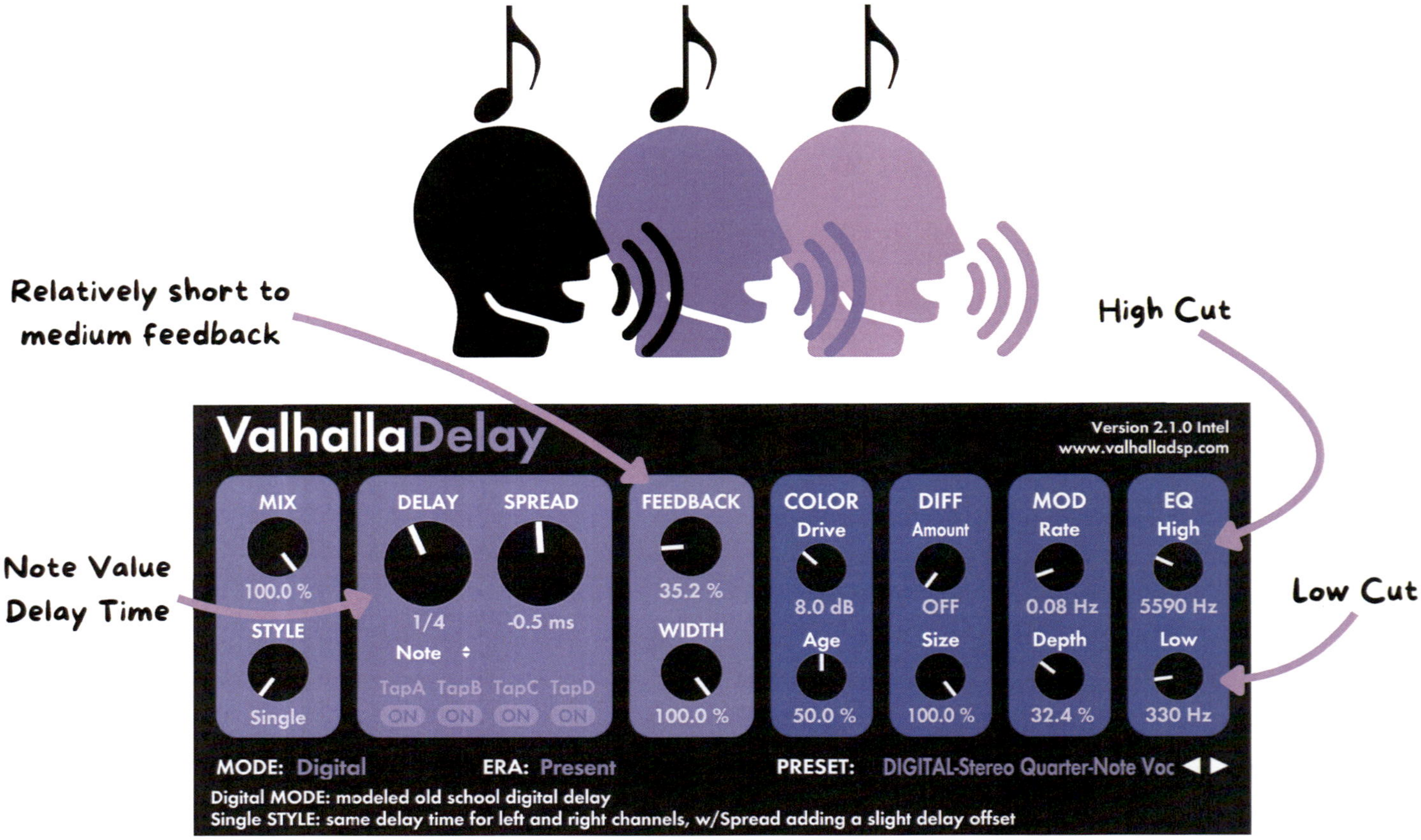

A timed or "tap" delay refers to a technique that interacts specifically within the rhythmic framework of a song. There are various ways to implement a timed delay, such as using an 8th note, 1/4 note, dotted 1/4, or even 1/2 note delay. The crucial factor is to select a rhythmic value that complements rather than distracts from the main focus of the song. Additionally, automating the level of this type of delay can help prevent it from overpowering the mix. This technique can be applied subtly or boldly for particular sections of a song. In conclusion, it's important to carefully observe how this effect interacts with the music and strike a balance between excitement and distraction.

Felt Delay

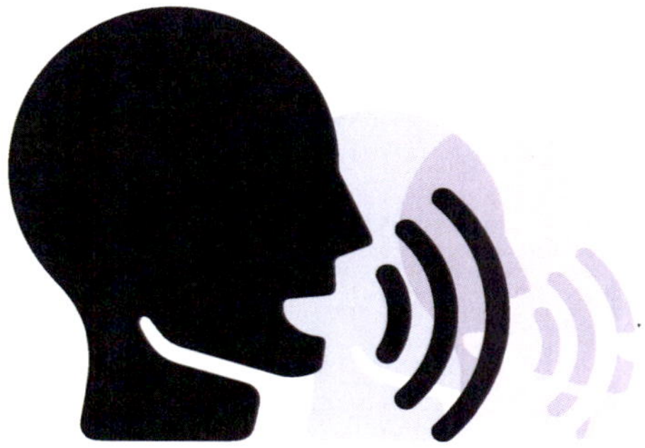

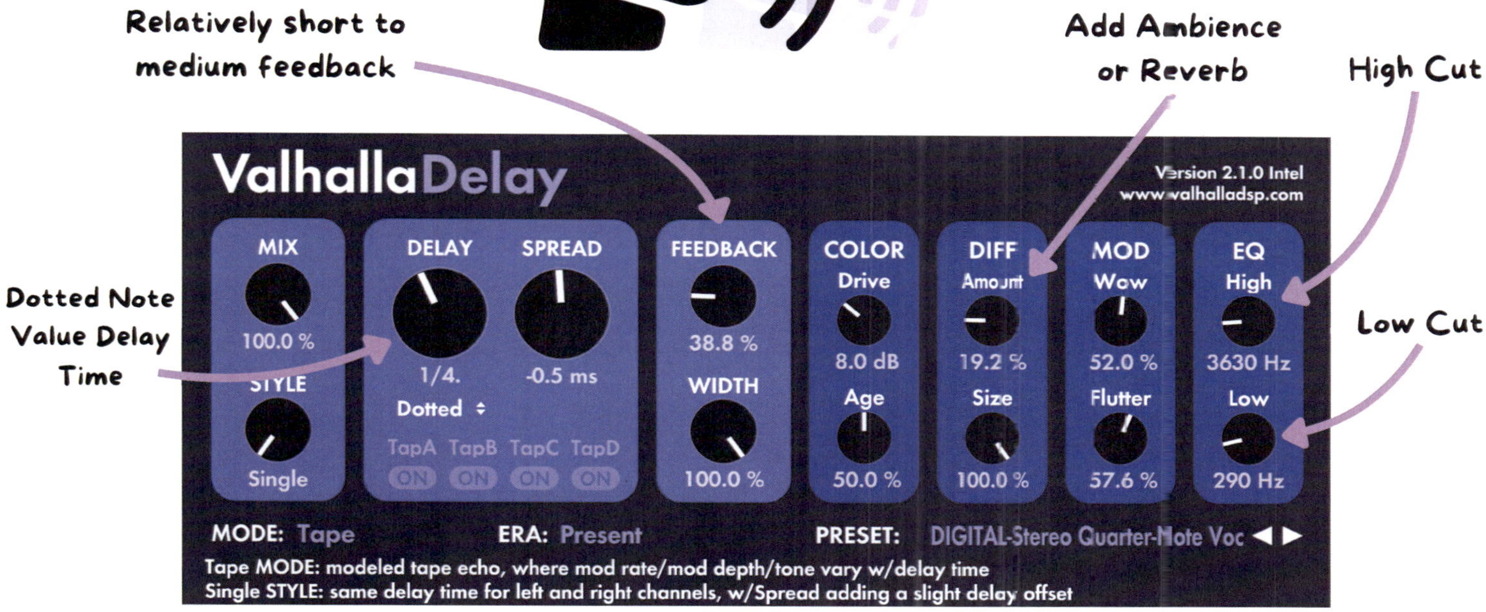

A felt delay refers to a delay that you perceive without actually hearing it. Essentially, aim to create a "haze" or subtle color that is blended within the mix. You can accomplish this through various methods, such as reducing extended high frequencies, selecting dotted note values, and incorporating reverb or ambient sounds. This technique is excellent for enhancing color and texture while keeping the effect hidden. In many respects, this method resembles a rhythmic reverb.

Recording Techniques

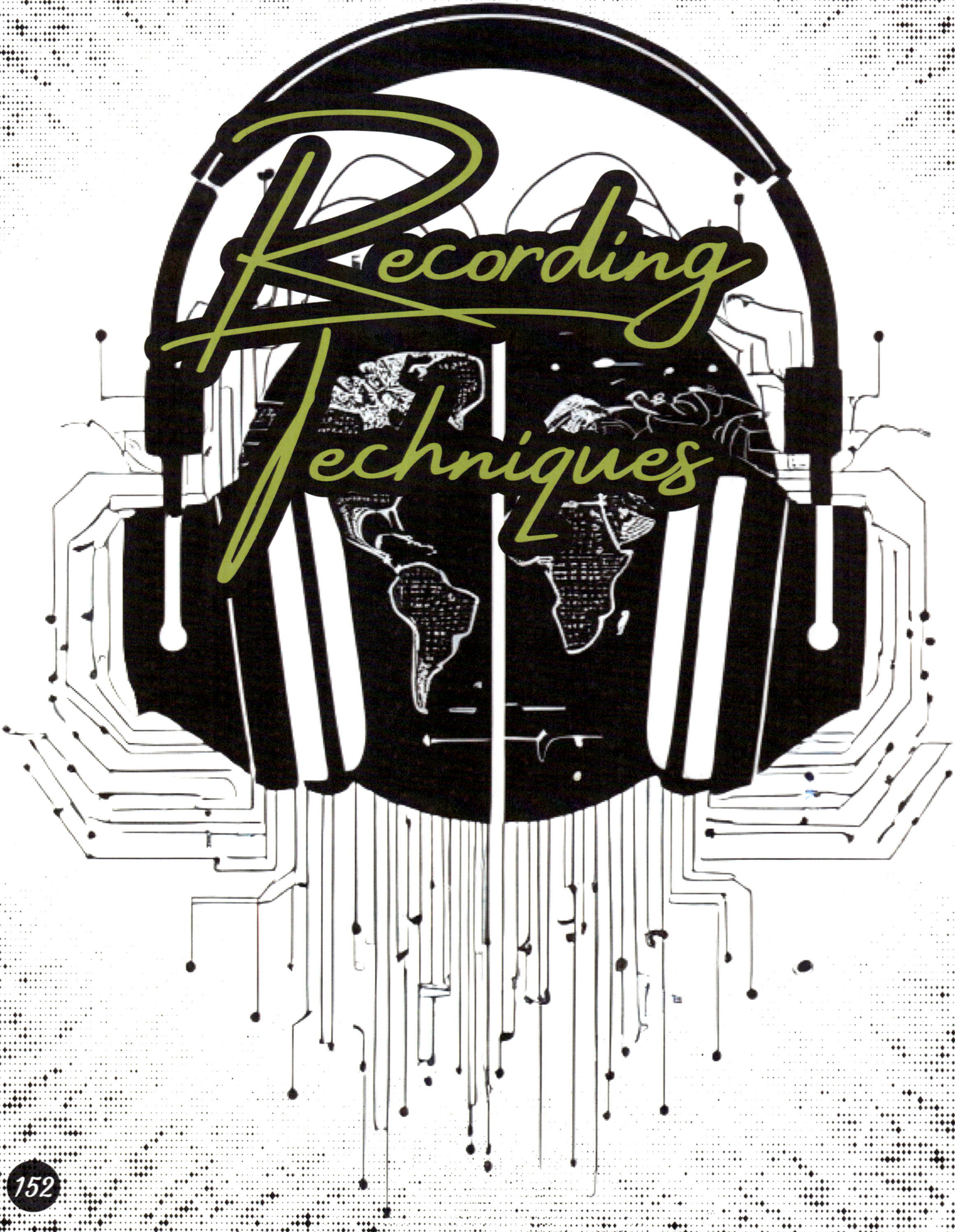

Preparing in the Pendulum Process

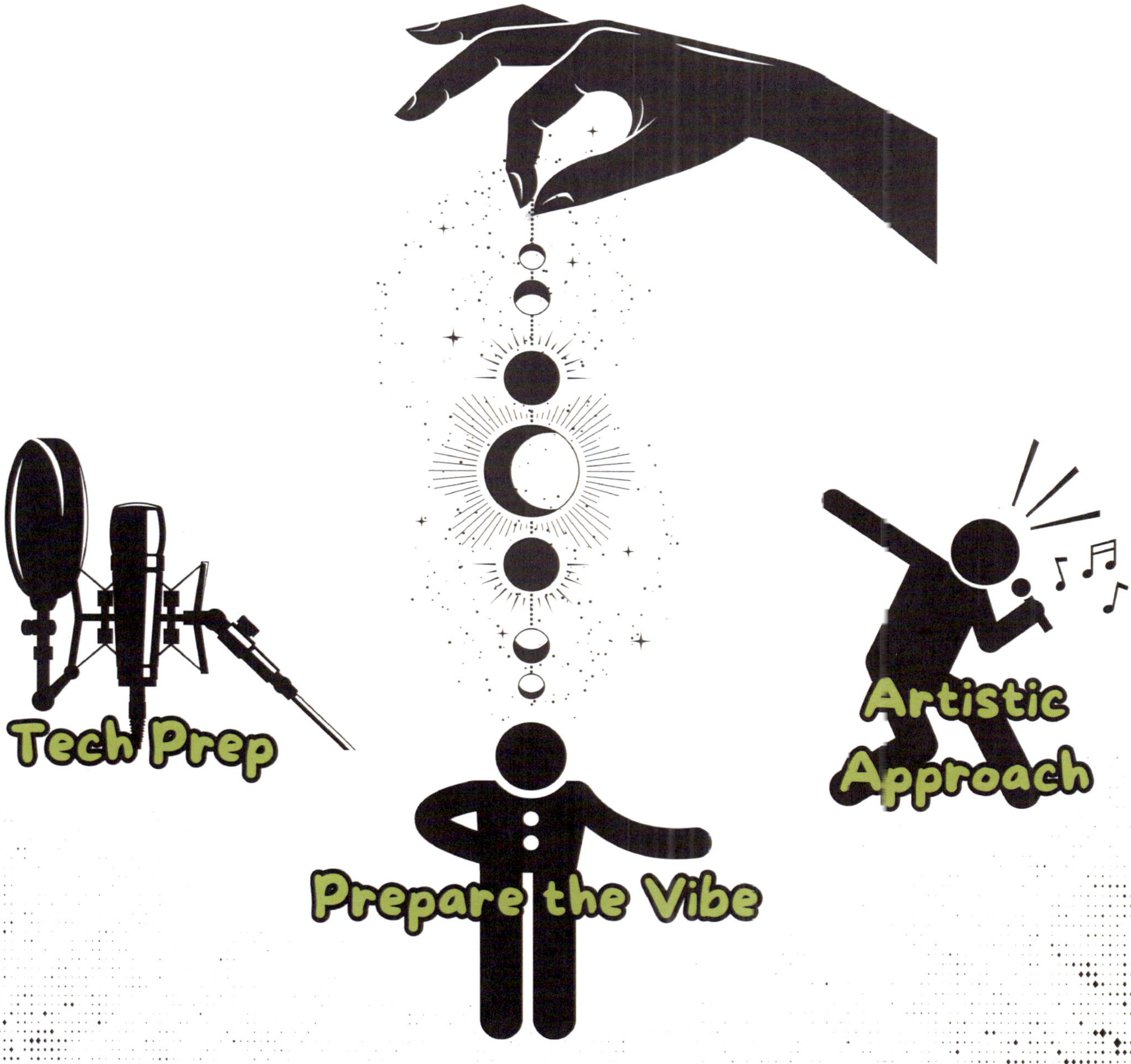

Producing in the Pendulum Process!

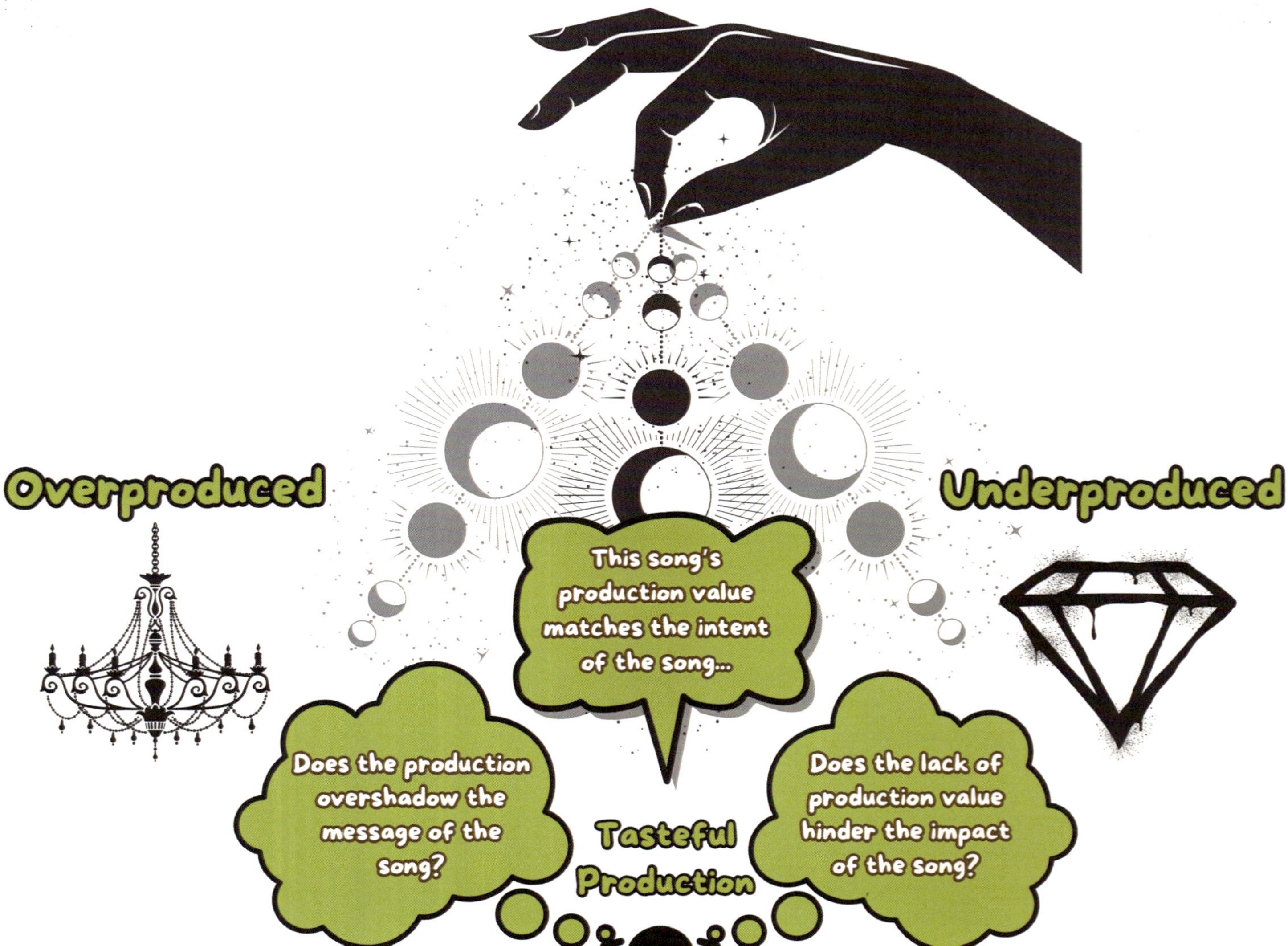

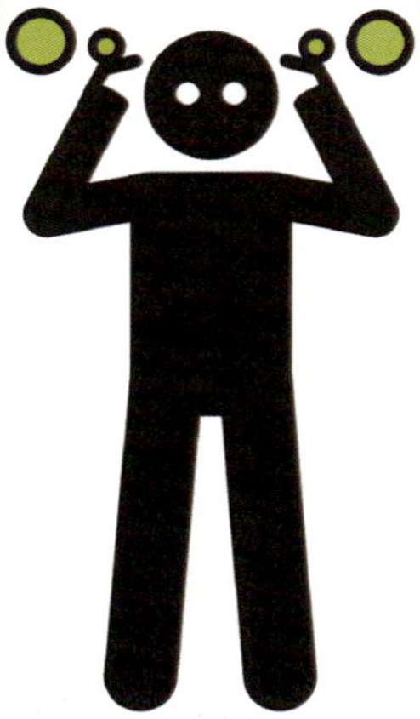

The fancy esthetic of a chandelier is similar to an "Overproduced" song. In this context, the aesthetic of the song becomes its central focus. While certain genres of music may require high amounts of production, if not tastefully managed, the result can feel lacking in humanity and authenticity.

Like a diamond in the rough, an underproduced song might initially seem to lack aesthetic appeal, yet with refinement, it can shine. If a song stays underproduced, it may hinder the overall impact of the song.

Creating in the Pendulum Process

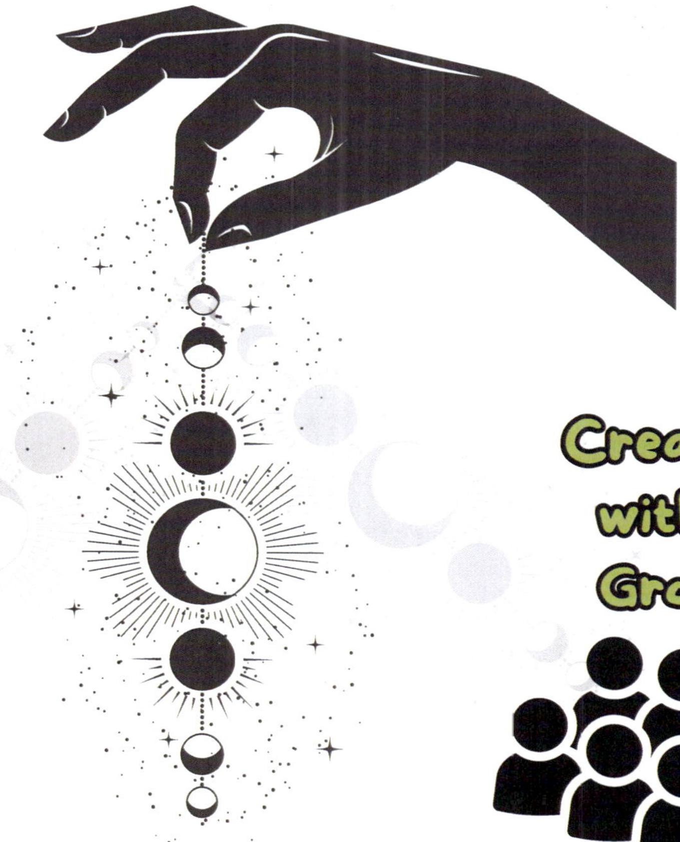

Creating Solo

When engaging in solo creation, individuals often adopt a more profound interpersonal approach towards the creative process. Songs tend to reflect deeper subject matter, and without the social pressure of a group, creatives tend to take more risks and show more vulnerability. This openness can lead to innovation, however the work load and pressure on a solo person comes with its obvious challenges.

Both individual and collaborative creation have their unique pros and cons. No matter which method you choose, a strong foundation of honesty and commitment is essential during your creative journey. By recognizing the natural tendencies of people, we can select the most effective environment for production.

Creating with a Group

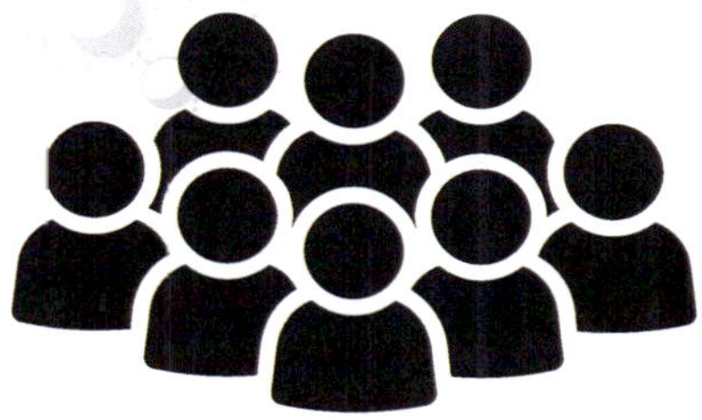

You can also create through collaboration. Such productions typically reflect the consensus of the group, resulting in a collective sound that lends itself well in social situations. While this method helps distribute the workload and reduces individual pressure, it can also cause original ideas to become diluted amidst differing opinions and social pressures.

Committing in the Pendulum Process

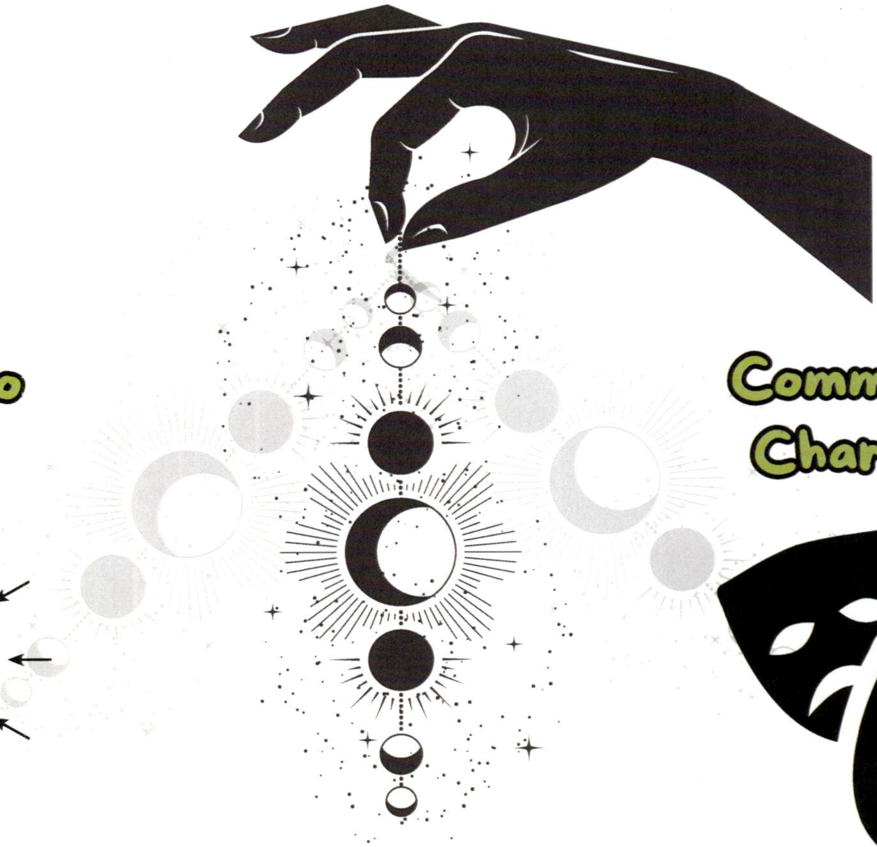

Commit to Yourself

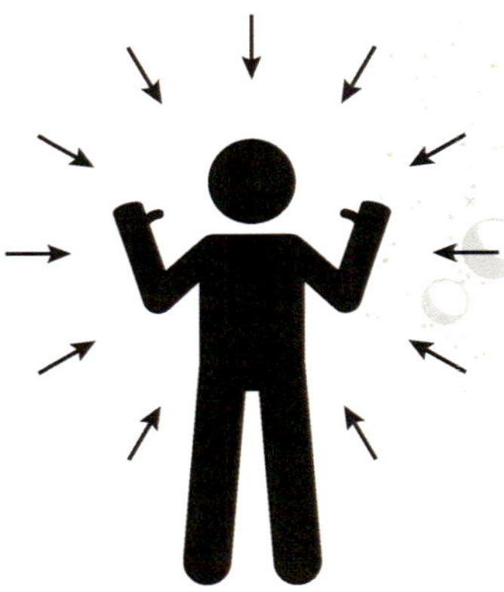

One of the most challenging tasks for an aspiring artist is to fully commit to one's authentic self. Self-expression can be a vulnerable endeavor, leading to an emotional rollercoaster. However, it's this leap of faith that is also one of the key factors in inspiring others.

Regardless of style or genre, commitment to a performance is one of the only traits shared by all influential artists. However, several challenges can hinder an artist's commitment. These challenges may include inadequate preparation, feelings of insecurity, excessive self-analysis, or physical limitations. Therefore, focus on minimizing distractions and prioritize "commitment to a performance" over the pursuit of "perfection." Ironically, many of the imperfections will smoothen out naturally.

Commit to a Character

Often, an artist creates a persona or alter ego to help commit to a performance. This choice often serves as a protective barrier between the audience and the artist's authentic self. Many times, this character represents the artist's core values while keeping personal details largely undisclosed.

"Anticipate the Project's Final Destination"

Strive to immerse yourself in the listener's context while creating. This understanding can effectively guide the creative process in a situationally tasteful direction. For example, if a song is going to be heard in a sports arena, imagine yourself cheering on your favorite sports team in a "loud / energetic" stadium, while creating the song. From a logistics perspective, this approach will also assist in addressing many technical and situational drawbacks beforehand. Imagine your music played in the following contexts....

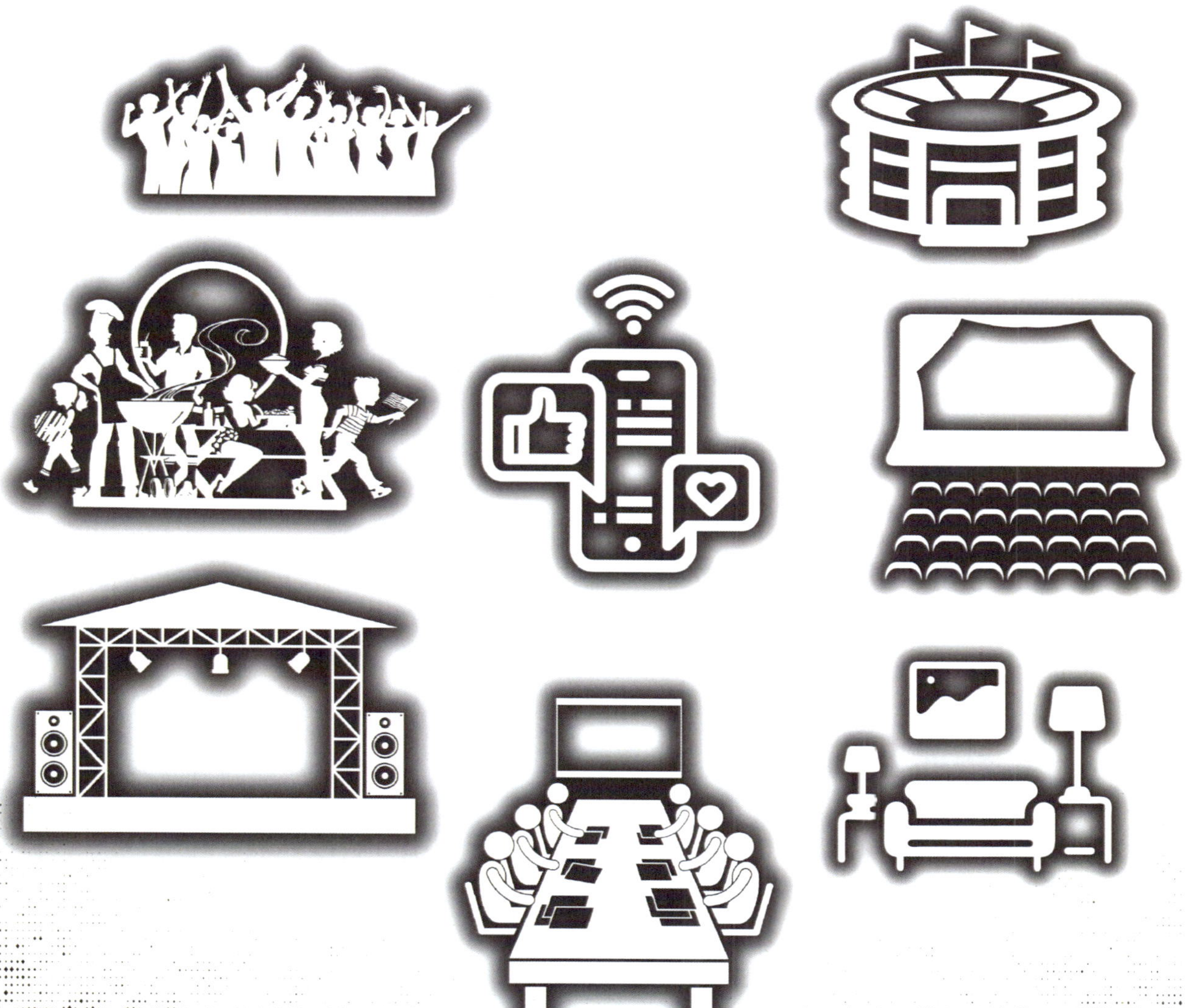

"Hosting the Party"

From the engineer's perspective, conducting a recording session is a lot like hosting a party...

As you continue to become more comfortable with your technical and artistic abilities, be sure to develop your hospitality skills. Find out what allows your client to feel comfortable.

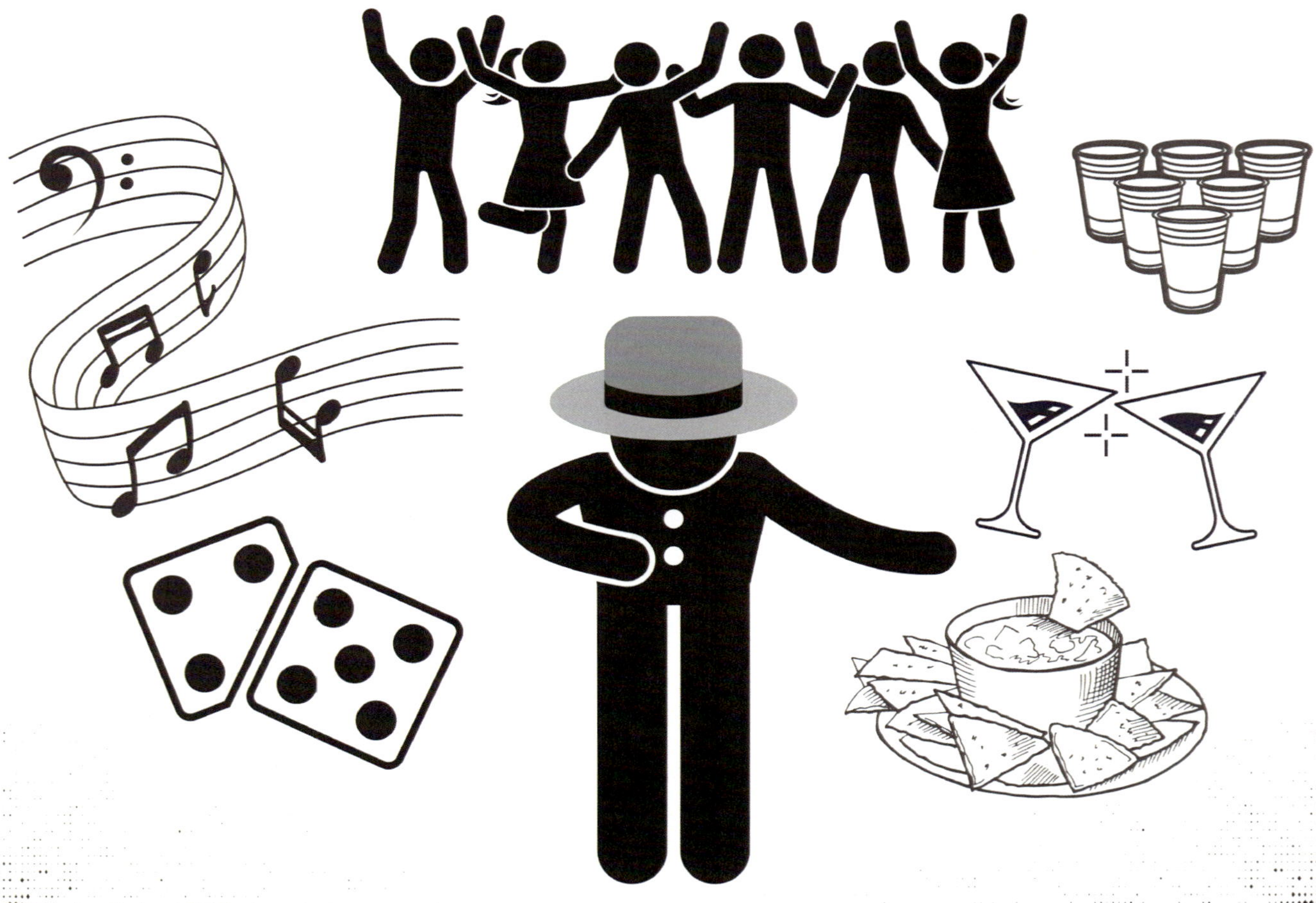

"Hosting the Session"

In addition to ensuring comfort, it's crucial to present yourself professionally. This involves being prepared both technically and in terms of hospitality. When a space looks ready to welcome someone, it fosters an immediate sense of trust, letting them know they are in good hands.

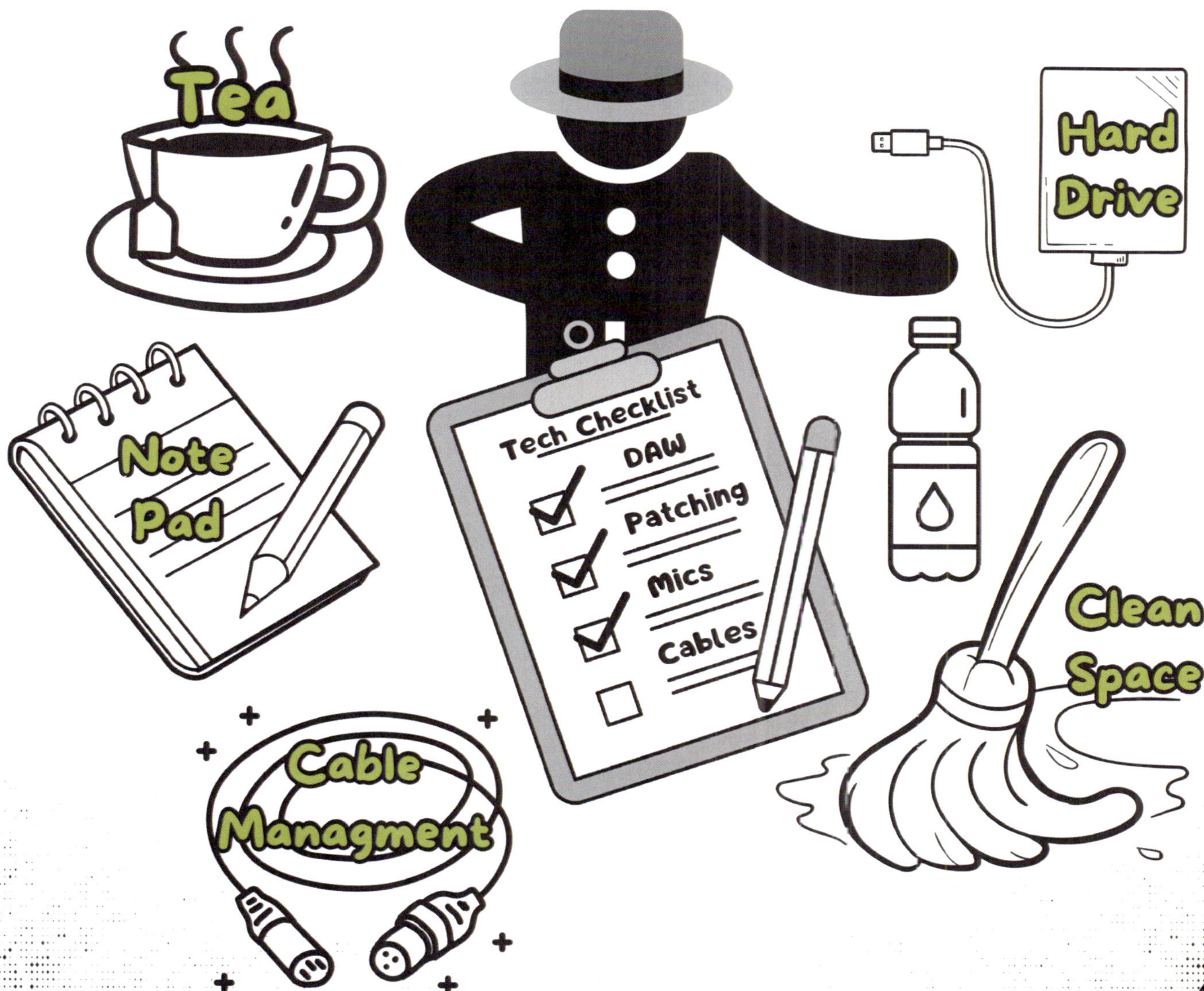

"Going Live"

There's a unique magic that happens when musicians perform together "LIVE." If the ensemble feels at ease while playing, tap into that energy to create an organic foundation for the recording. This chemistry is the group's signature and cannot be artificially created.

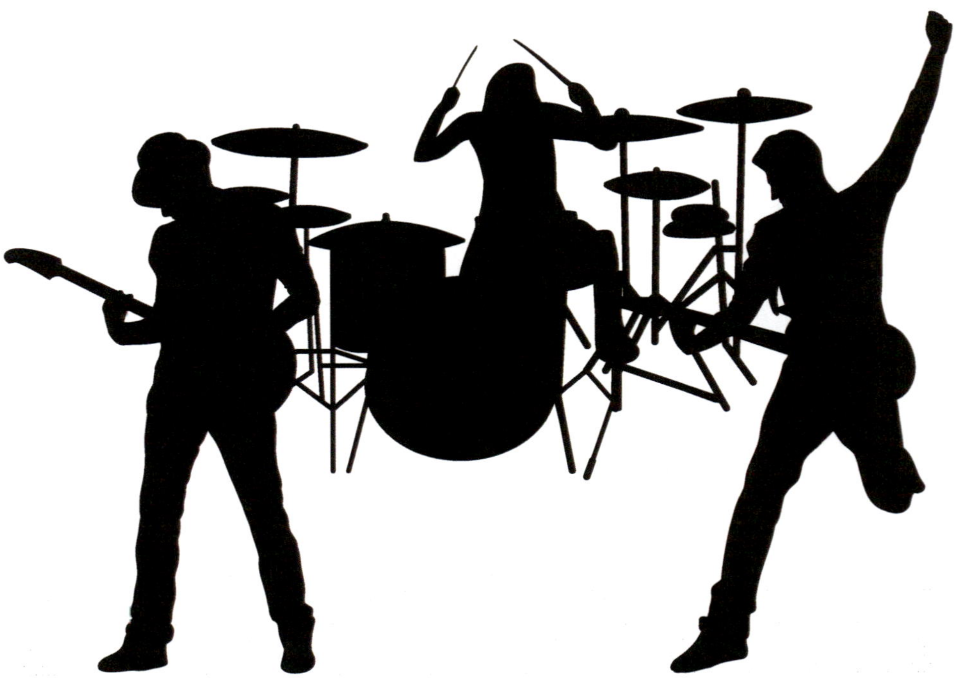

Computer Playback Settings for Recording

During the recording process, one of the first technical settings to verify in your DAW is the playback "buffer size." It's important to select a maximum of 128 samples. This ensures that the delay or "latency" during playback won't distract the recording artist. While a lower buffer size, such as 64 or 32 samples can further reduce latency, it may also impact your computer's performance based on the session size. Therefore, opting for 128 samples strikes a balance, allowing your computer to utilize additional resources while minimizing latency for recording.

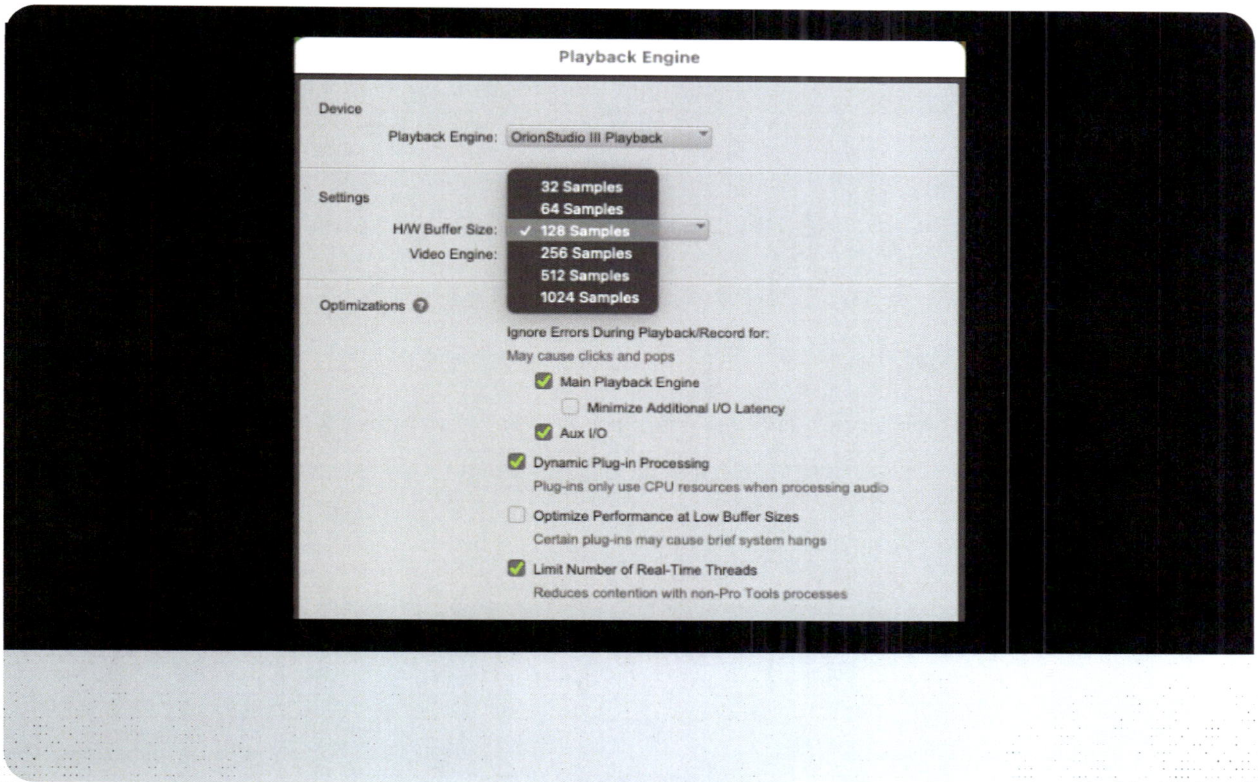

A Great Performance Starts with a Great Monitor Mix!

Regardless of whether an artist is performing solo or with a band, their ability to hear both themselves and the surrounding music is essential. Nothing ruins the flow of creativity faster than an uninspiring headphone mix during recording. Each artist has their own unique listening preferences, focusing on specific elements of the music. However, there are strategies we can use to avoid complications when creating monitor mixes, especially as the number of artists and mixes increase. Instead of starting from scratch, begin by establishing a basic mix that allows each performer to hear themselves clearly (at "Unity"), while keeping everyone else at -5dB or -10dB. Many artists will find this mix to be a good starting point, while making adjustments from this point is often minimal.

Turn to the next page to explore this technique in detail...

Monitor Mix Technique

First: Turn all the mic pre's down as much as possible. Then, set every input channel and master fader to "Unity" (0dB)

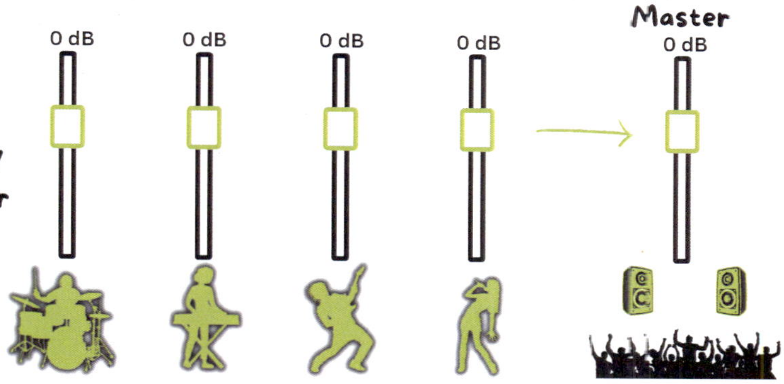

Main Mix

Second: Set faders so each artist hears themselves at "Unity" and everyone else at "-5dB" or "-10dB" (This depends on how loud an artist may want themselves vs. everybody else)

Third: While listening to the Main Mix, achieve a balanced mix without attenuating the faders. This technique is essentially balancing the source signal before it reaches any fader distribution. At this point, your fader positions should look the way they sound, achieving "Relative Fader Position"

Singer's Mix

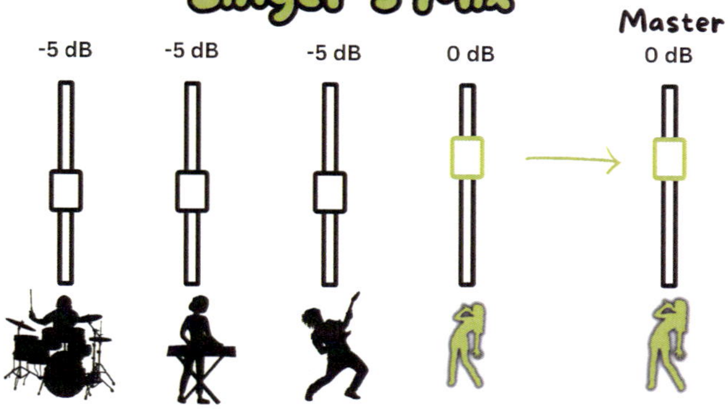

Guitar's Mix

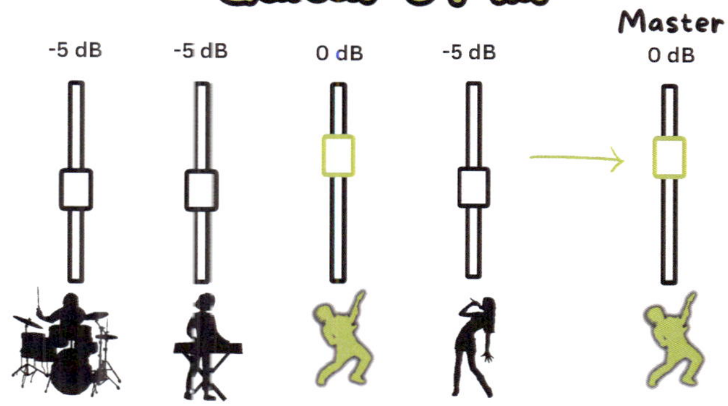

Key's Mix

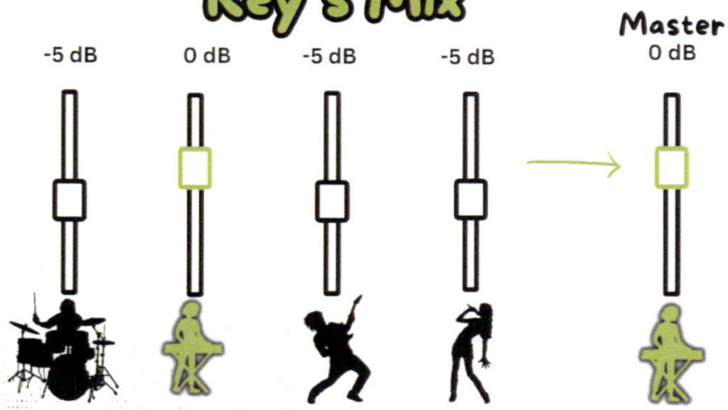

Drummer's Mix

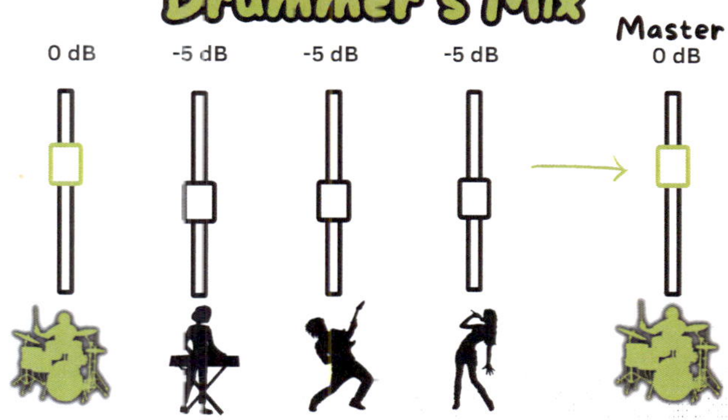

"Overdubbing"

The method of recording or layering new audio tracks over previously existing audio...

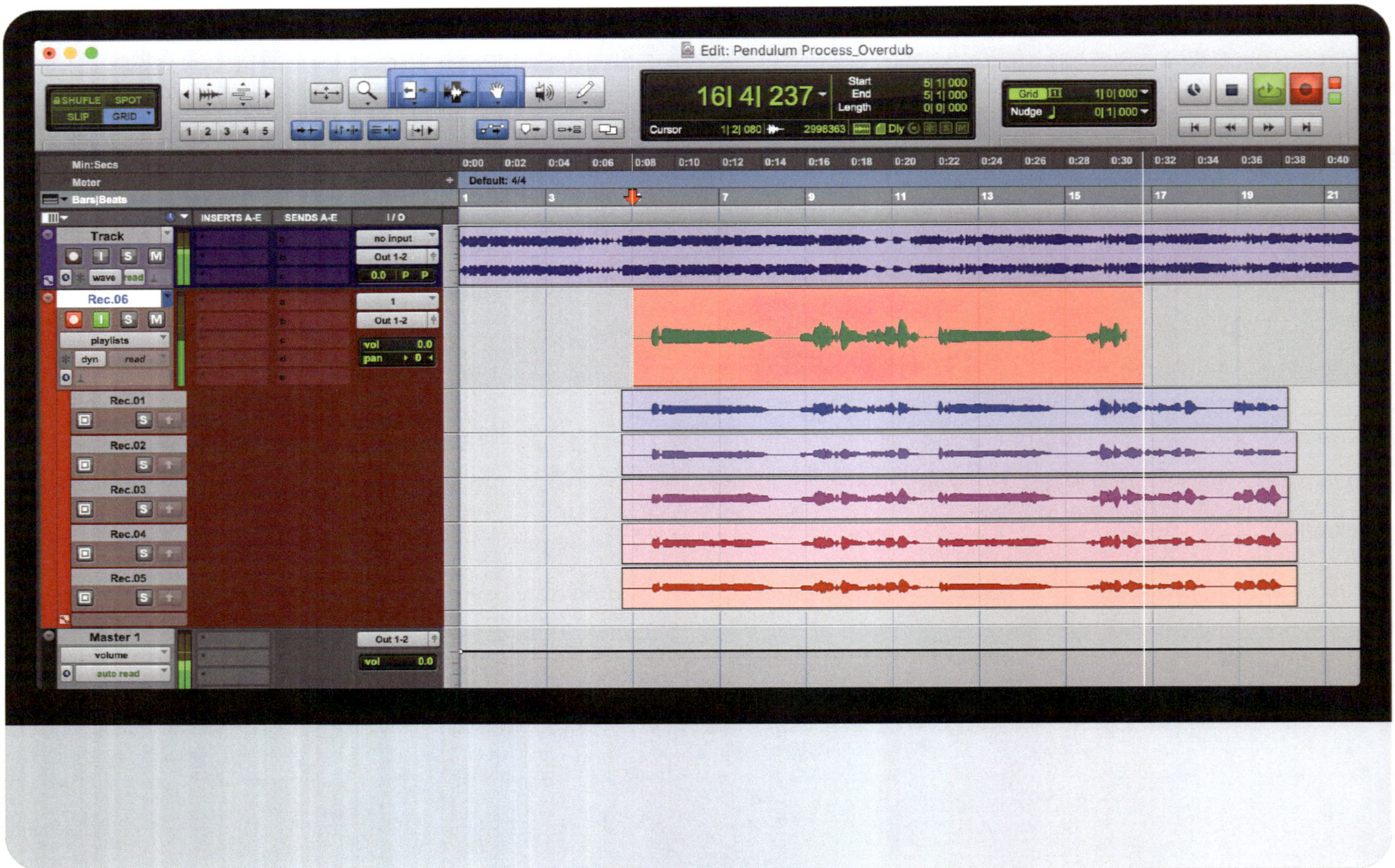

With the growing popularity of producing instrumentals or beats "in the box" using a computer, overdubbing has emerged as one of the most prevalent techniques for recording. Understanding the nuances of effective overdubbing and recording techniques is essential, given that many of the most important elements "like vocals" are recorded in this fashion.

"Playlist Recording"

Tip* Think of an audio track as a folder that holds all the recordings specific to that track.

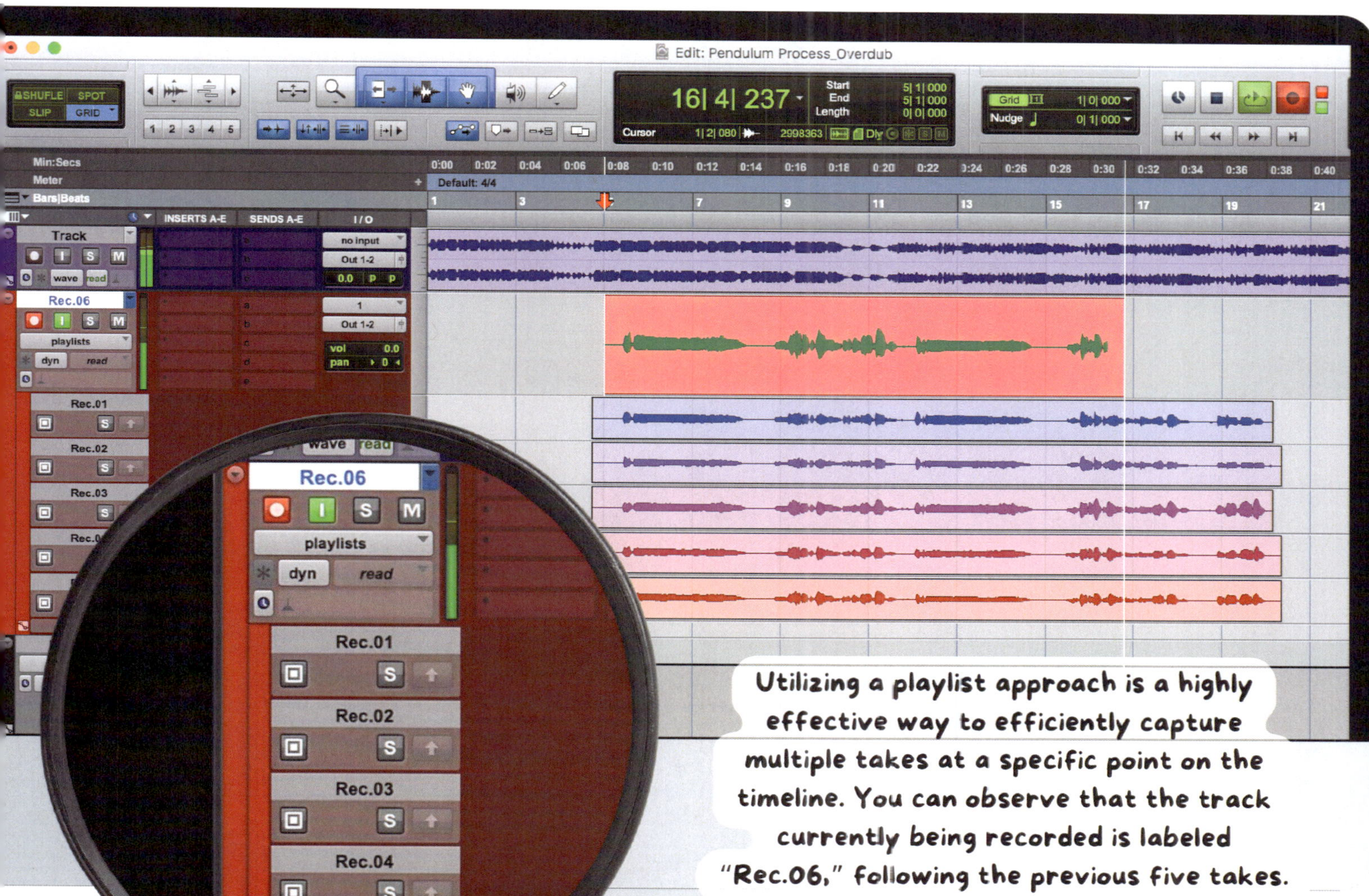

Utilizing a playlist approach is a highly effective way to efficiently capture multiple takes at a specific point on the timeline. You can observe that the track currently being recorded is labeled "Rec.06," following the previous five takes.

Playlists are primarily used during the overdub process for individual instruments or vocals, but they can also be applied to ensemble recordings. The key requirement is that the session must follow a consistent timing guide. This can be achieved using a click track or a previously recorded track that establishes the timing for the piece of music.

"New Playlist Take" Shortcut*

^ control + \

"Playlist Comping"

When producing a "Comp," the finest takes from each segment can be seamlessly combined to create a compilation or "All Star" track.

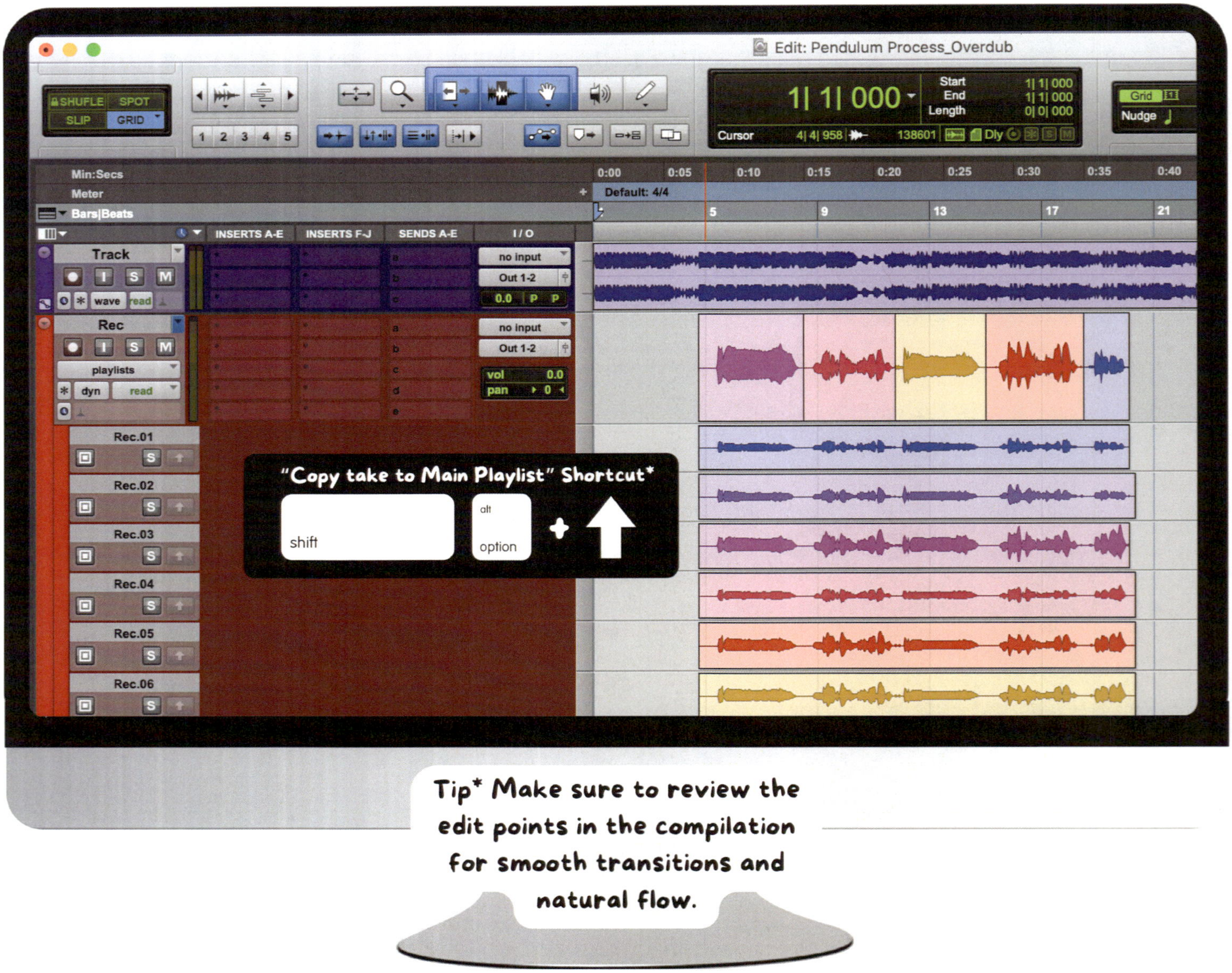

Tip* Make sure to review the edit points in the compilation for smooth transitions and natural flow.

Since all recordings are captured at the same timestamp, you can seamlessly combine the best elements from each take. You'll notice that the "comp" or "main playlist" track features multiple colors, reflecting pieces from each recording. Additionally, on the left side of the track, there is a separate solo button for each take, enabling you to listen to each recording in the context of the song.

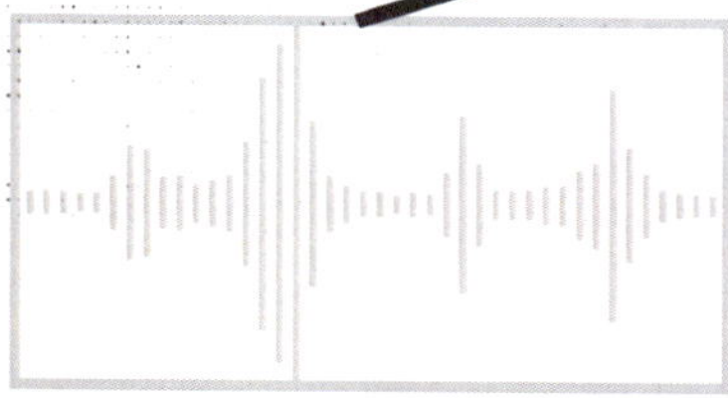

Copy + Drag Technique

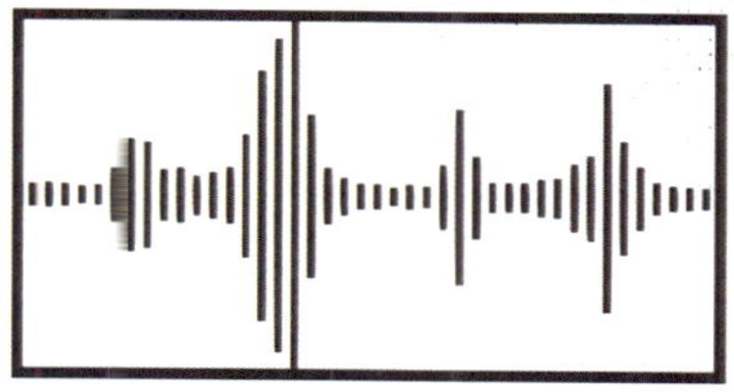

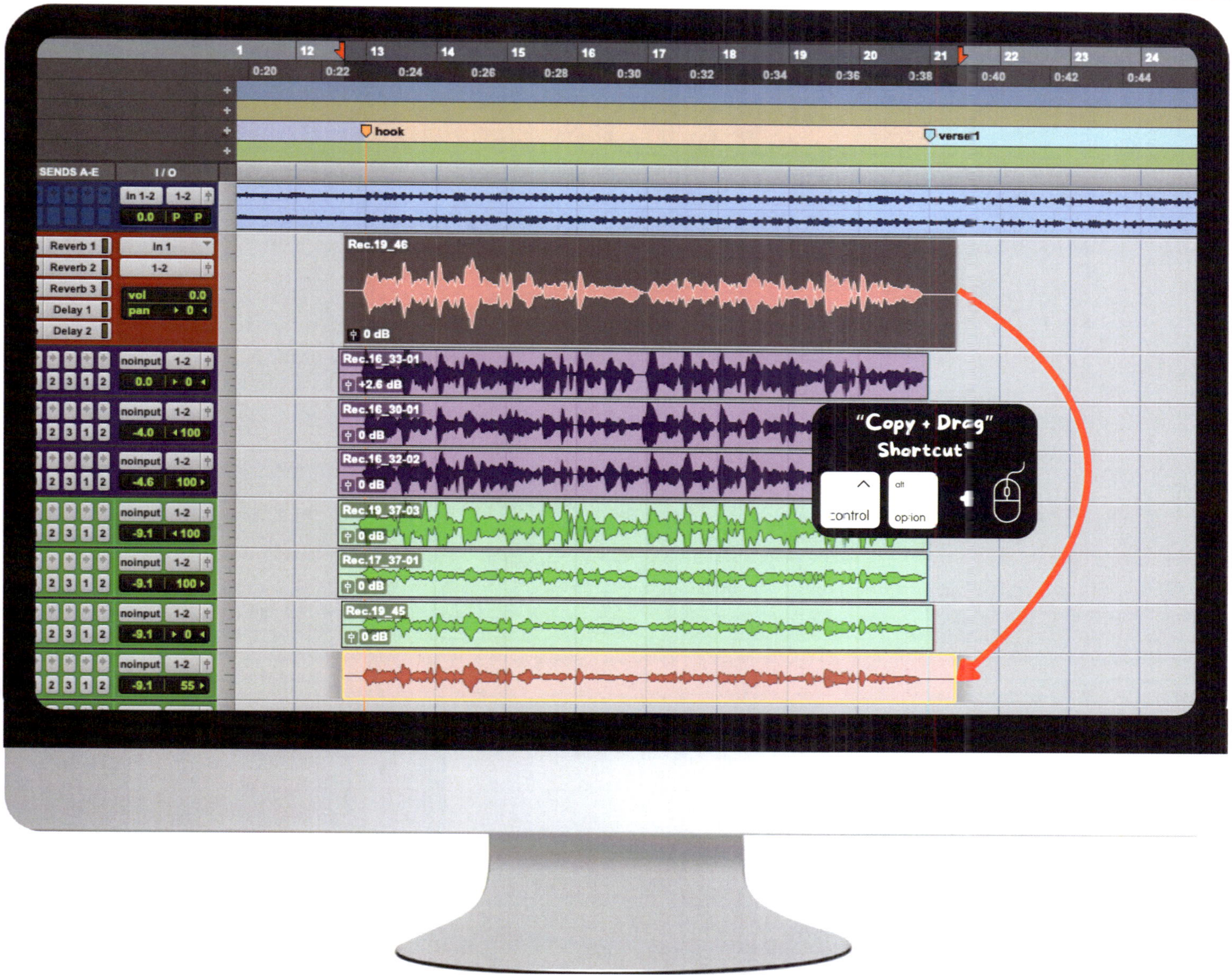

A highly effective method is to use the "copy + drag" technique for overdubbing while recording. This involves setting up a dedicated recording track above and a destination track below. Instead of treating the editing process as an afterthought, you can record until you capture a great take, then simply "copy + drag" the track into a destination track such as "Lead Vocal" or "Harmony Vocal." This technique also enables you to maintain an inventory of the entire overdub session within a single track. Additionally, this method can facilitating easy playback listening and communication while an artist is "in the booth".

Vocal Overdub Setup

Save time and energy by creating a template with the desired tracks, labels, colors, inserts, FX, and routing before the recording begins.

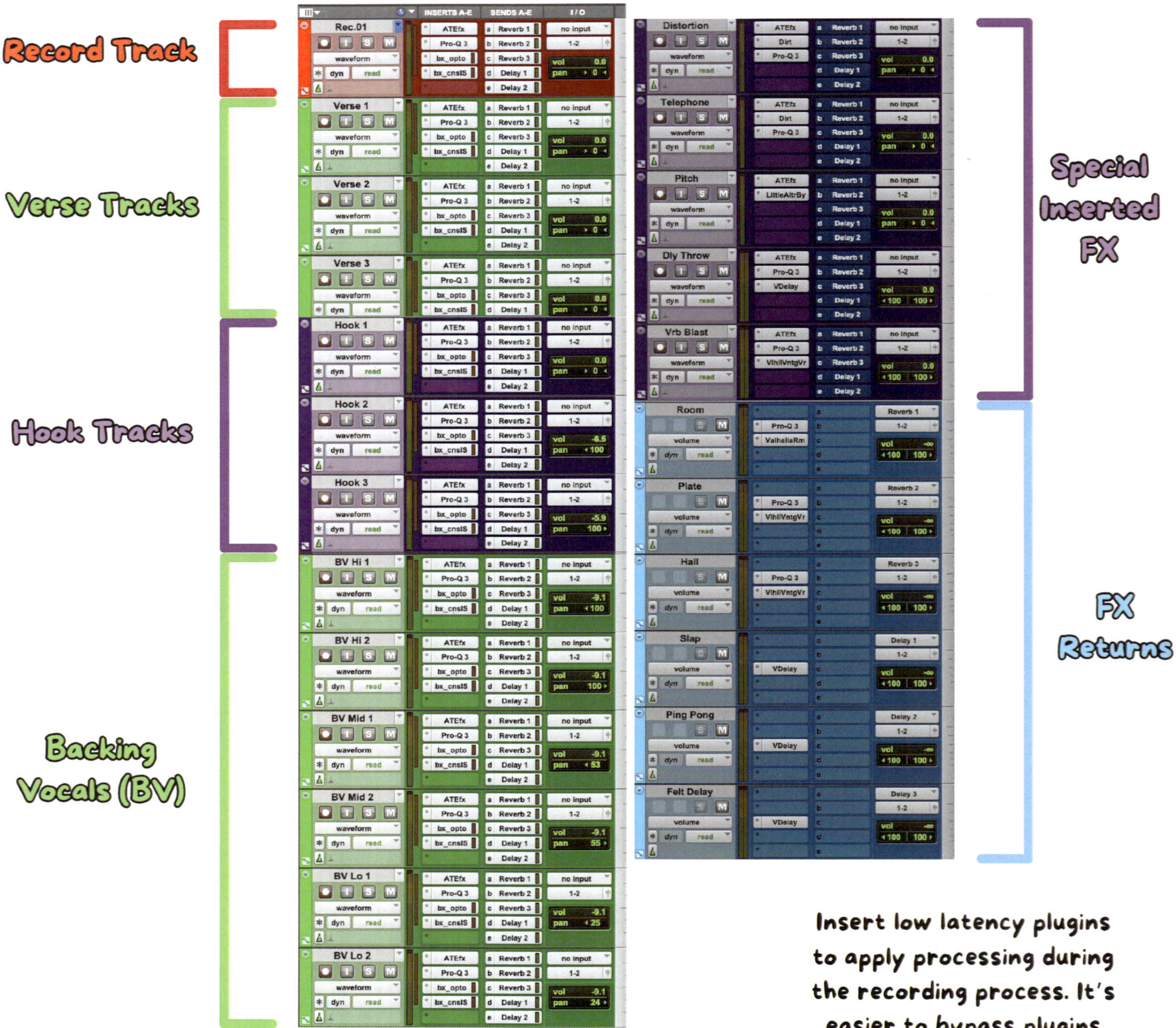

Insert low latency plugins to apply processing during the recording process. It's easier to bypass plugins than it is to apply them during a session.

Route all inputs and outputs of channels and create sends for FX.

Note* There is a theme of 3...
You don't have to use all 3 recorded tracks, but it is a great start for allowing variation in your production.

Record Track

This track will function as your dedicated recording track, providing inventory of all the recorded "takes".

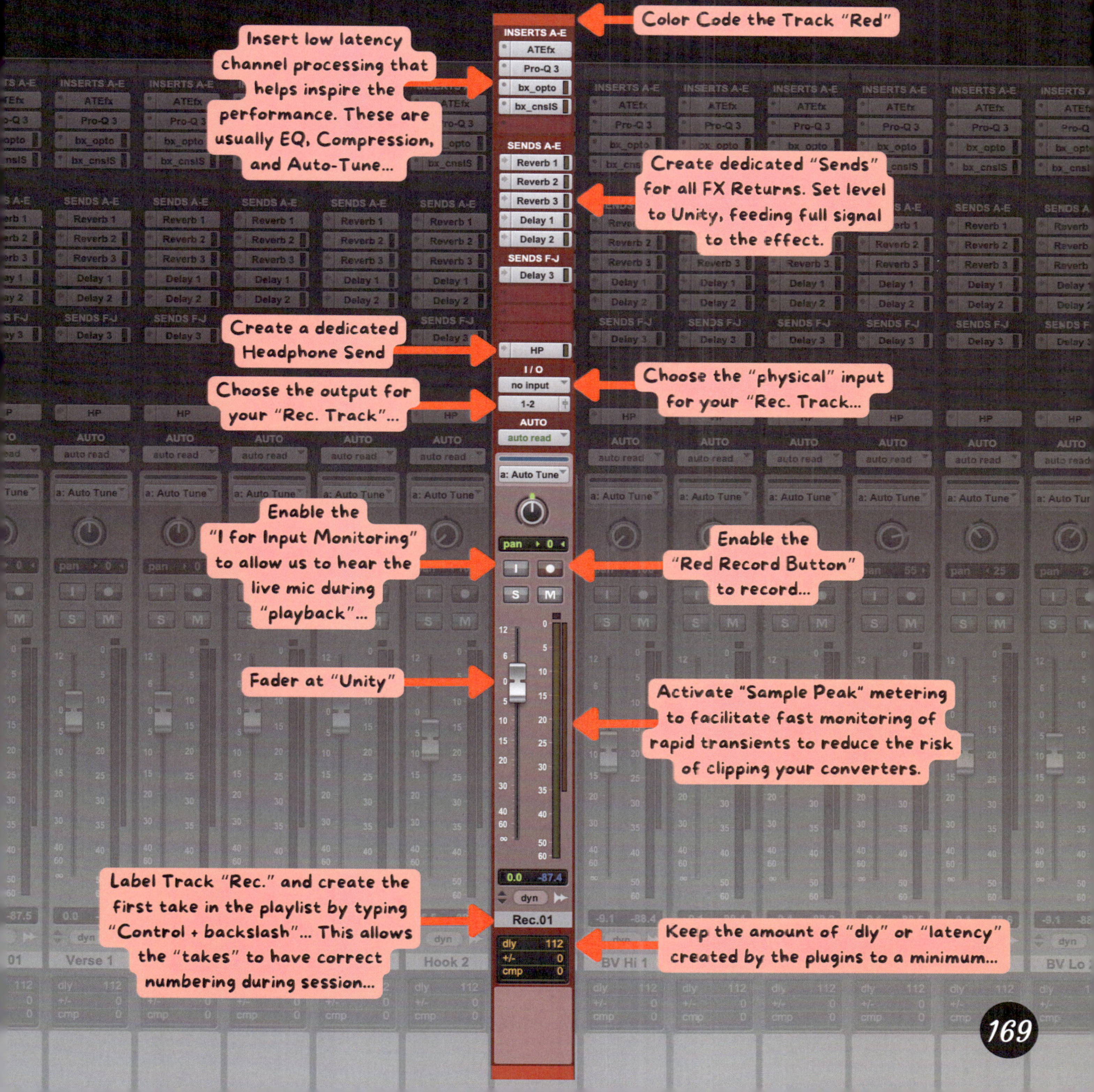

Verses

Utilizing three separate verse tracks provides significant flexibility when crafting your verses. You can apply distinct processing techniques to each verse, seamlessly add doubles or triples, or stagger the lines to accommodate overlapping phrases.

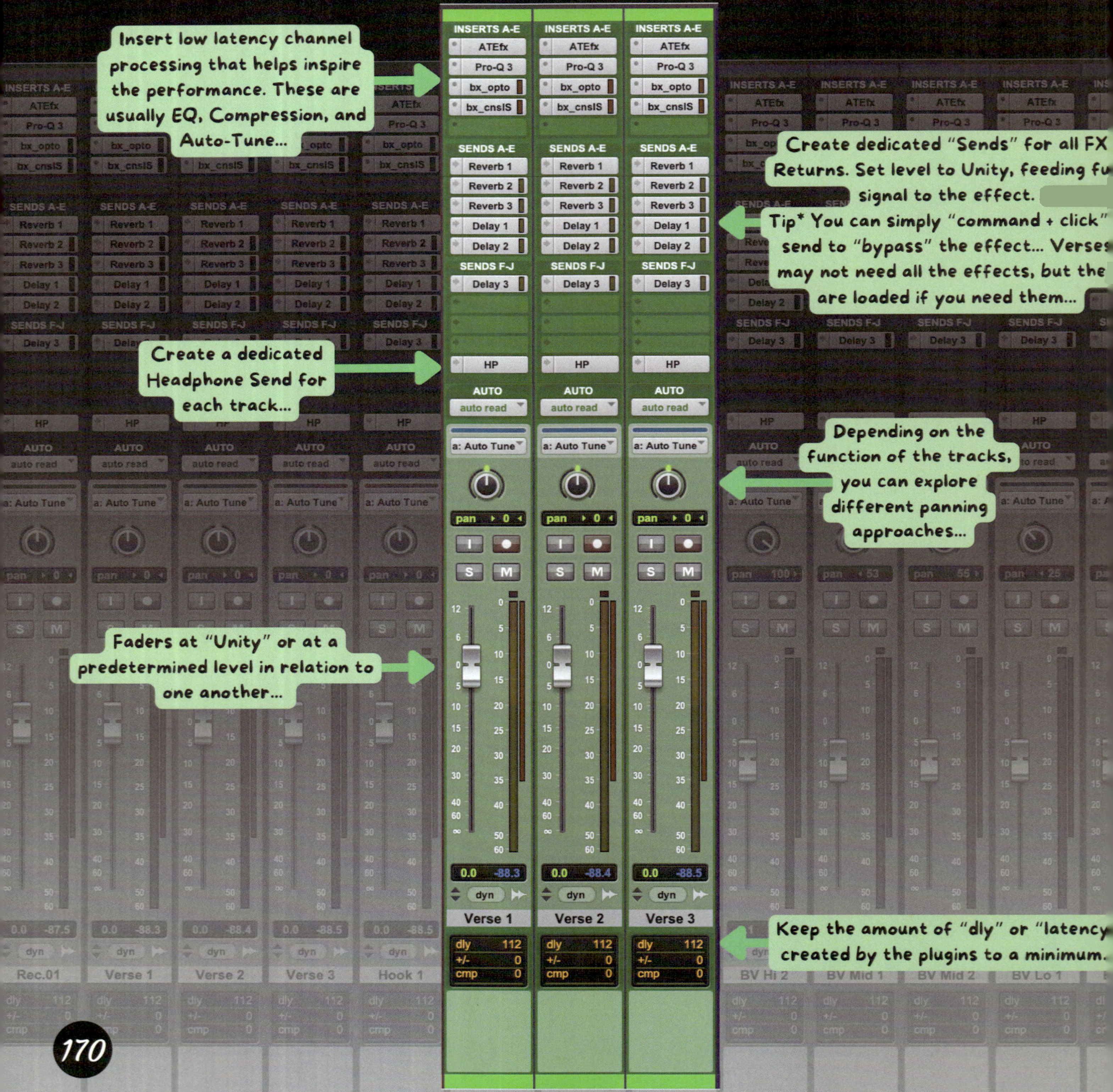

Hooks

Similar to the 3 verses, record 3 hook tracks. A great lead vocal technique is the LCR approach. This is where 3 of the same vocal are panned to the left, center, and right in the stereo image. The amount of panning and levels of the tracks can vary, but this approach provides lots of flexibility.

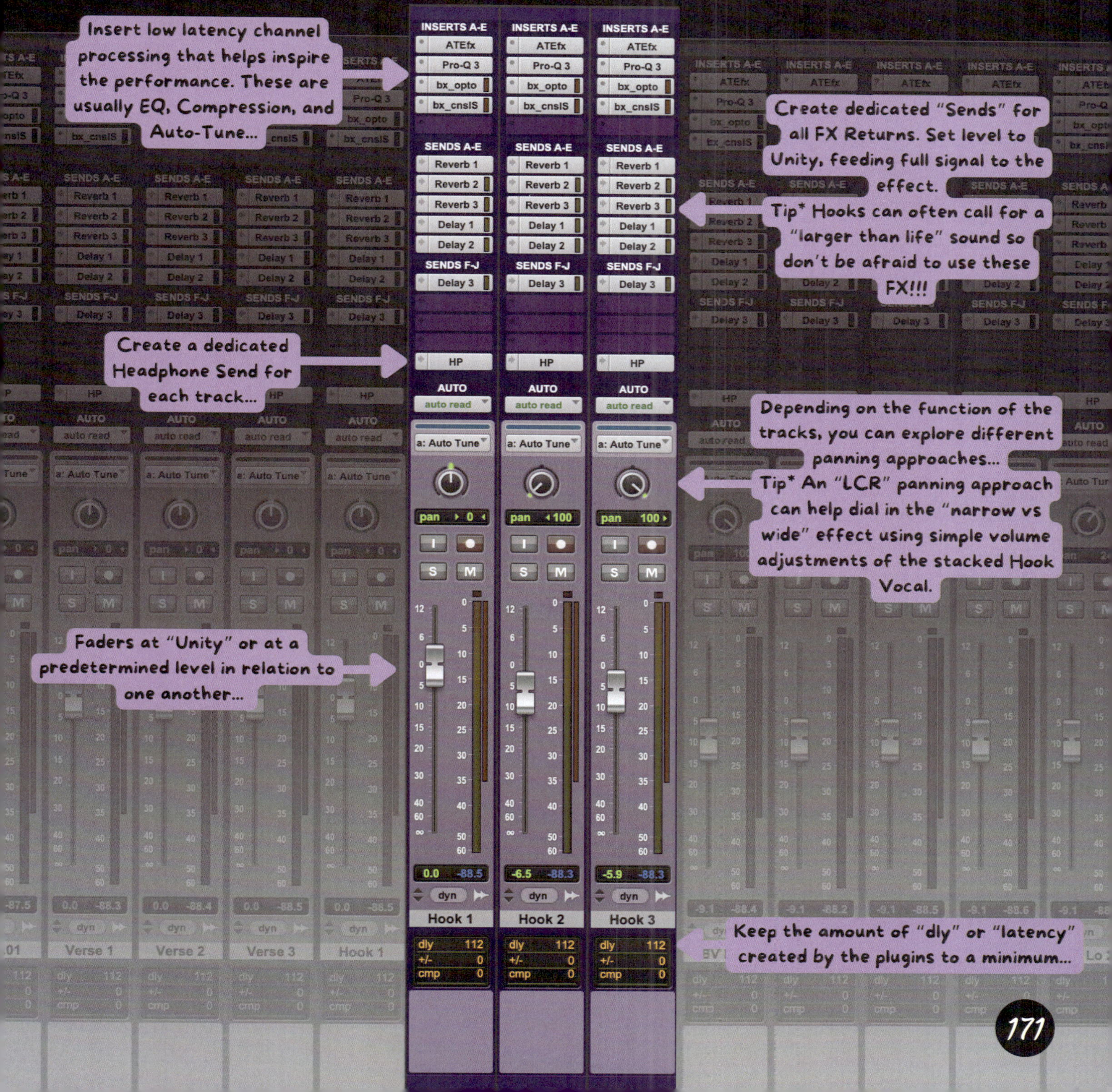

Backing Vocals

Backing vocals or "BV's" can be implemented in a number of ways. Sometimes a few simple "Ad Lib" tracks is all you need to support a lead vocal. Other times, BV's can take on a more complex texture, stacking large amounts of tracks. An example of this is capturing a three-part harmony with doubled Highs (Hi), Middles (Mid), and Lows (Lo). This can serve as an excellent foundation for layering harmonies. You can further develop this idea by incorporating additional layers as needed.

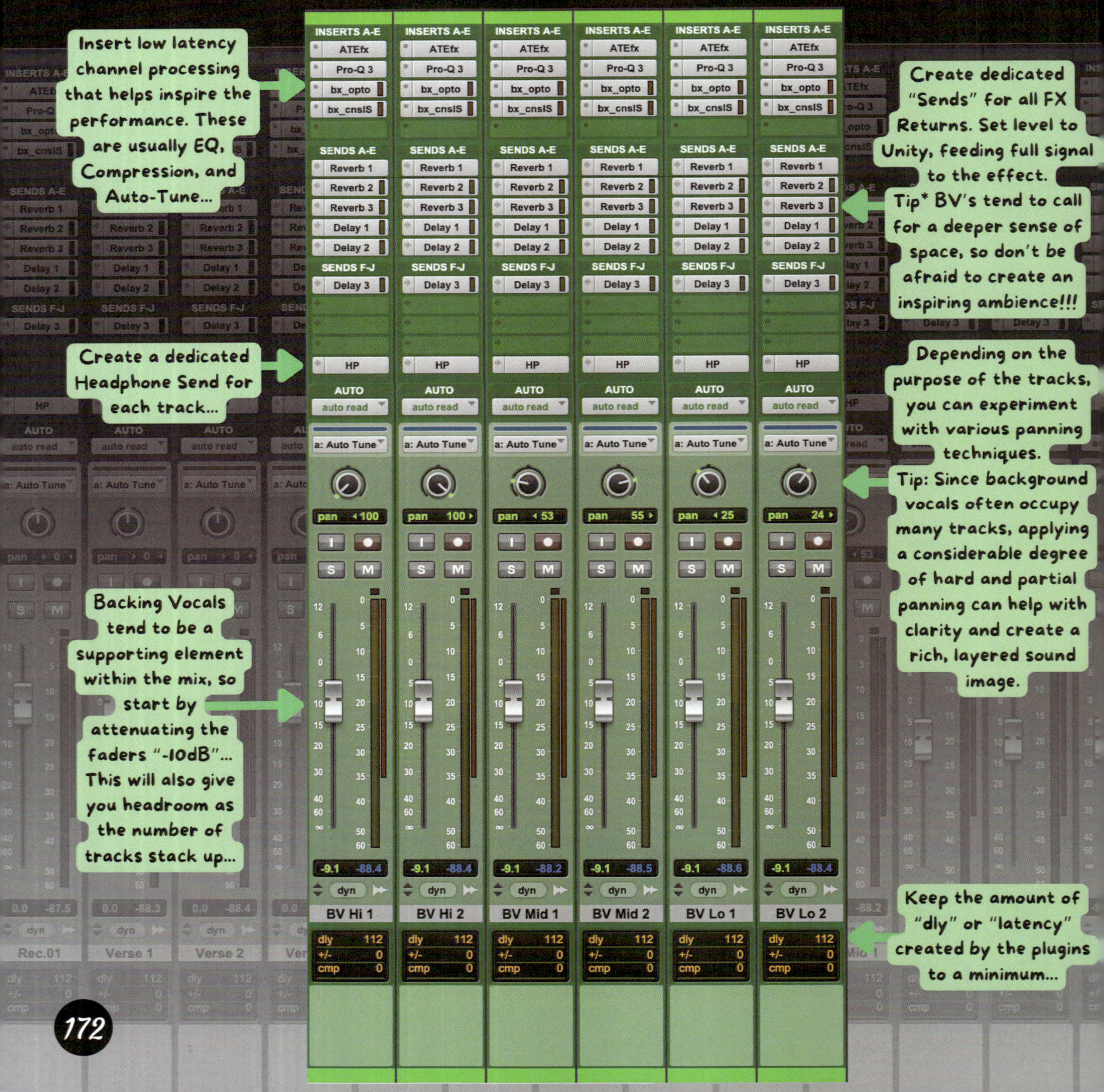

Special Inserted FX

Depending on the music genre, there are numerous special effects that can elevate vocal performances. Common effects to consider include Distortion, Pitch Shifting, Chorus, Telephone, and Long Echoes. Although the options are limitless, it's crucial to have a range of special effects "ready to go" to maintain creative momentum. Keep in mind that these tracks are just audio tracks equipped with pre-loaded effects. This means you only need to copy or drag audio into the track to hear the effect.

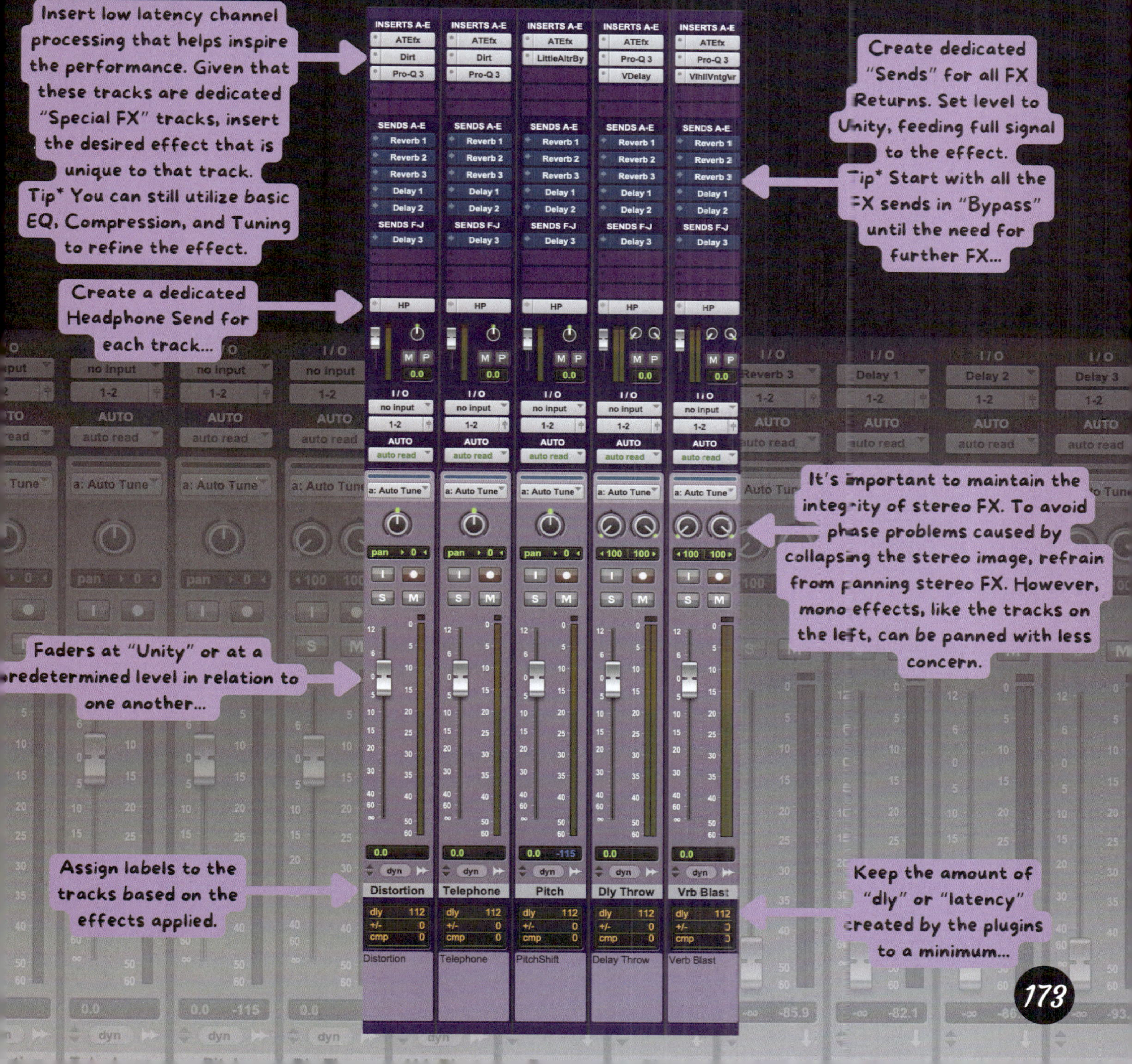

FX Returns

Load the FX Returns or "Aux Inputs in Pro Tools" with a few flavors of Reverb and Delay. Remember you don't have to use them all, but have them ready to go when inspiration strikes. You can always tweak afterwards.

Insert low latency channel processing that helps inspire the performance. Given that these tracks are dedicated "FX Returns", insert the desired effect that is unique to that return.

Tip: For Reverbs and Delays, position the plugin two or three slots down from the top. This method enables you to EQ or process the send before it reaches the effect.

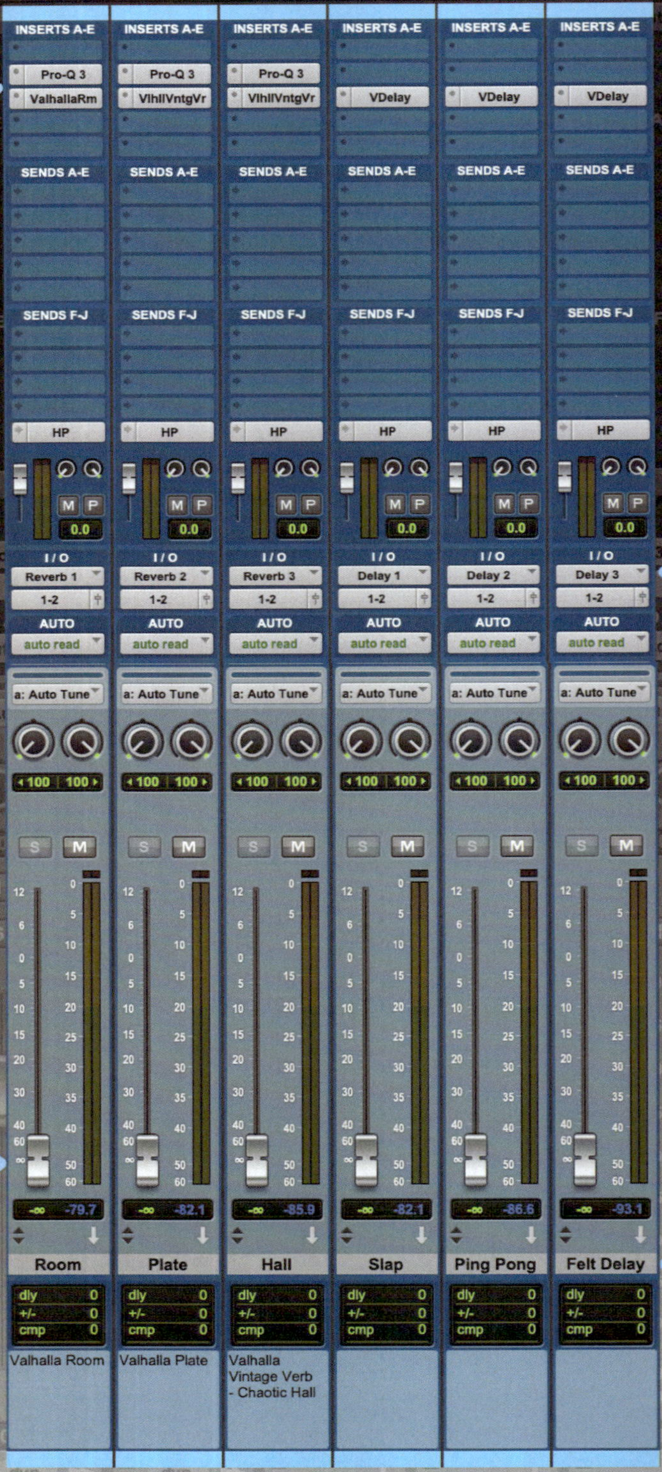

Create a dedicated Headphone Send for each track...

Route the Bus used for the effect (Ex. Reverb 1) to the Input of the Aux Input channel...

It's important to maintain the integrity of stereo FX. To avoid phase problems caused by collapsing the stereo image, refrain from panning stereo FX. However, mono effects, like the tracks on the left, can be panned with less concern.

Begin with the fader levels set to the "off" position. Gradually increase the effect as required. Keep in mind that the send levels directing the effect are transmitted at full "unity."

Assign labels to the tracks based on the effects applied.

"Recording a Band or Ensemble"

The initial step in recording a band or ensemble is to create all the audio tracks. As the number of tracks grows, the importance of organization also rises. Consider color coding your tracks according to the musician or instrument group. There is no right or wrong way to organize your tracks, but establishing a strategy and maintaining consistency in your workflow will be beneficial.

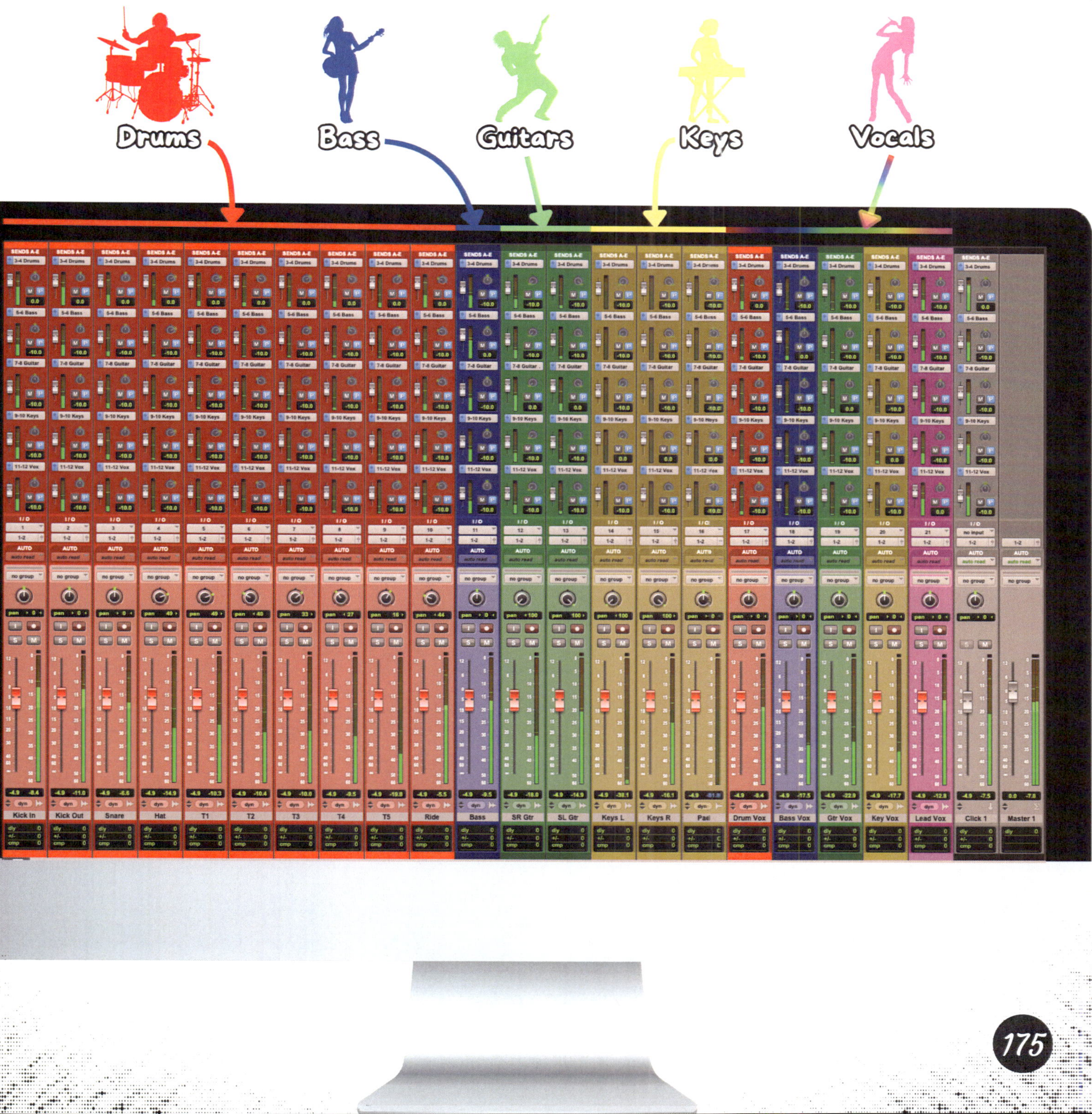

Linear Recording "Approach"

One of the easiest ways to record a band or ensemble is by utilizing a linear recording approach. Notice how all 4 takes are recorded "back to back" on the timeline. This creates a simple view of all the "takes". To help your workflow during the editing process, try color coding each take.

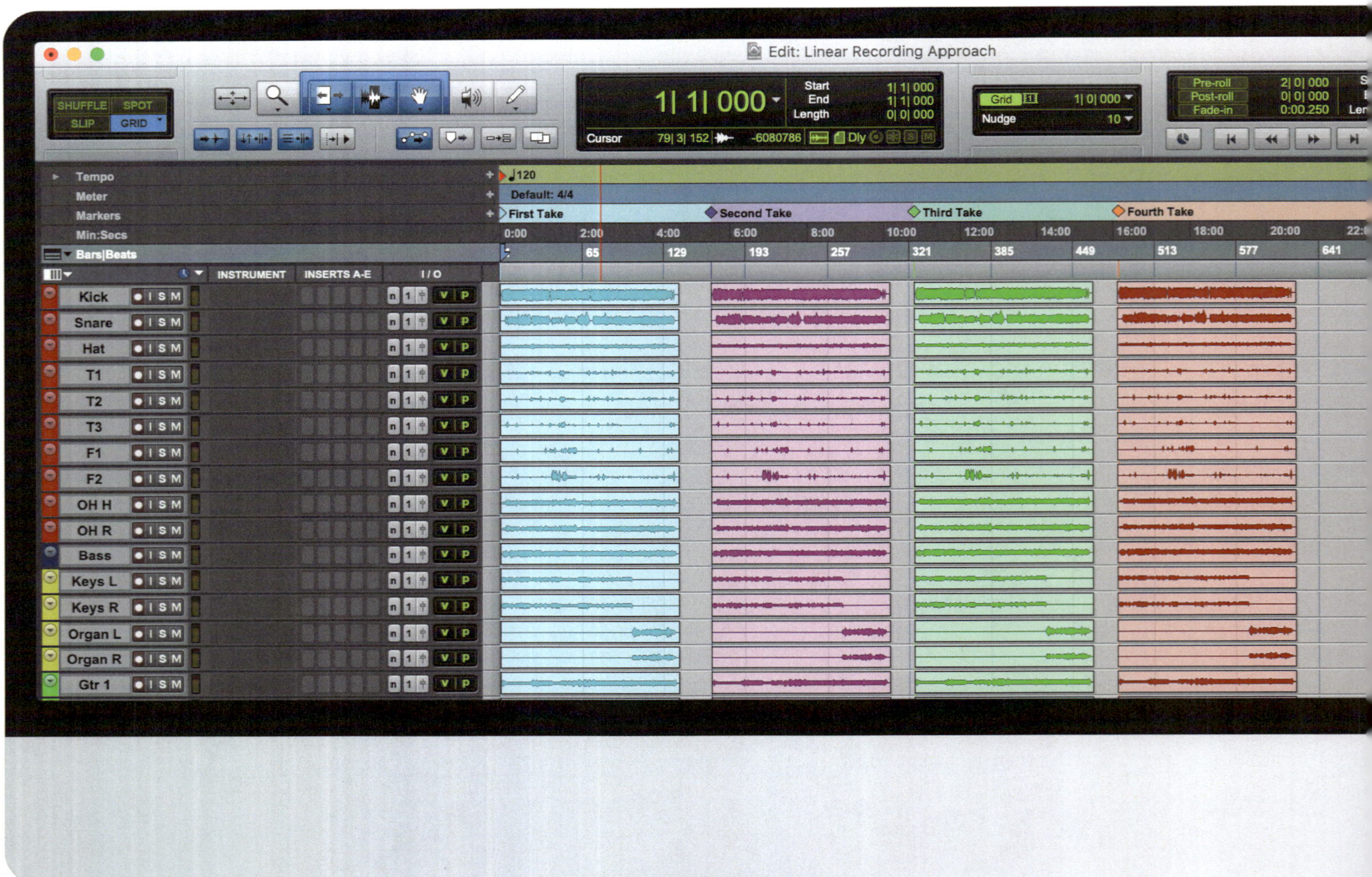

In situations where a click track is not used, be sure to revert to this technique, as every take will vary in tempo and feel. This is a great starting point to capture the initial recording and feel of a group of musicians.

Don't Forget! If overdubs will follow the initial recording, make sure you overdub the right take...

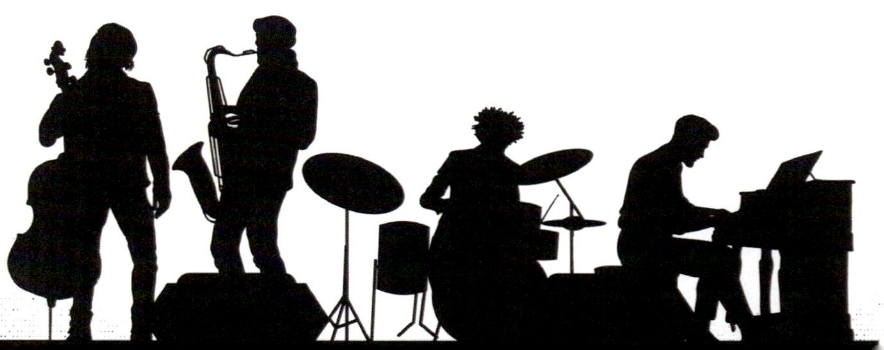

Linear Recording "Approach for Songs"

A linear recording approach can also be a great way to capture several songs in one session. This ensures a consistent sound from song to song, while maintaining a larger point of view with a project.

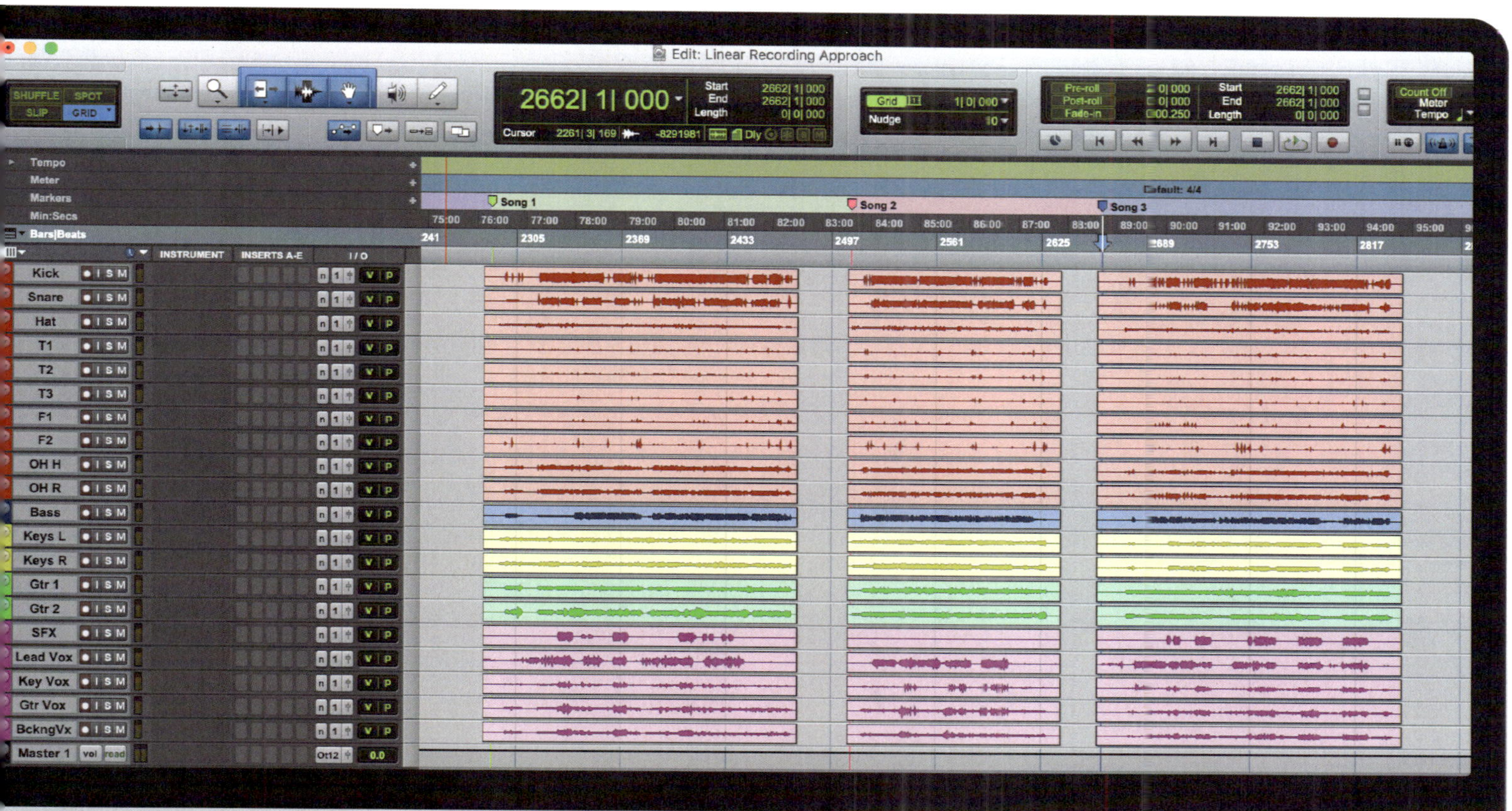

This approach works best for initial recordings, simple sessions, and projects with simple to moderate complexity. Given the diversity or complexity of the songs, this approach may not suit your needs. Be aware that this is not the only approach for recording bands or ensembles, but it's simple and gets the production out of the way of the musicians.

Doing a "Save As" for the session can allow you to create drastic changes to the session while keeping the initial linear recording in tact.

Linear Recording "The Comp"

Creating a composite take or "comp"

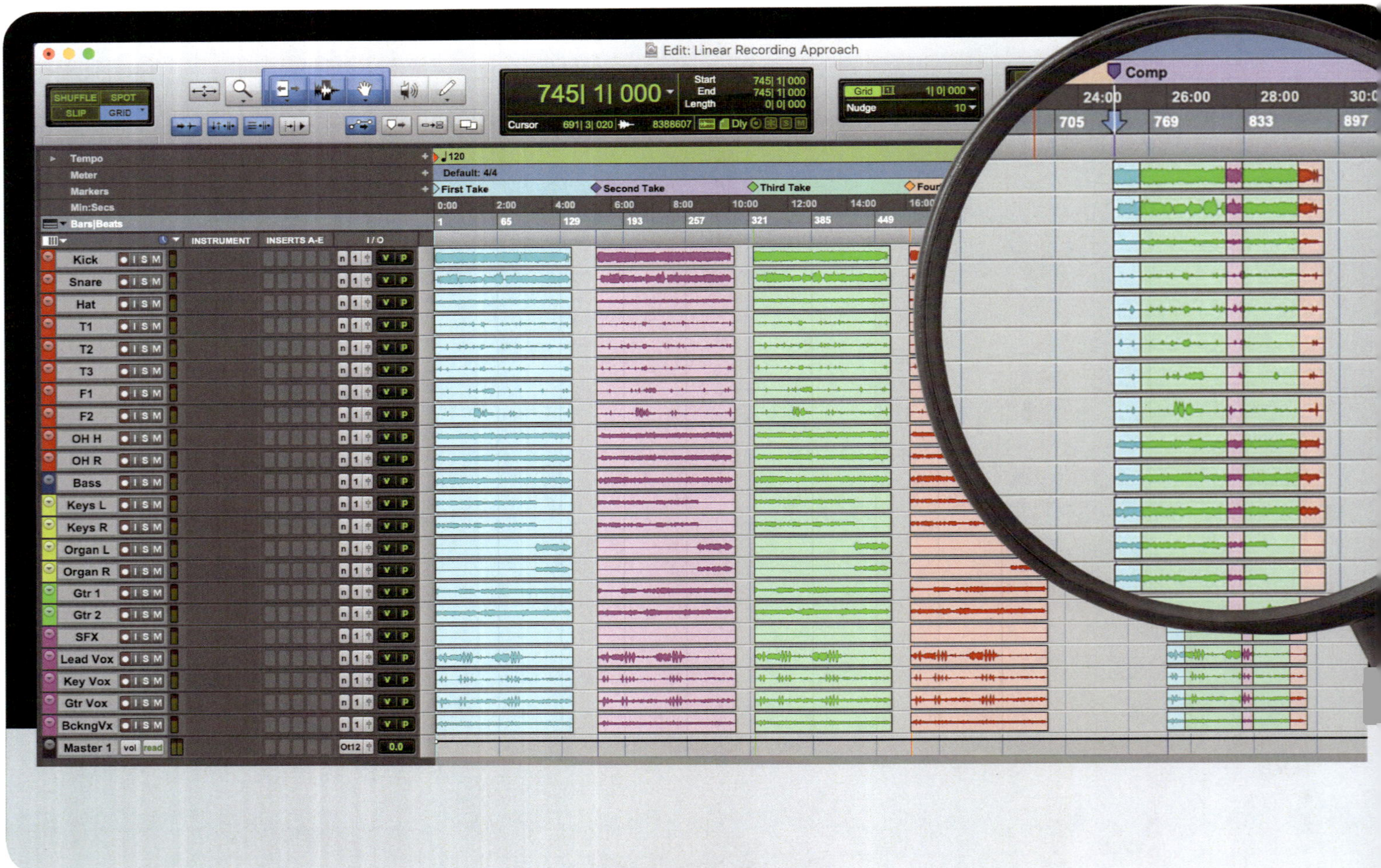

With a linear recording approach, we still have the ability to "comp" the best sections of each take. Notice the dedicated "comp" on the far right side of the timeline. Color coding makes this process easy to see, while creating an edit group can prevent clips from shifting out of sync. Pay close attention to the edit points and crossfades comp'ing the takes. Sustained instruments can cause unnatural ambience changes, so try to choose chord changes or rhythmic attacks for your edit points. With difficult transitions, you may need to apply unique crossfades for different tracks. As long as you can't hear the edit points, your "comp" is good to go!

Headphone Mixing "From DAW"

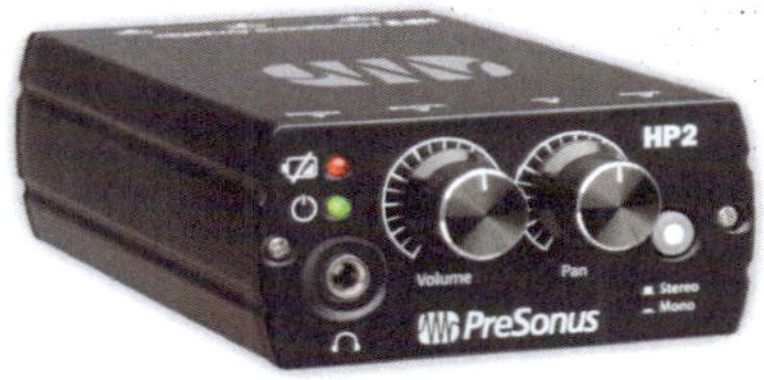

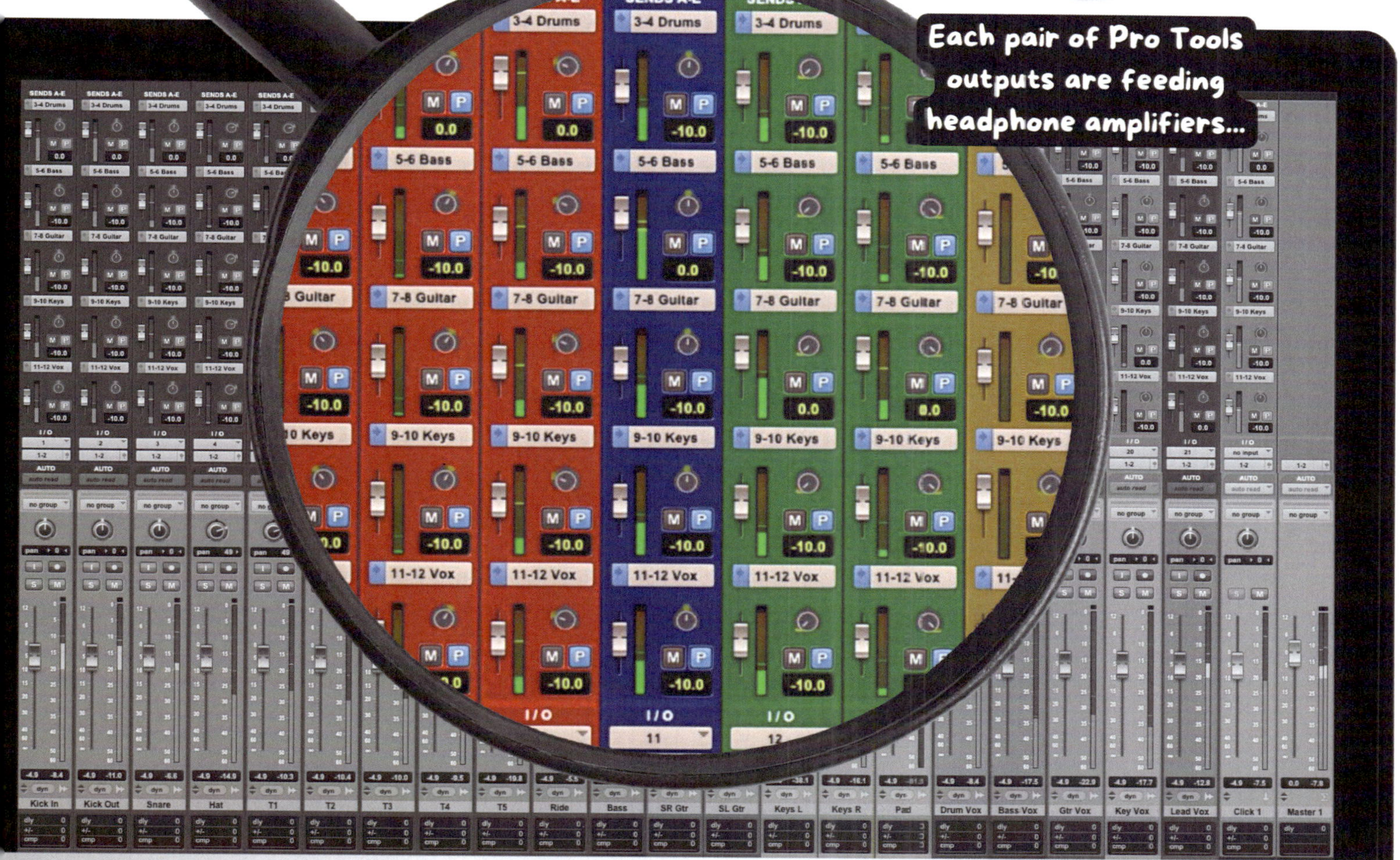

In the above example, you can see the headphone method in action, where the Pro Tools engineer oversees the mix for each musician. Moving from left to right, you'll notice that each instrument (displayed in different colors) has its own send to a specific output feeding the performer's headphone mix. The "relative fader position" you see serves as an excellent starting point for your session. As a reminder, refer back to the "Monitor Mixing Technique" discussed earlier in this chapter. Additionally, the "blue P" on each send indicates "pre-fader" sends to the mixes. In this setup, each band member receives a mix that is independent of the main mix, ensuring that monitor mixes remain stable even if the engineer adjusts the volume of the main mix during the session.

Headphone Mixing "Self Monitoring"

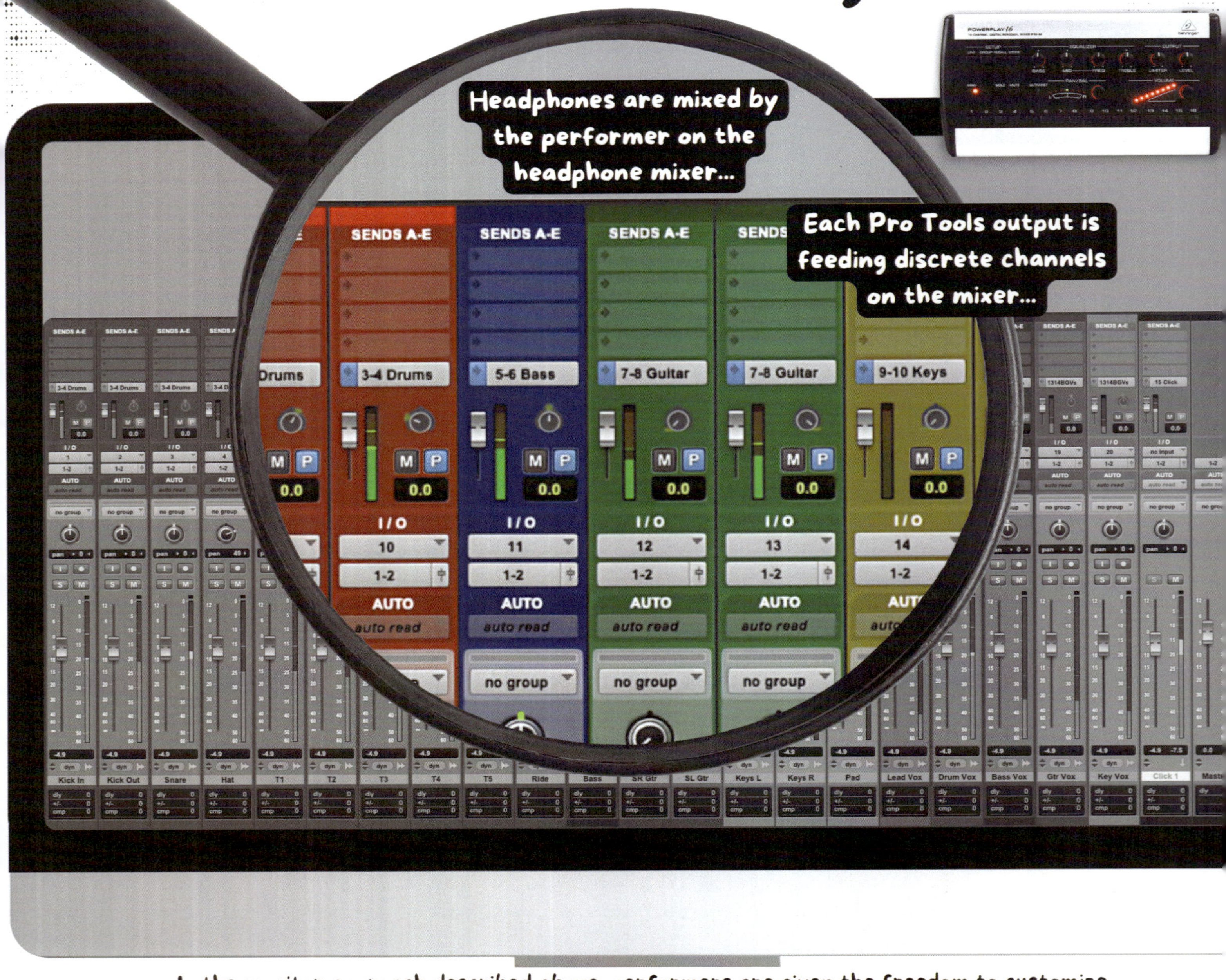

In the monitor approach described above, performers are given the freedom to customize their own mixes. Rather than having outputs serve as separate headphone mixes, they are utilized to send discrete audio channels or sub-groups to headphone mixers operated by performers. Each track is sent at "unity" and "pre-fader," giving musicians access to the complete signal. There are various types of "self-monitoring" headphone mixers available, such as the Behringer P16, which supports 16 audio channels. As the engineer, you can determine which tracks are routed to specific channels on the mixer. For instance, you might send instruments through a stereo pair of channels while routing vocals through a different pair, then reserve discrete channels for a "talk back" or "click track." Be aware that this method calls for fundamental mixing and technical abilities from the performer. Depending on their level of competency, this can lead to either happiness or frustration for the performer.

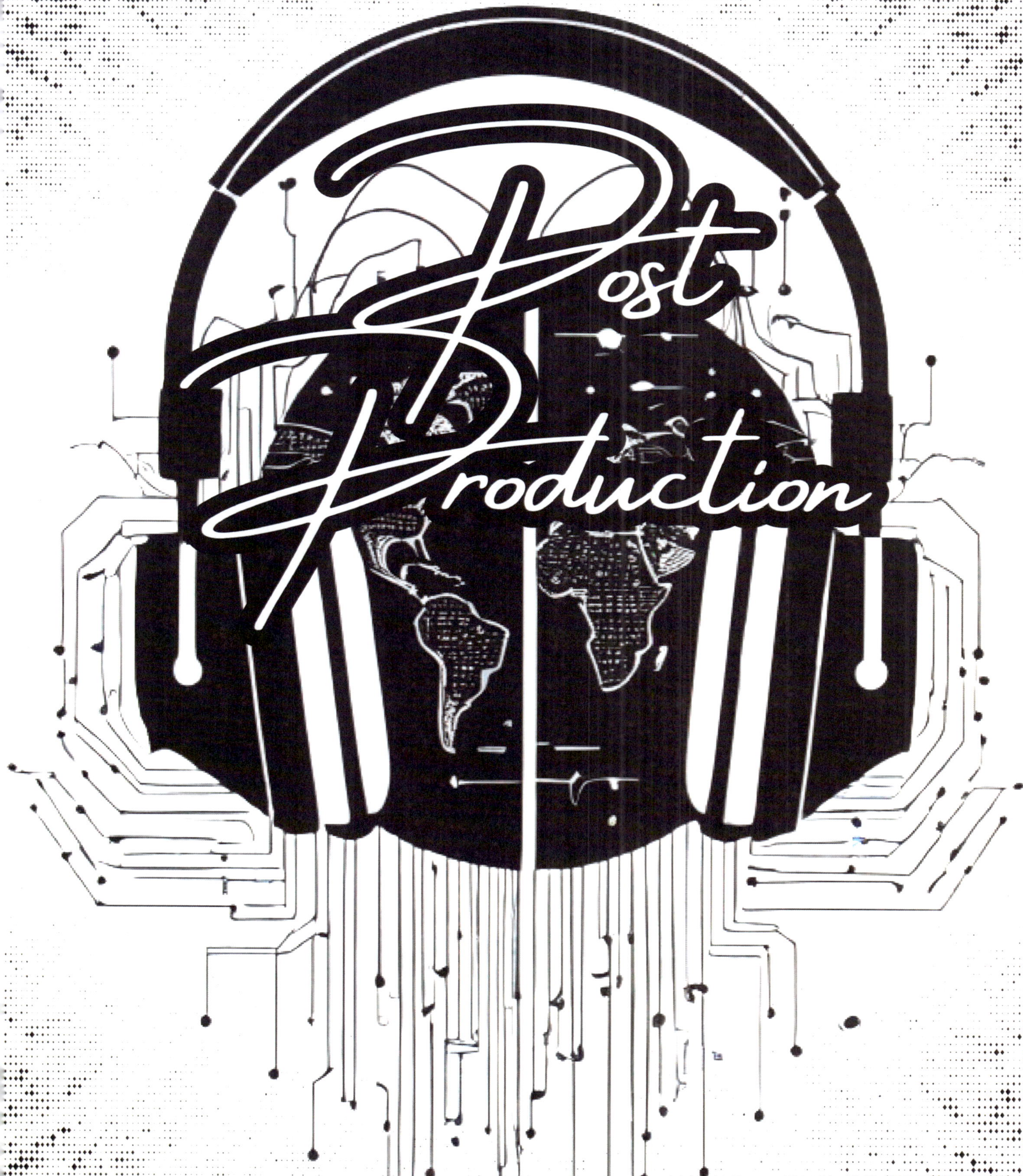
Post
Production

"Mix Preparation"

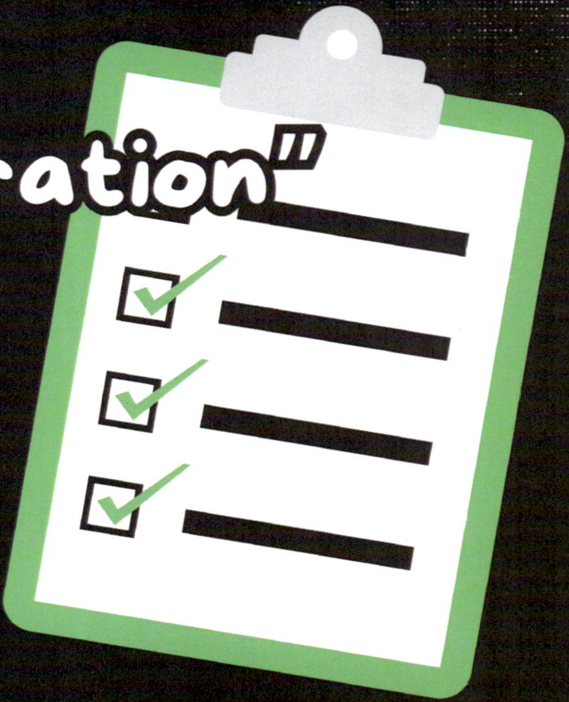

Make sure:

Tracks are labeled...

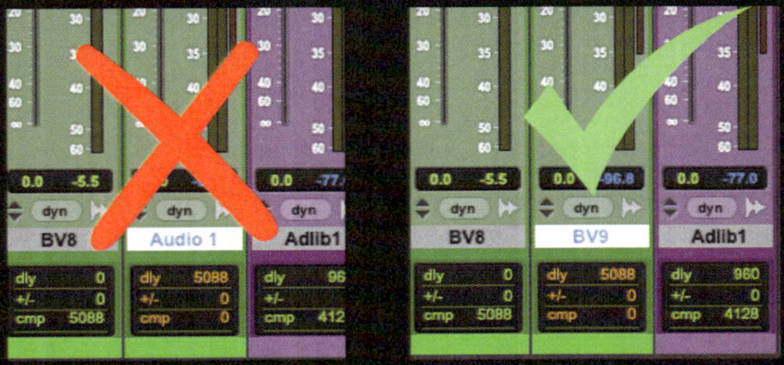

All clips have good level and Fades...

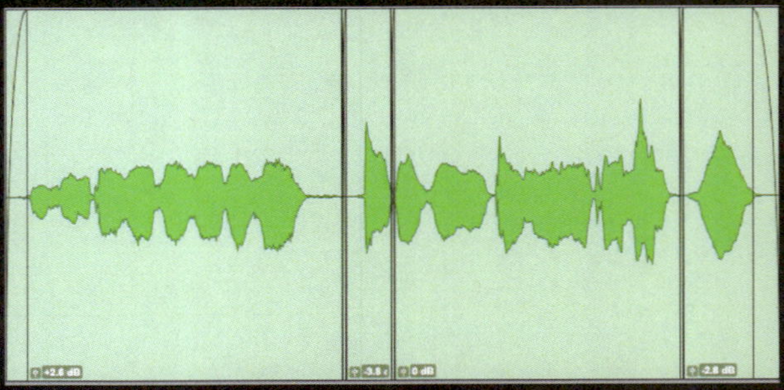

All corrective editing (Pitch and Timing) is done...

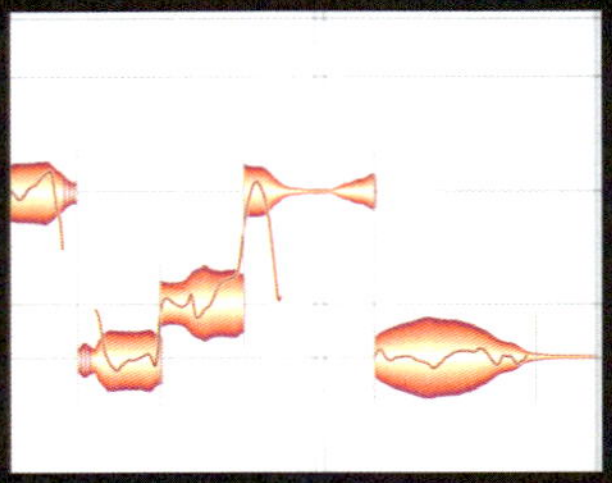

Color Code and Organize Track Order...

FX sends/returns ready to go...

Tracks and Groups have correct routing...

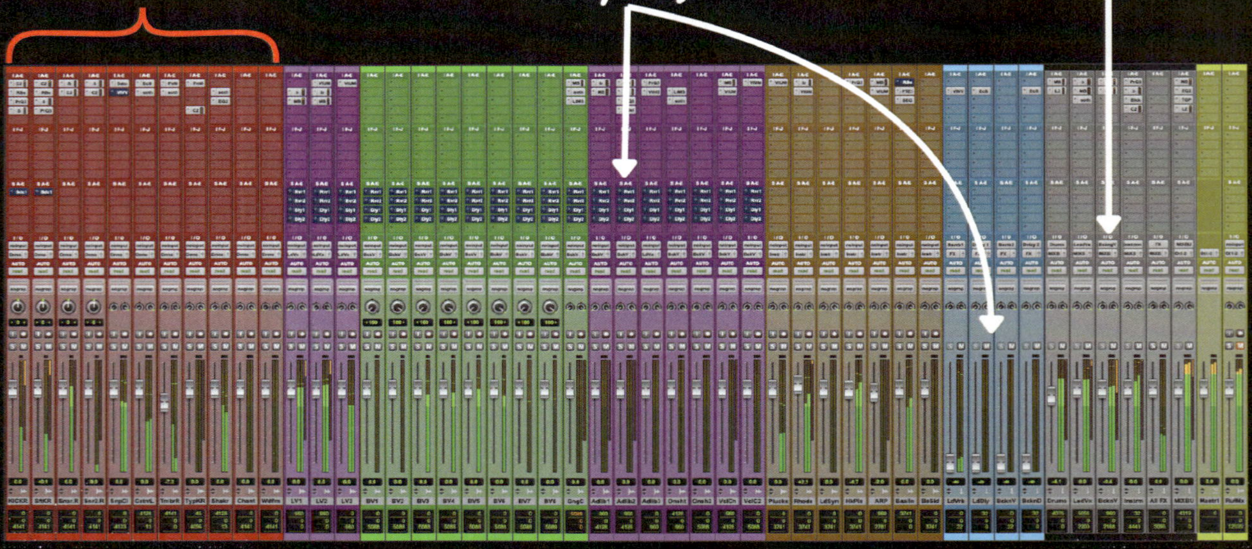

Computer Playback Settings for Post Production

During the post-production phase, one of the first technical settings to verify in your DAW is the playback "buffer size." To optimize your computer's resources, it's recommended to choose a "1024 sample" buffer size. This is a detail that can easily be overlooked when shifting from the recording stage to post-production. If you forget, your DAW may trigger a "system overload" error. Don't worry; simply increase your buffer size and continue mixing. Remember to revert to a lower buffer size if you need to record again. Depending on plugin latency, you might need to freeze or commit tracks before adjusting the buffer.

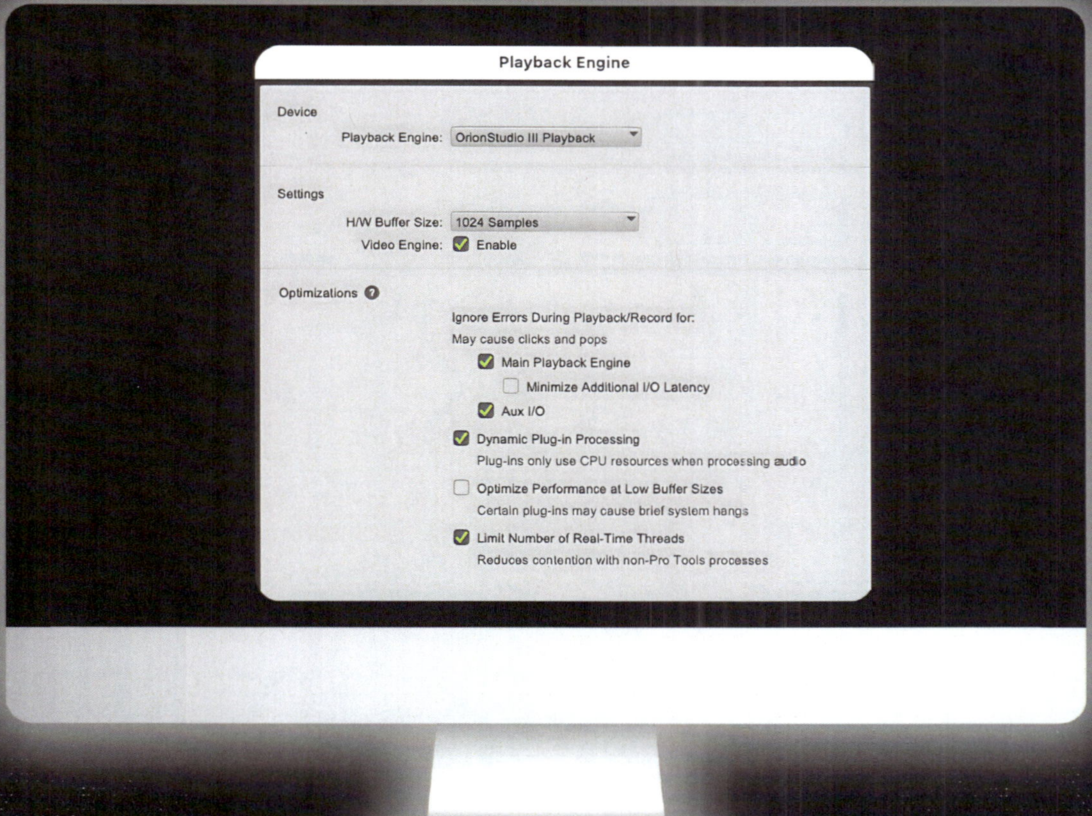

"Cleaning Clips"

Making a conscious effort to trim ambient noise and unnecessary sound from your recordings will dramatically improve your signal to noise ratio and computer performance. Remember, your computer is still crunching the numbers on low level noise. Allow your computer to focus on crunching the numbers on the important audio.

Applying fades to "All" clips is also a necessity. It can prevent unwanted pops, ticks, or sudden spikes in audio. Again, allow your computer time to crunch the numbers. A sudden spike of data from an unfaded clip can also create artifacts to the rendered "bounce" of your session.

Trim Left/Right

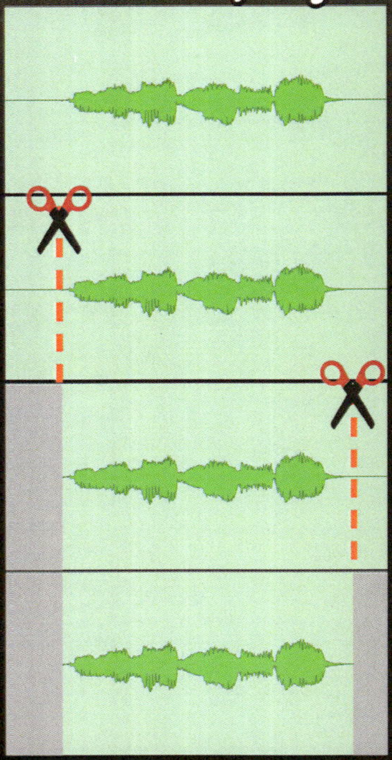

Let's clean this clip!

Trim Left... Click the edit point and Press "A"

Trim Right... Click the edit point and Press "S"

The clip's left and right silence is now trimmed!

Fade In/Out

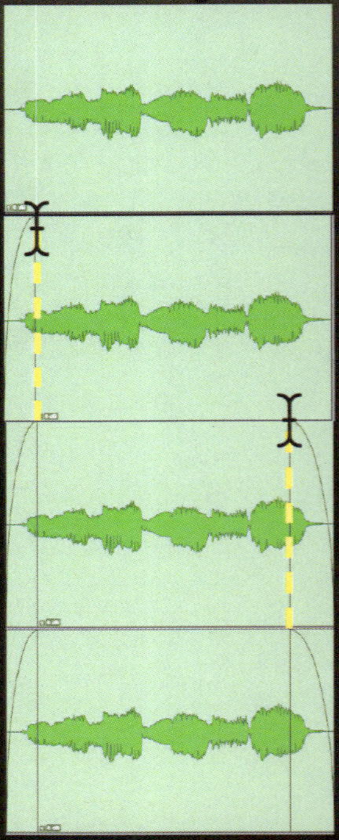

Let's create fades for this clip!

Fade In... Click the edit point and Press "D"

Fade Out... Click the edit point and Press "G"

Your clip now has both a fade in and a fade out!

Isolate a section

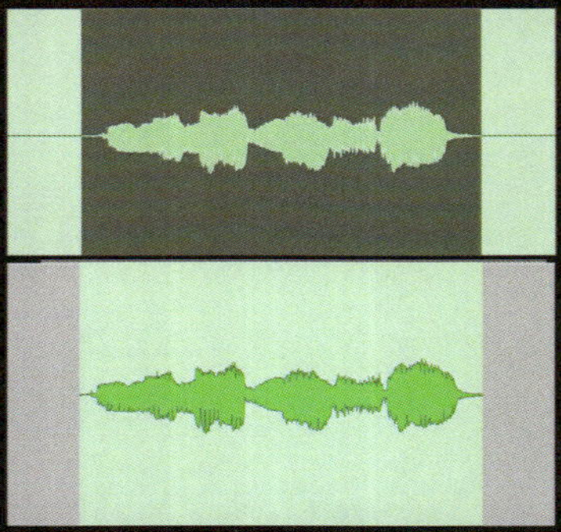

Highlight the area you wish to isolate!

Press... ⌘ command +T.
Now you have an isolated section

Cross Fade

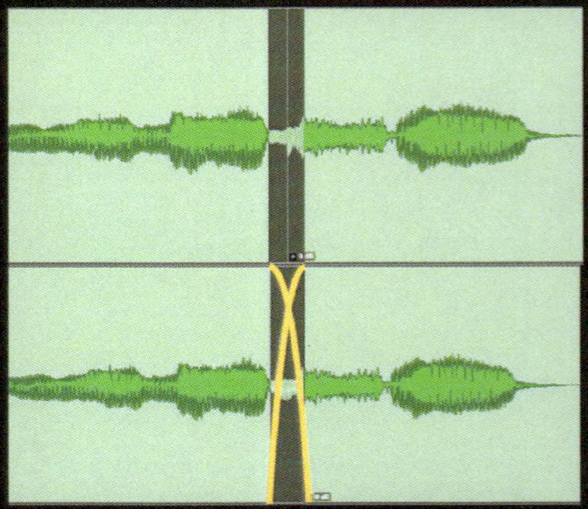

When two clips connect, apply a crossfade! Highlight the area you wish to apply the crossfade...

Press... ⌘ command +F.
Now you have applied a crossfade between two clips...

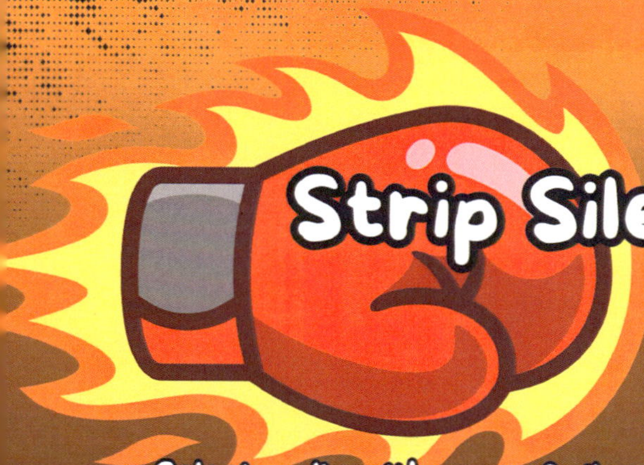

Strip Silence to Batch Fade Combo!

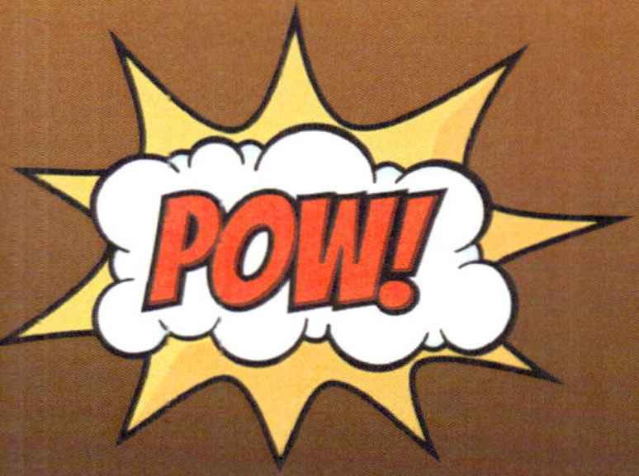

Select a clip with gaps of silence...

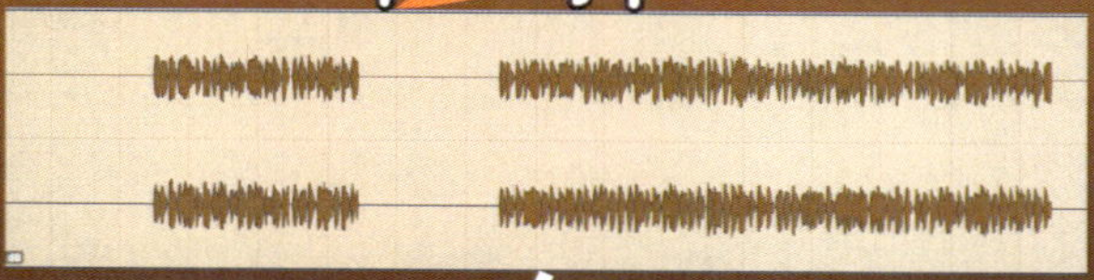

Press... ⌘ command +U
The Strip silence parameters will pop up...

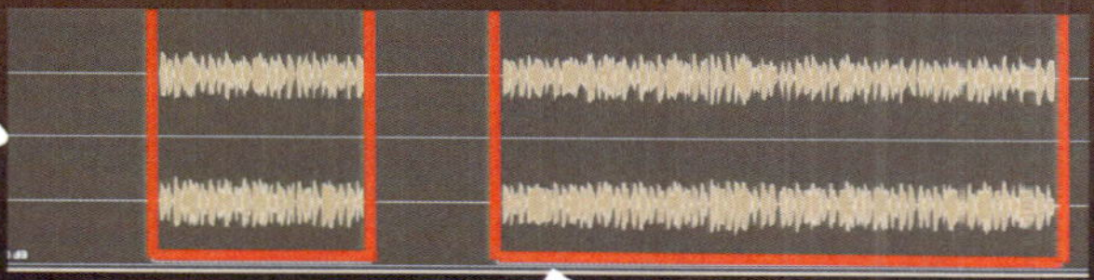

Adjust the threshold until only the gaps are separated... Adjust the start and end times to leave a little space...

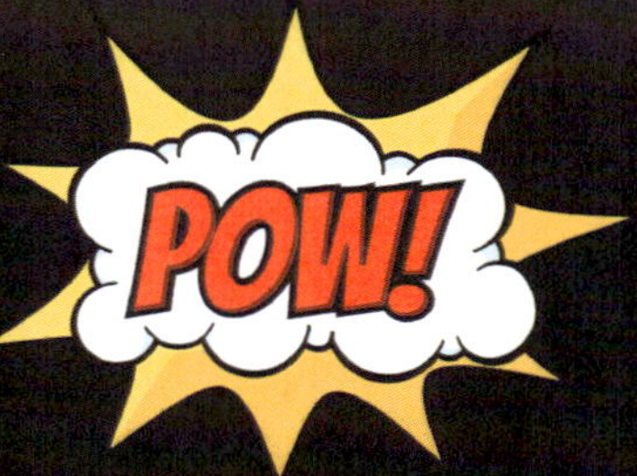

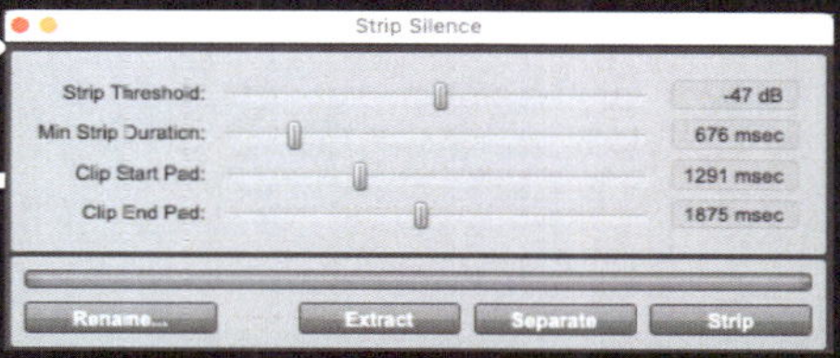

Select "Strip" when parameters are ready...

Notice! The silence is deleted and both clips are still highlighted...

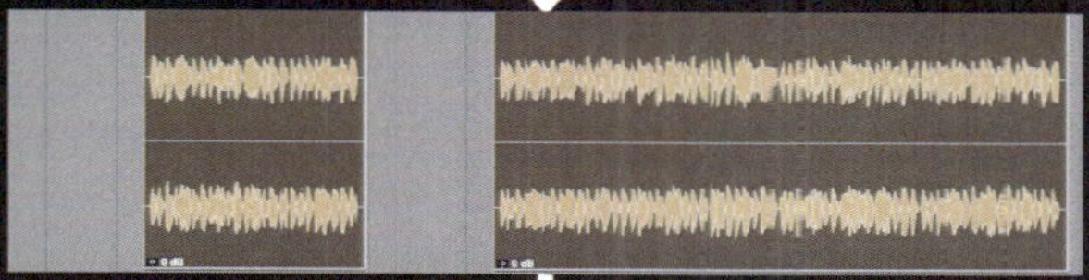

Press... ⌘ command +F
The Batch Fade parameters will pop up...

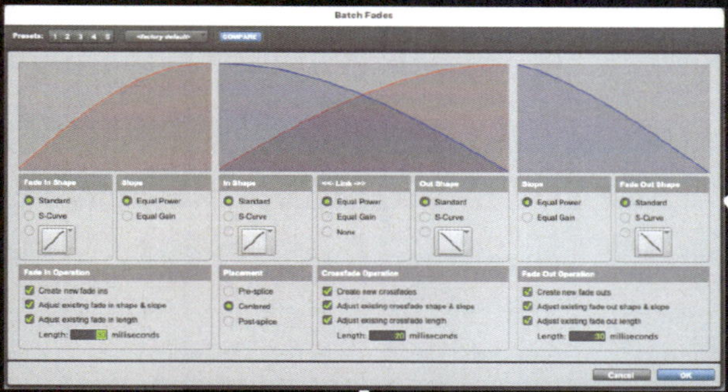

Set the desired "In", "Out", and Crossfade (If needed)...

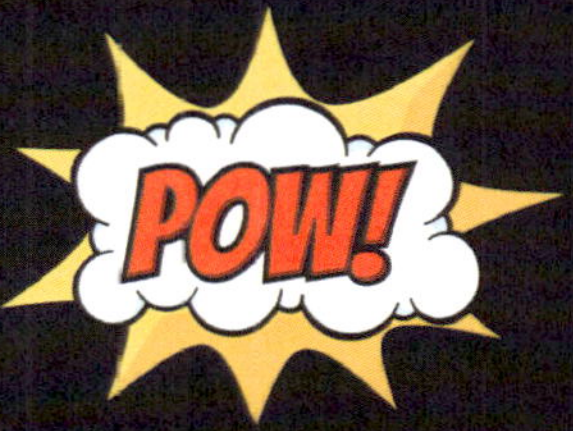

Select "OK" and both clips will share the same fade parameters...

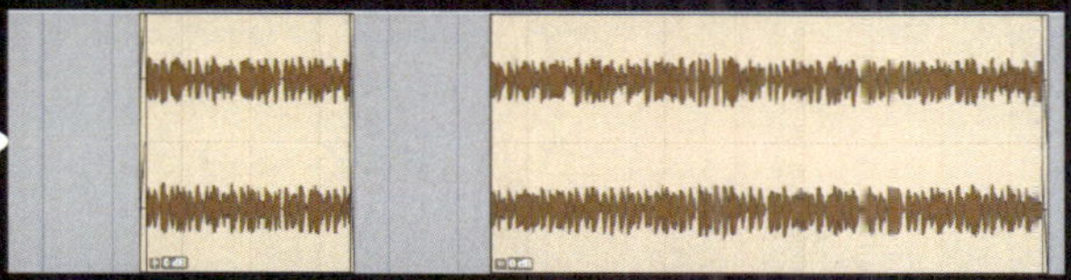

Have a cupcake!

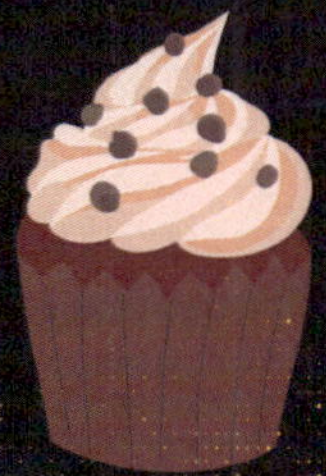

Clip Gain

Original Clip: At times, the volume levels in a recording can fluctuate significantly. Using clip gain is a practical way to balance the dynamics of a performance directly at the waveform level.

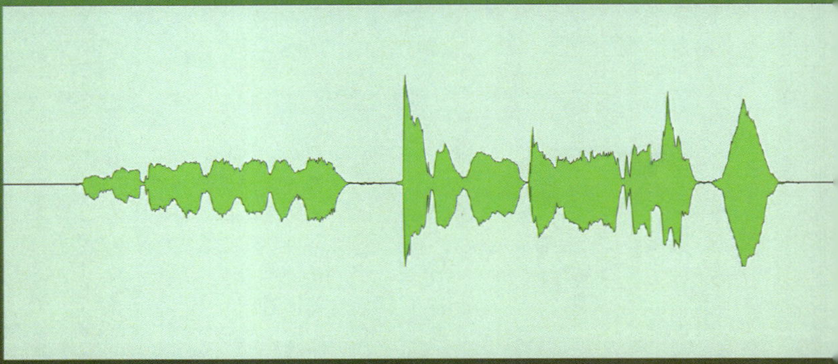

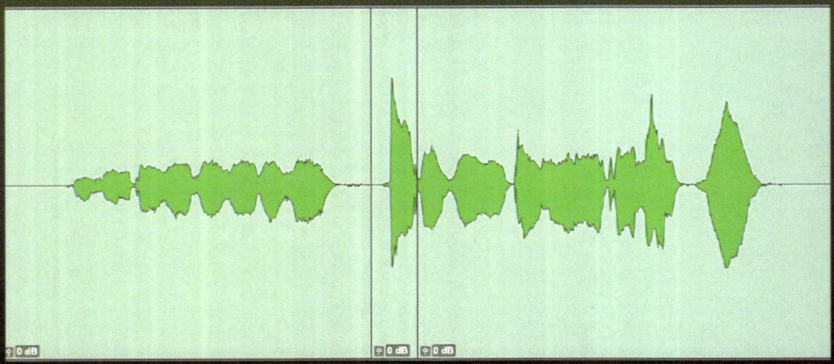

Break Clip: Breaking the waveform into smaller segments will allow yourself to balance the gain of the clip. Notice the break right at the sharp transient in this clip.

Adjust Clip Gain: Each clip segment has its own distinct volume level. Observe the -3.8dB attenuation applied to the loud transient and the 2.6dB boost on the first segment.

Tip* You may also utilize the "clip gain line" feature, which allows for drawing clip gain levels with the smart tool in Pro Tools.

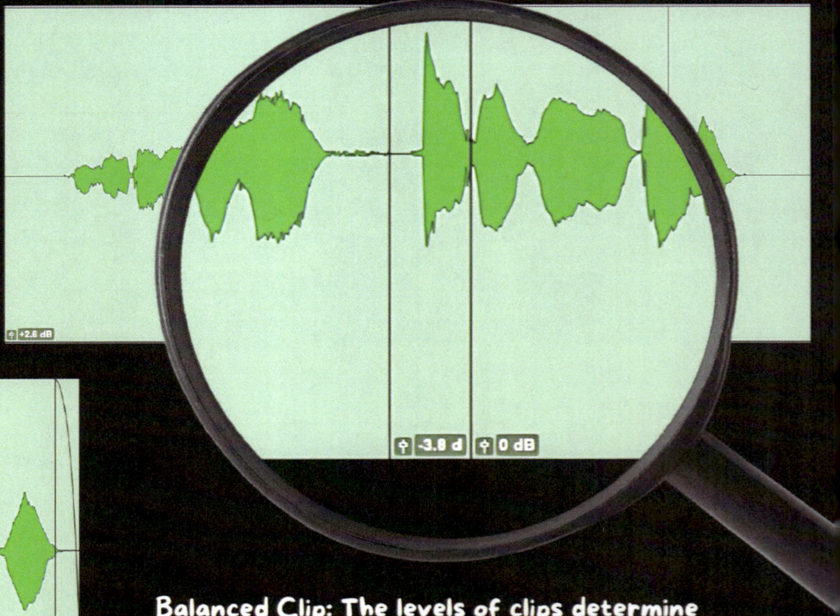

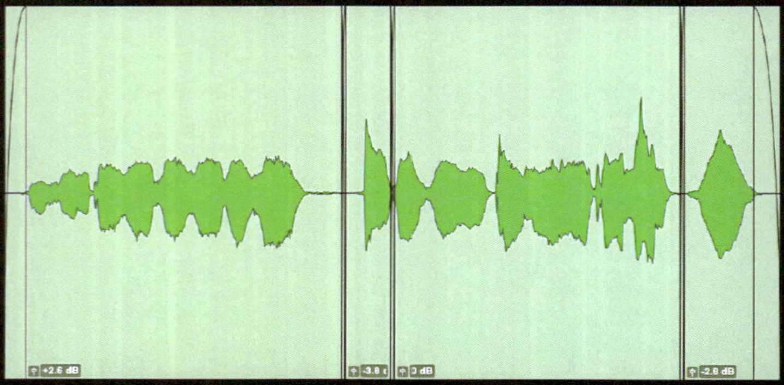

Clip Gain Up/Down Shortcut

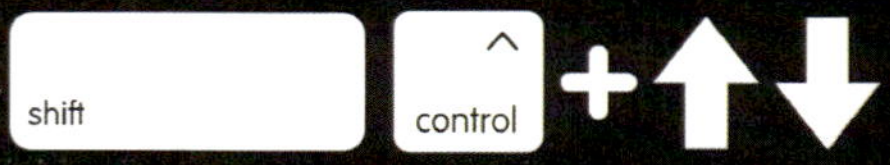

Balanced Clip: The levels of clips determine the initial signal entering the channel strip and bus processing within the DAW. Adjusting the clip in a DAW is similar to modifying the gain on a mixer. Now that the clips are balanced, this adjustment will improve the performance of level-dependent dynamic processors like compressors, enabling them to operate more efficiently.

"Tuning and Pitch Correcting"

Pitch correcting can be approached in a couple ways. We all know "Auto-Tune", which is an automatic "one size fits all" approach. However, there are pitch correction approaches that provide greater precision while maintaining a more natural and "transparent" sound. Software like Melodyne enables adjustments to the finer nuances of tuning, including vibrato, pitch drift, and formant shifting, alongside overall pitch correction.

In many ways, pitch correcting is a delicate process. You can take a conservative approach, focusing only on that "one wacky note", or you can craft a perfectly "in tune" performance, nuancing every syllable. Tuning calls for a keen musical ear and strong situational awareness. Even if it's possible to sculpt a perfectly "in tune" performance, should you? Perfect isn't always better. In fact, many would argue that the imperfections are what make a performance great. This is a controversial topic, but the reality is it takes more than good pitch to inspire someone.

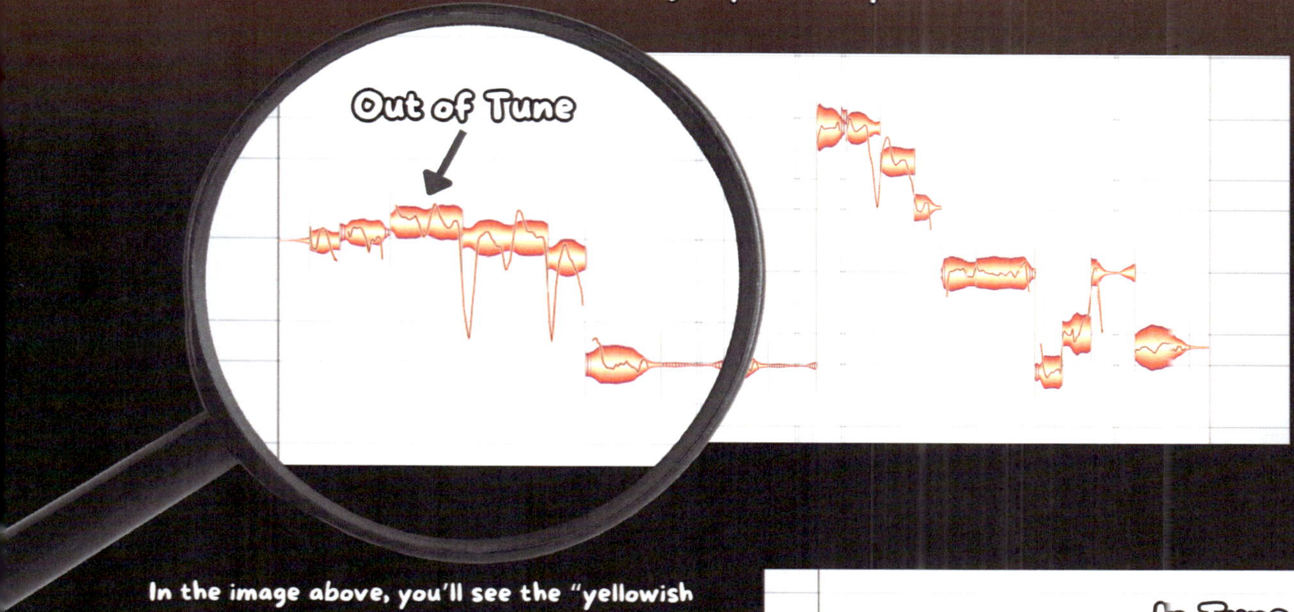

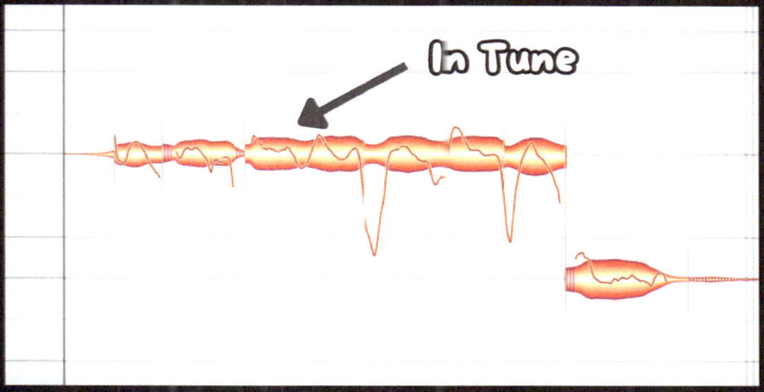

In the image above, you'll see the "yellowish blobs" symbolizing the fundamental notes within Melodyne's edit view. The "squiggly lines" illustrate the pitch contour, while the horizontal lines indicate the reference pitch grid. You may also observe that the middle part of the phrase rises into an out-of-tune "sharp" position. Now, turn your focus to the corrected pitches on the right. Notice how the fundamental notes align perfectly with the pitch grid. Additionally, the pitch contour varies "up and down," which is perfectly normal. Use your artistic judgment to decide if the notes require further adjustments.

"Time and Alignment Editing"

First: After listening to a drum recording, you notice the first hit of this drum hit sounds early in the context of the song...

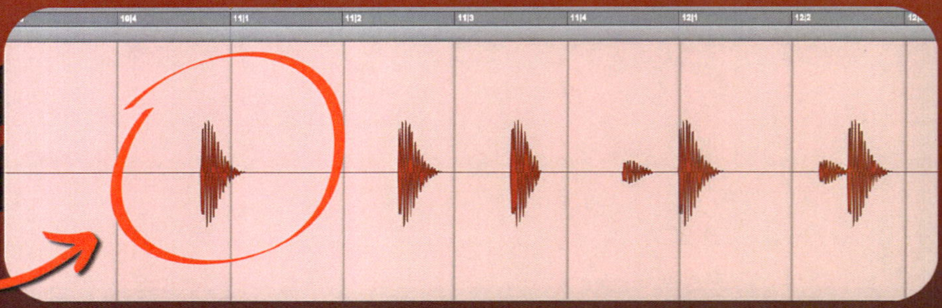

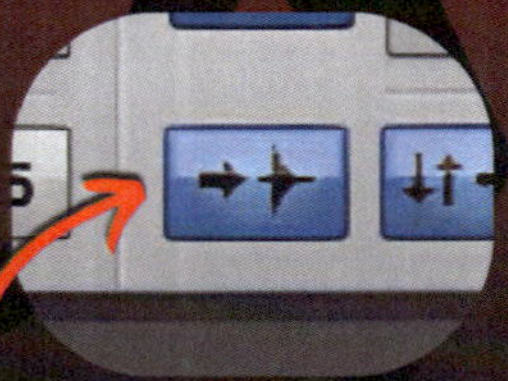

Tip* The "tab to transient" function in Pro Tools is great for locating the beginning of a transient...

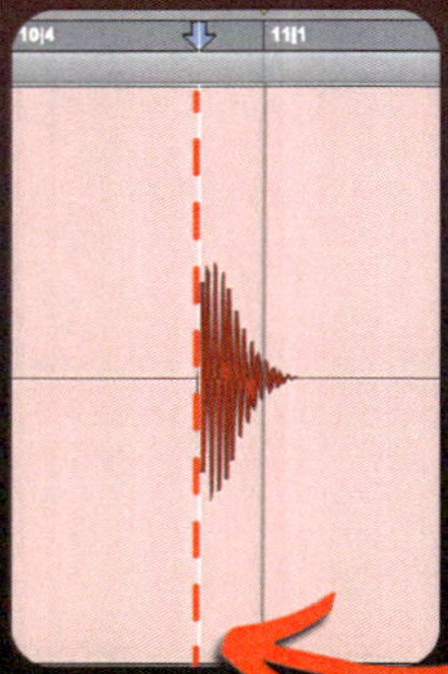

Second: Identify the beginning of the "out of time" transient...

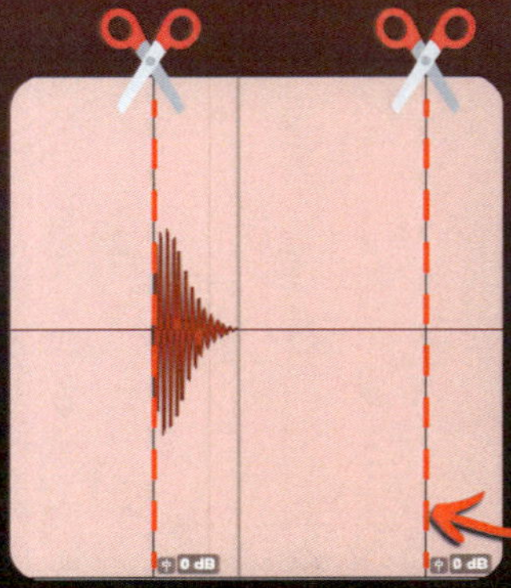

Third: Isolate th "out of time" note by "breaking" the clip...

Fourth: Modify or "nudge" the timing of the clip until it aligns with the grid. Not every circumstance calls for "perfect" beat alignment, so use your best judgment on what sounds best in context...

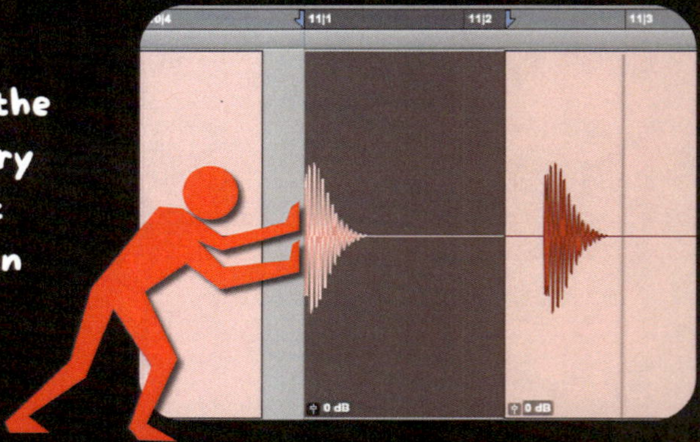

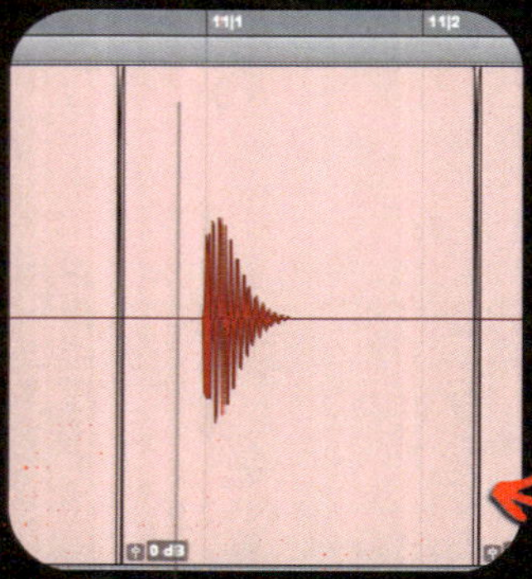

Fifth: Apply crossfades to merge the broken clips. This will ensure seamless transitions at the edit points. For best results, adjust the length and location of these fades until the edit sounds most natural.

"Top Down Mixing"

A "Top Down Mixing" approach refers to the mindset of focusing on the mix as a whole, while addressing the individual tracks in relation to the "Big Picture". This approach keeps the priority of the mix workflow on how tracks sum together, rather than the refinement of tracks from an individual perspective. With that said, if the need for individual attention arrises, do so while keeping the big picture in mind.

Keep in mind to apply all "Main Mix" processing to the Mix Bus rather than the Main Output Master fader.

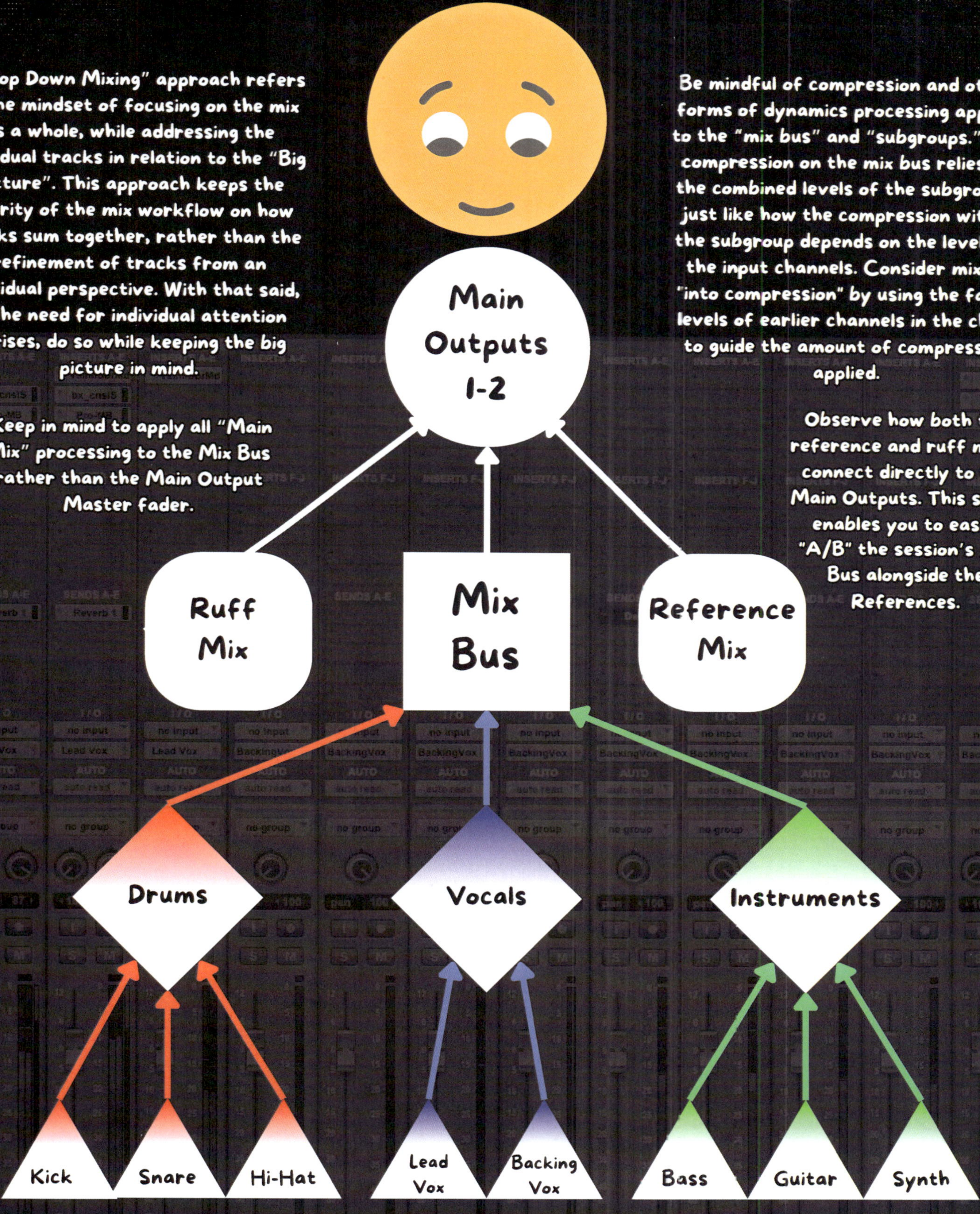

Be mindful of compression and other forms of dynamics processing applied to the "mix bus" and "subgroups." The compression on the mix bus relies on the combined levels of the subgroups, just like how the compression within the subgroup depends on the levels of the input channels. Consider mixing "into compression" by using the fader levels of earlier channels in the chain to guide the amount of compression applied.

Observe how both the reference and ruff mixes connect directly to the Main Outputs. This setup enables you to easily "A/B" the session's Mix Bus alongside the References.

Building the Top Down Mix

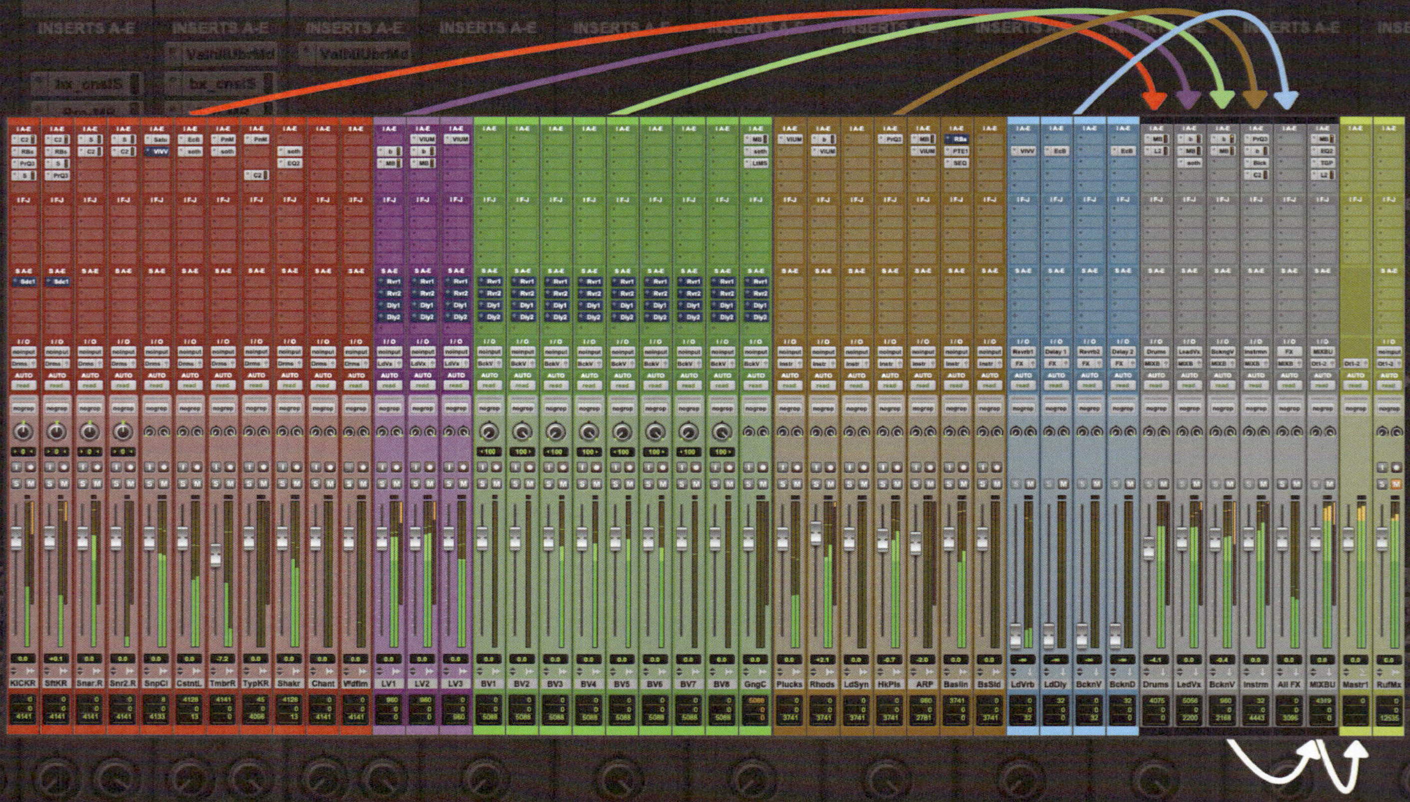

Tip* The volume of the individual channel faders will effect the level feeding the FX (especially compression) on the Subgroup... Think of it like "driving" your levels into the group FX...When you have a nice "Mini Mix" within the subgroup, altering the volume of the subgroup won't change the integrity of the sound, however the individual channels will...

The Mix Bus "Top of the Mix"

Observe how all the outputs of the subgroups feed the "Mix Bus" before the mix bus sends to the main outputs 1-2. Think of the Mix Bus as the final stage of processing before it feeds the main output. This routing method enables the master fader to manage a broader view of the entire session, rather than focusing solely on the primary mix. In essence, this approach prepares us for seamless "A/B" comparisons of reference mixes and system tuning.

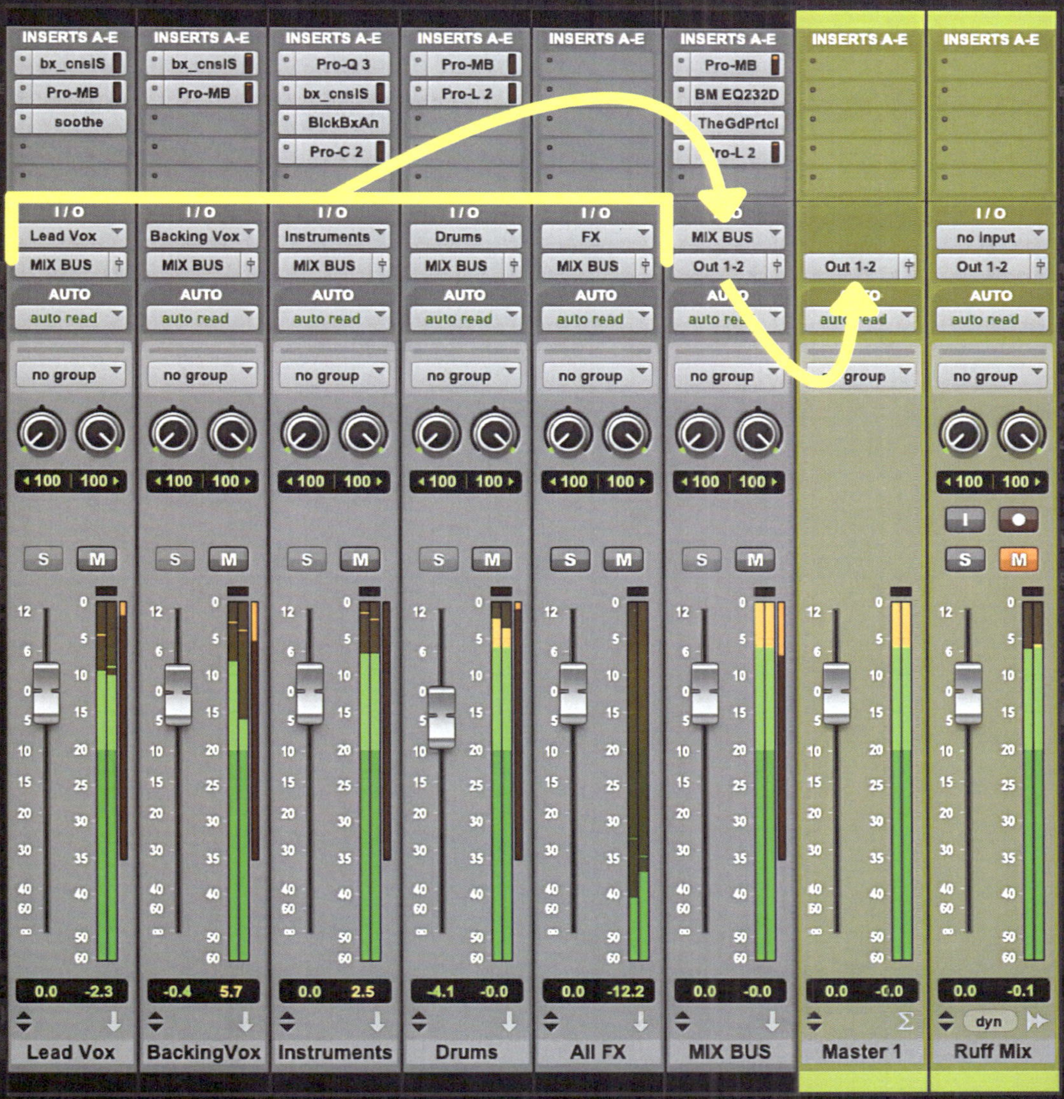

Reference Mix

As mentioned before, the use of a dedicated mix bus allows for seamless A/B comparison of a reference mix. Notice how the reference mix is routed directly to the main ouputs 1-2, just like the mix bus. Remember, the master fader should not have inserted plugins, as it would would effect not only the main mix but the reference.

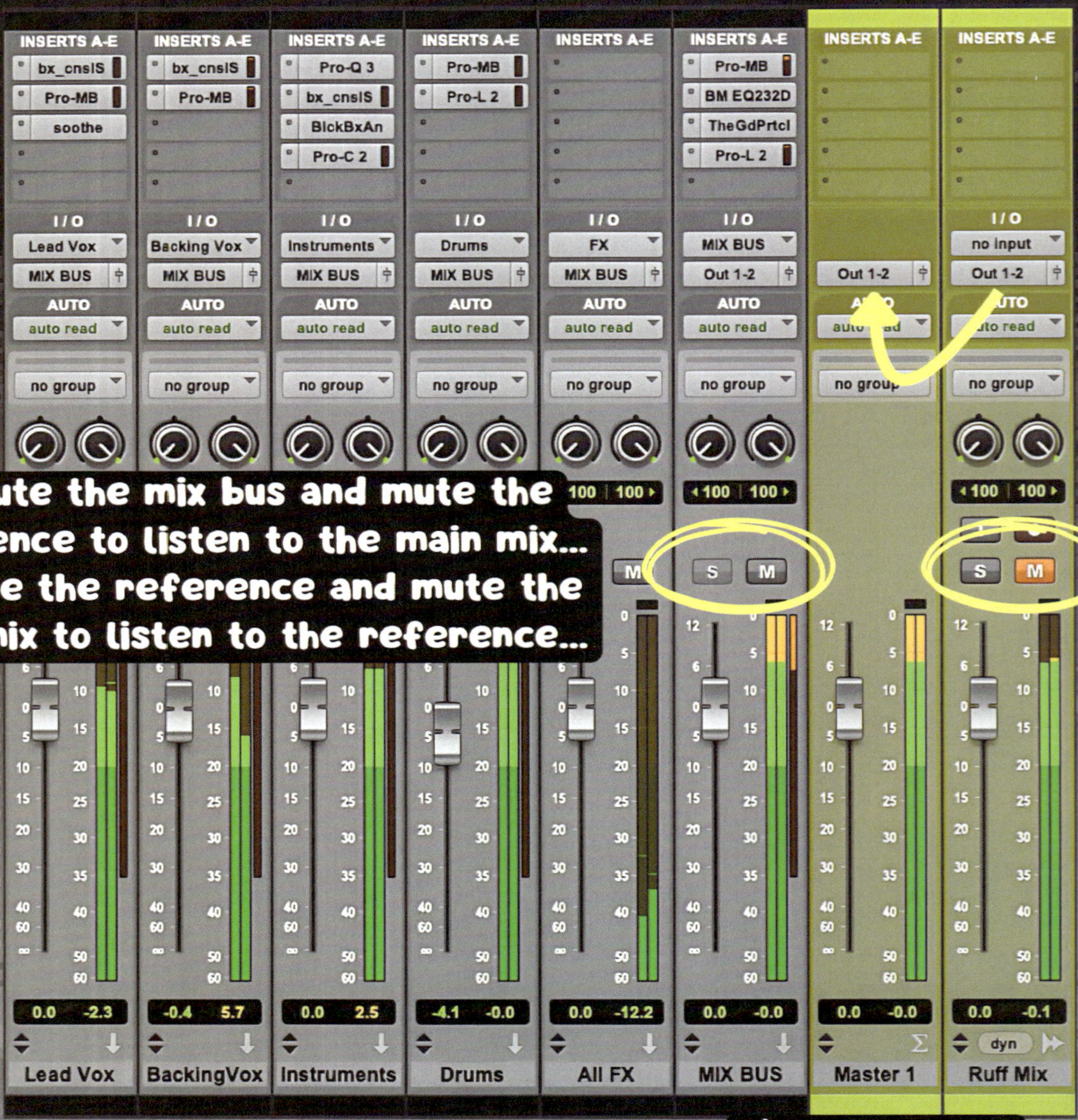

A/B Toggle Shortcut
Select the names of 2 tracks
Shift + M

Subgroup

By creating a "Subgroup" with a stereo "Aux Input", you can process all the "Backing Vocals" together within a single channel... This gives you efficient level management for both individual and group volume adjustments... Think of a Subgroup like a "Mini Mix" within the mix... Notice how both the "output" of the individual channels and "input" of the group channel are assigned to the "Backing Vox" Bus... This will allow the individual channels to flow into the group channel...

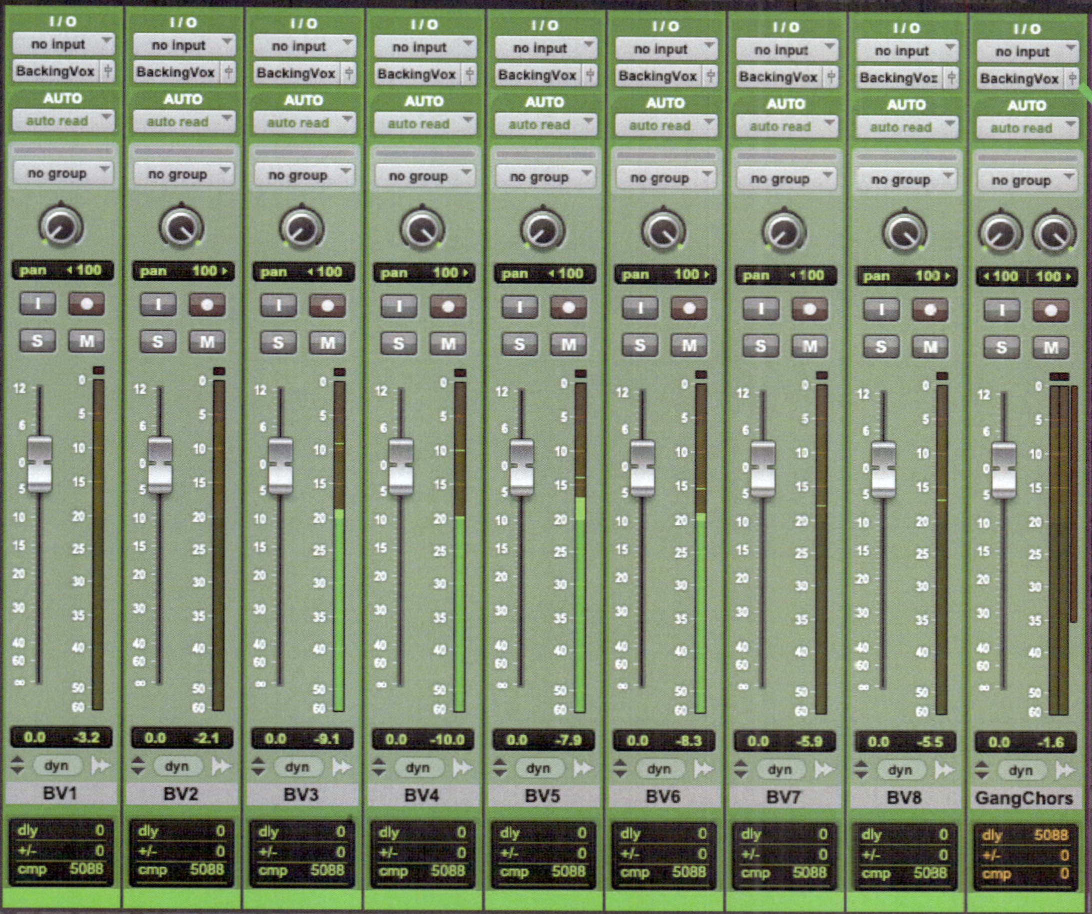

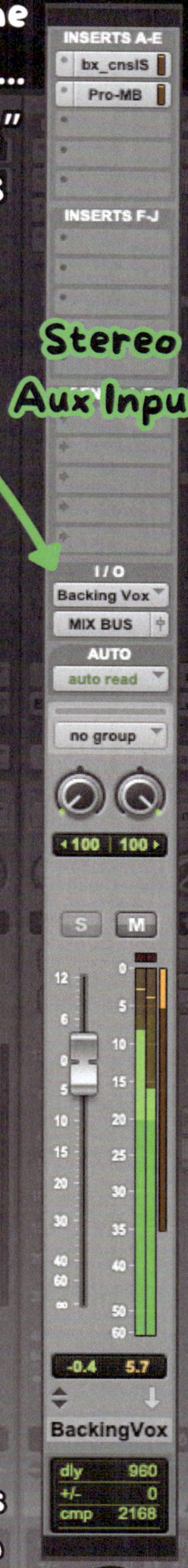

Tip: The individual channel fader volumes will influence the level sent to the FX, particularly compression, on the Subgroup. You can think of it as "driving" your levels into the group FX. When you create a balanced "Mini Mix" within the subgroup, adjusting the subgroup's volume won't compromise the sound's integrity. This is because the subgroup fader does not alter the signal feeding the subgroup's plug-ins. In short, you won't lose your "nice compression" sound when turning down your subgroup fader, unlike the channel faders feeding the subgroup.

Mastering

Mastering is the concluding stage of the production process, where a mix receives its final refinement and processing before delivery. Often confused with mixing which addresses the summing of individual tracks within a song, mastering starts with a single rendered audio file of the mix itself. EQ, compression, and target loudness all play a vital role in creating a universally polished song that sounds great on various types of speaker systems. It's also common to master a number of songs together, creating a sense of cohesiveness within an album project. Lastly, it's not a bad idea to hand off the mastering process to someone with a fresh pair of unbiased ears. By this point in the process, you have already spent countless hours on the project and your judgment has acclimated to what you are used to hearing. If you feel fresh enough, give mastering a try, but there is no shame in handing it off to a second pair of ears.

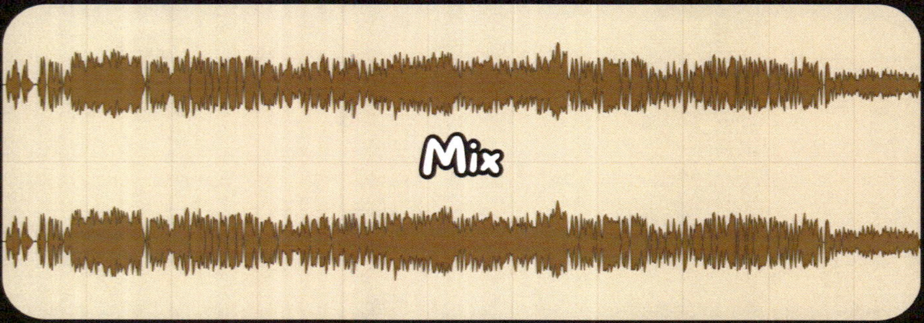

Even if a finished mix appears balanced and sonically correct for the music, it might still fall short in delivering the fullness, consistency, and loudness expected for a final product.

When a mix is mastered, additional EQ, compression, and limiting is applied to achieve the desired polish and loudness. Notice the density and consistency of the waveform. The more compression applied to the master, the more dense the waveform will appear.

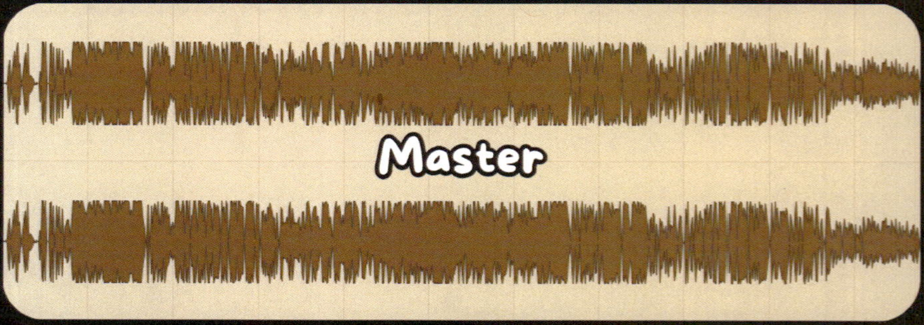

"Exporting Files"

Aost recording sessions consist f various fragmented clips that together create the complete rack. When exporting files, it's nportant to consolidate tracks to ensure they remain in sync with the song. If clips are xported without consolidation, these "floating clips" will have no reference of time.

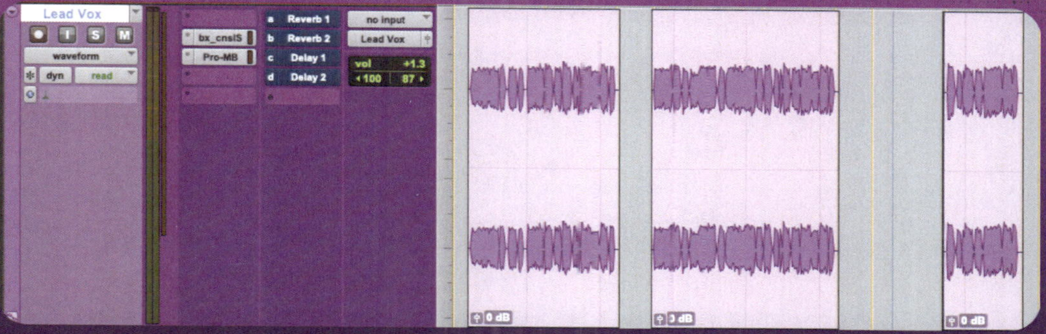

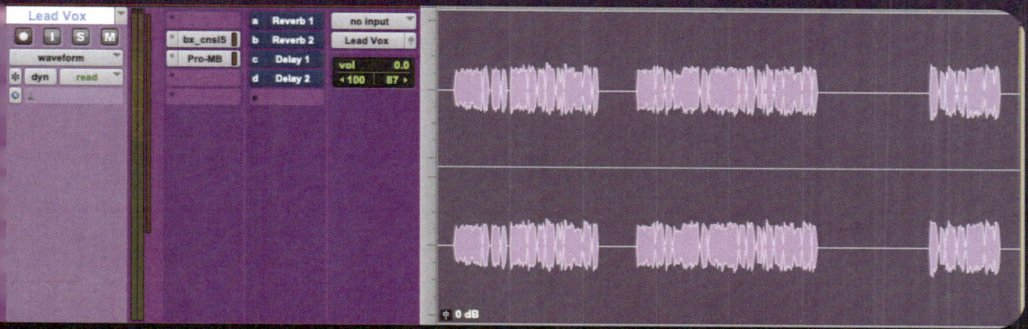

To "Consolidate" fragmented clips into a single clip on a track, start by selecting the area of the clip you wish to consolidate, then press "Shift+Option 3." This action will generate a new consolidated clip. Keep in mind, this approach will not render the inserted plugins on the track. Lastly, be sure that all consolidated files have a common "start" and "end" time. The recipient of these files will thank you!

To export the files from the Pro Tools session, first highlight the desired clips in the edit window. Your'll notice those same clips are highlighted within the clip list. Right click one of the grey clips and choose "Export Clips as Files...". Be sure to create a dedicated destination folder for the exported files. Again, this process of exporting files does not render the processing inserted on the tracks.

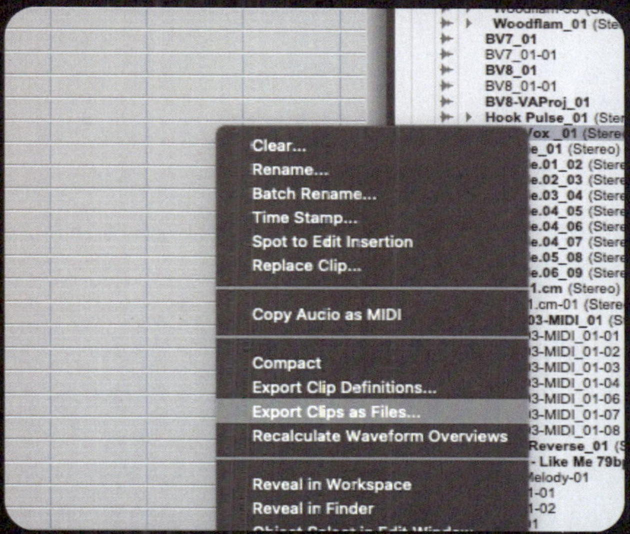

Stems

Disclaimer: Before "bouncing stems," it's wise to consult with the client. Rendering individual track stems can be approached in a number of valid ways. Theoretically, if all the stems are played together, it should sound like the mixed song, but stems can also be used for a "work in progress" mix.

Creating "Stems" involves generating a consolidated audio file of a specific track within a session, which typically incorporates inserted processing. However, depending on the situation, stems can also be rendered without processing to provide more mixing flexibility for the recipient. A few common examples of stems would an instrumental stem, vocal stem, drum stem, guitar stem, ambience stem, FX stem, or an a cappella stem.

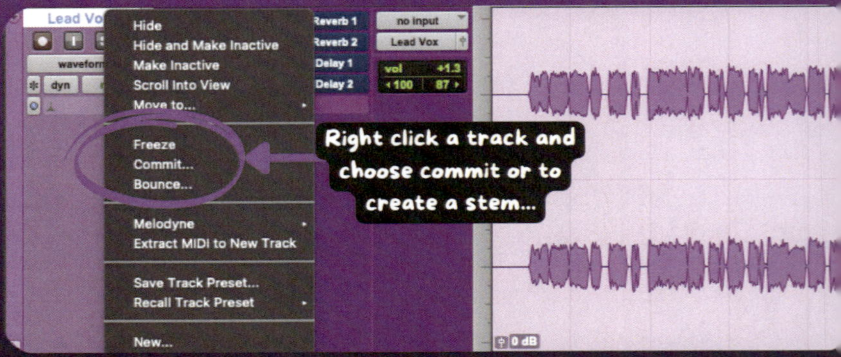

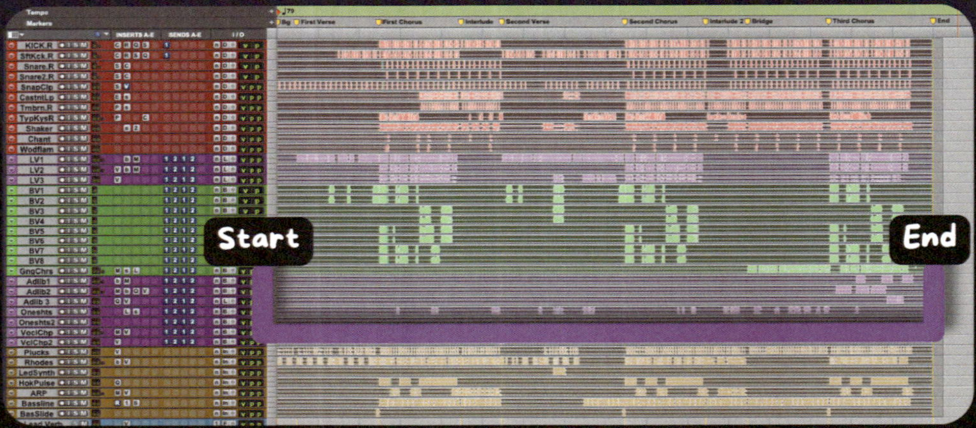

Make sure that all the rendered "Stems" have a common start and end point. This will assure all tracks remain in sync if the stems are transferred to a different session.

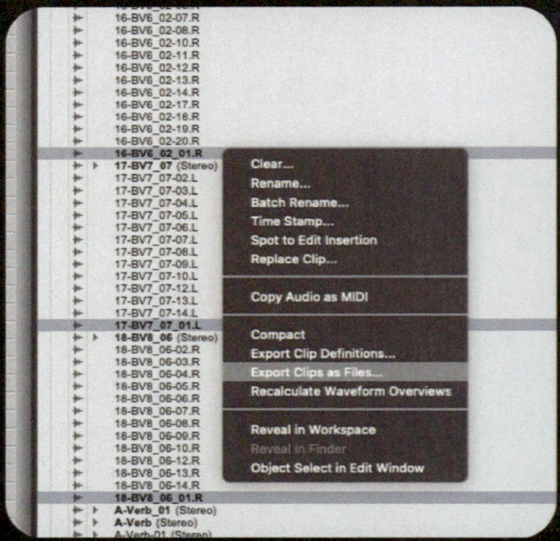

To export the files from the Pro Tools session, first highlight all the rendered stem clips and view the clip list on the right side of the edit window. The selected clips will be highlighted in dark grey. Right click one of the grey clips and choose "Export Clips as Files...". Be sure to create a dedicated destination folder with the "Stems" in the title.

Tip: Use the following labels for your stem files.

Song Title_Track_Stem_bpm

Ex: Cool Song_Lead Vox_Stem_120bpm

Concluding in The Pendulum Process...

Thank you for being a part of The Pendulum Process. In this guide, we've focused on one essential aspect of music production: "Tech." It is my hope that you not only feel well-informed but also inspired to discover your unique role within this vastly diverse world of music production. Remember to revisit the elements that inspired you to pursue music in the first place, and strive to fall in love with the process, not the end result.

-Samuel Friend

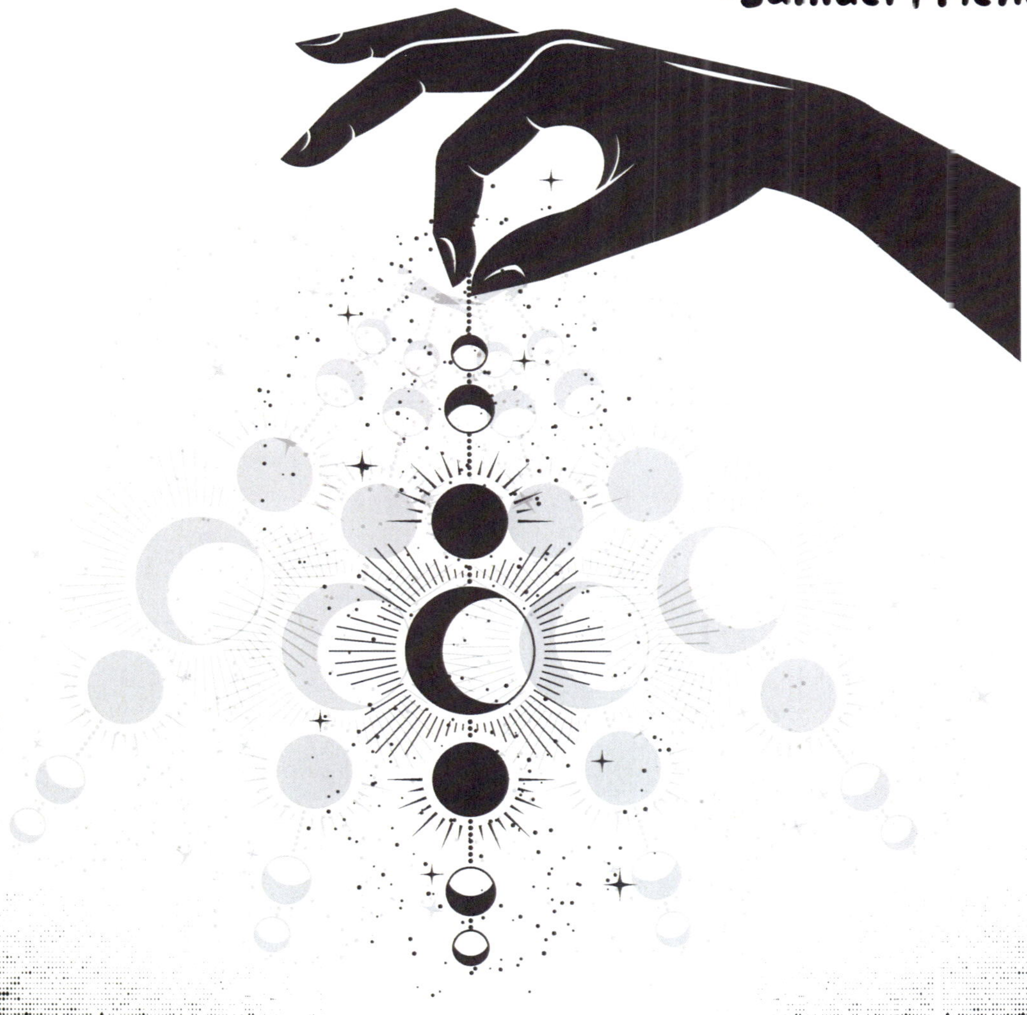

www.ingramcontent.com/pod-product-compliance
Lightning Source LLC
LaVergne TN
LVRC080247220926
848482LV00042B/345

* 9 7 9 8 9 9 2 6 1 3 7 1 1 *